Resources, Climate and Sustainable Development

Series Editors

Malin Song, Anhui University of Finance and Economic, Bengbu, China

Hooi Hooi Lean, School of Social Sciences, Universiti Sains Malaysia, Gelugor, Malaysia

The book series are dedicated to consolidating projects, case studies, and practical initiatives within the realms of natural resource utilization, climate change, energy and environmental economics, and sustainable development efforts. These contributions collectively tackle global challenges of paramount significance. Within the broader context of the global commitment to a green transition, nations have forged collaborative efforts aimed at achieving green transition and sustainable development objectives and fostering a more environmentally sustainable way of life on a global scale. Consequently, there arises a pressing demand for the provision of cutting-edge information and comprehensive case studies to serve as valuable resources for advancing energy transition, natural resource utilization and sustainable development efforts across diverse domains, which is this comprehensive objective that this book series endeavors to fulfill. We extend an open invitation for the submission of monographs, contributed volumes tailored to an academic and professional readership, as well as peer-reviewed conference proceedings. Relevant subject matter encompasses a wide range of areas, including but not confined to efficient utilization of nature resources, energy system revolution, low-carbon economic transformation, individual lifestyle choices and emission quota allocation, familial strategies for energy transition, corporate performance within the sphere of energy transition, recent energy transition policies alongside noteworthy pioneering cases, as well as the development and application of advanced models for simulating the dynamics of energy transition, natural resource utilization and sustainable development efforts.

Jianlin Wang · Malin Song
Editors

The Green Path of Chinese Transportation Enterprises

Springer

Editors
Jianlin Wang
Low-Carbon Development and Carbon Finance Laboratory
Anhui University of Finance and Economics
Bengbu, China

Malin Song
Collaborative Innovation Center for Ecological Economy and Management
Anhui University of Finance and Economics
Bengbu, China

ISSN 3005-0502 ISSN 3005-0510 (electronic)
Resources, Climate and Sustainable Development
ISBN 978-981-95-5288-7 ISBN 978-981-95-5289-4 (eBook)
https://doi.org/10.1007/978-981-95-5289-4

Translation from the Chinese Simplified language edition: "no" by Jianlin Wang and Malin Song, © Author 2025. Published by not published. All Rights Reserved.

Title of the original Chinese edition: 中国交通企业的绿色之路

The original submitted manuscript has been translated into English. The translation was done using artificial intelligence. A subsequent revision was performed by the author(s) to further refine the work and to ensure that the translation is appropriate concerning content and scientific correctness. It may, however, read stylistically different from a conventional translation.

This Springer imprint is published by the registered company Springer Nature Singapore Pte Ltd.
The registered company address is: 152 Beach Road, #21-01/04 Gateway East, Singapore 189721, Singapore

Preface

We stand at a critical juncture in history, where the imperatives of climate change, technological disruption, and sustainable development are converging to redefine the future of global industries. Among these, transportation—the vital artery that has connected economies and shaped civilizations for millennia—faces an unprecedented mandate for transformation. Its journey from a facilitator of growth to a leader in green and intelligent innovation will profoundly influence the trajectory of human progress. As the world's largest developing nation, China's pursuit of a modern, low-carbon transportation system is not merely a domestic priority but a matter of global significance. It is a cornerstone for achieving the nation's "Dual Carbon" goals (peaking carbon emissions and achieving carbon neutrality) and a fundamental pillar for building a new, sustainable development paradigm. The choices made here will resonate through global supply chains, international climate efforts, and the blueprint for other emerging economies.

This book, "Green Pathways for Transportation Enterprises: A China Perspective," emerges from this urgent context. It seeks to move beyond theoretical discourse and provide a systematic, empirical examination of the green transition across the entire spectrum of China's transportation sector. By delving into the unique challenges, innovations, and strategic pathways of sub-sectors from high-speed rail and automated ports to urban transit and sustainable aviation, this work aims to uncover the underlying environmental-economic logic of this great transformation. It is our hope that the analysis within these pages will serve as a valuable resource for policymakers, industry leaders, academics, and students alike, offering practical insights and a comparative framework for understanding one of the most critical transitions of our time.

The global dialogue on sustainable transportation is enriched by diverse perspectives. By placing China's experience in conversation with global practices—such as the infrastructure modernization efforts in nations like Romania—this book underscores that there is no single path to a green future. Instead, it highlights the universal themes of technological leapfrogging, policy-driven innovation, and the immense potential for international cooperation. We invite you to explore the pages that follow, not as a final word, but as a contribution to an ongoing, vital global conversation about building the efficient, clean, and intelligent transportation systems that our shared future demands.

Jianlin Wang
Low-Carbon Development and Carbon Finance Laboratory
Anhui University of Finance and Economics
Bengbu, China

Acknowledgement

We would like to show our appreciation for the support of the National Natural Science Foundation of China (No. 72574001, 71934001), Anhui Provincial University Research Plan Project (2024AH050012).

Introduction

Throughout the history of human civilization, transportation has always served as the "major artery" of economic and social development, playing a central role in connecting economic veins and shaping the forms of civilization. As the world enters the third decade of the twenty-first century, climate change and sustainable development have become global common imperatives. The transportation sector, being a key source of carbon emissions, faces a green transformation that is not only crucial for the future of the industry itself but also profoundly impacts the sustainable development path of human civilization. As the world's largest developing country, China's modernization and greening of its transportation system are both a critical area for achieving its "Dual Carbon" goals and an important support for constructing a new development paradigm. This book focuses on the green development pathways for transportation enterprises, systematically reviewing the current status, challenges, and trends across various sub-sectors such as road, rail, waterway, civil aviation, pipeline transport, and urban rail transit. Through cross-industry comparative analysis, it aims to reveal the internal logic and practical pathways of the transportation industry's transformation from an environmental-economic perspective, exploring a green transportation development paradigm that aligns with China's national conditions while drawing on international insights.

As China's economy shifts from high-speed growth to high-quality development, the traditional transportation development model faces efficiency bottlenecks and environmental carrying capacity constraints. On the one hand, rising urbanization rates and consumption upgrades are driving passenger and freight transport demands toward greater efficiency, personalization, and green transformation. On the other hand, the regional economic coordination development strategy imposes higher requirements on the comprehensive transportation system. Against this backdrop, transportation enterprises need to achieve green and intelligent upgrades in infrastructure construction, transport service provision, and industrial ecosystem development, supporting economic restructuring with lower resource consumption and reduced environmental costs. The rapid development of high-speed rail, for instance, has not only shortened regional spatiotemporal distances but also boasts energy consumption per passenger-kilometer that is only one-fifth that of road transport and one-twelfth that of air transport, making it a typical representative of green transportation. In-depth research on the environmental-economic effects of transportation enterprises helps resolve the conflict between economic development and resource-environmental constraints, maximizing both the ecological and economic benefits of transportation infrastructure investments.

This book takes enterprises from major sub-sectors of China's transportation industry as its research objects, covering road transport, rail transport, urban rail transit, waterway transport, civil aviation, pipeline transport, expressways, ports, airports, pipeline facility operations, express delivery and postal services, automobile manufacturing, shipbuilding, and more. It analyzes these sectors from dimensions such as policy environment, market size, technological innovation, environmental impact, and international competition. Chapter 1 introduces the green development path of China's road transport enterprises, analyzing achievements in policy support, market expansion, structural optimization, and technological progress, while also pointing out issues such as the imbalance between transport capacity and volume and insufficient informatization. Chapter 2 focuses on the current development status and challenges of China's rail transport enterprises, delving into their operational conditions, technological innovations, and market competition, showcasing practices in green development and service optimization through case studies of representative enterprises. Chapter 3 elaborates on the current status and future trends of China's urban rail transit enterprises, analyzing their core competitiveness from dimensions such as business scope, corporate structure, and financial status, while also identifying problems in finance, operations, technology, and external factors. Chapter 4 discusses the development status and trends of China's waterway transport enterprises, outlining the market size, enterprise status, and technological innovations in the sector, analyzing internal and external environmental challenges, and proposing recommendations for promoting green transformation and sustainable development. Chapter 5 analyzes the development status and market environment of China's civil aviation enterprises, examining their economic situation, competitive landscape, challenges, and opportunities from both global and Chinese perspectives. Chapter 6 studies the development and challenges of China's pipeline transport enterprises, detailing their business distribution and current status in areas such as oil and gas pipelines and chemical transport, and analyzing market competition and technological challenges. Chapter 7 focuses on the intelligent and green development trends of China's expressway enterprises, analyzing challenges and opportunities from aspects such as policy systems, market size, corporate structure, and technological development. Chapter 8 discusses the position and role of China's port enterprises in global trade, outlining their development status, infrastructure upgrades, and achievements in smart port construction. Chapter 9 analyzes the historical transformation and global contribution of China's airport enterprises, exploring operational management challenges from the perspective of their historical development and current status globally and in China. Chapter 10 studies the business models and development strategies of China's pipeline facility operation enterprises, analyzing their market competitive position, challenges, and opportunities, and proposing corresponding strategic recommendations. Chapter 11 discusses market competition and service innovation in China's express delivery and postal enterprises, analyzing market size, competitive landscape, and practices in technological innovation, service optimization, and market expansion. Chapter 12 focuses on technological innovation and international competitiveness of China's automobile manufacturing enterprises, comprehensively analyzing their development status, policy support, market size, and technological level, and discussing the international competitive environment and future trends. Chapter 13 introduces the global competitiveness of China's shipbuilding enterprises, showcasing technological breakthroughs and market expansion in high-end shipbuilding through case studies of leading enterprises. Chapter 14 focuses on the rail transit vehicle manufacturing industry, detailing innovation development in areas such as high-speed rail and metro, and market expansion along the "Belt and Road". Chapter 15 systematically analyzes the development status of the civil aviation equipment manufacturing industry, focusing on the driving effect of the domestic large aircraft project on the industrial chain and the construction of a technological innovation system. Chapter 16 studies the pipeline transport equipment manufacturing industry, revealing the transformation and upgrading path of equipment manufacturing under the energy transition through case studies of typical enterprises. Chapter 17 comprehensively reviews service innovation models in transportation enterprises, including emerging formats such as shared mobility and smart logistics. Chapter 18 specifically studies the construction of intelligent transportation systems, showcasing China's cutting-edge exploration in areas from vehicle-road coordination to urban brain from multiple dimensions. Chapter 19 focuses on the application of green low-carbon technologies in the transportation sector, including practices such as new energy vehicle promotion and carbon emission reduction pathways.

Looking at the global transportation development landscape, different countries have formed different development models due to variations in historical evolution, geographical conditions, and development stages. Romania, located in Southeastern Europe as an EU member state and an important country in the Danube River Basin, has a railway network of approximately 10,611 kilometers, with a density higher than the EU average, yet its modernization level remains somewhat insufficient—with an electrification rate of only about 38% [1]. In comparison, China has built the world's largest high-speed rail network, with 47,000 kilometers of high-speed rail out of a total of 162,000 kilometers of railway [2], and an electrification rate of 75.2%, demonstrating a paradigm of latecomer countries achieving catch-up through technological leaps. In waterway transport, Romania relies on the Danube-Black Sea canal system to achieve a river-sea intermodal network. The Port of Constanța, as the largest port on the Black Sea, handles 120 million tons of cargo annually, ranking first on the Black Sea [3]. Currently, with EU support, it is actively upgrading its automated terminal facilities. In contrast, ports like Shanghai and Ningbo-Zhoushan consistently rank among the top globally in container throughput, with their massive port clusters and efficient operational capabilities being the core support for China's role as the "world's factory" integrated into global supply chains [4]. In civil aviation, Romanian airlines participate in the EU's "Clean Aviation" program to develop bio-jet fuel technology pathways; China Eastern Airlines took the lead in achieving commercial flights using domestically produced Sustainable Aviation Fuel (SAF), and the C919 large aircraft project integrates green design throughout the entire lifecycle of the domestic aircraft model.

These differences reflect the strategic choices in transportation development between the two countries. Romania, as a key logistics node in Southeastern Europe connecting the EU with the Black Sea, Caucasus, and Central Asia, and an important stop on the southern route of the China-Europe Railway Express, focuses on regional network integration, which is also a key means for Romania to leverage its advantages. China leads globally in the application of technology and construction speed in areas such as high-speed rail, automated ports, and smart logistics platforms, demonstrating strong system integration and engineering implementation capabilities, as its transport network primarily serves a vast domestic market and export demands. China faces challenges related to managing the complexity of an ultra-large-scale network and regional development imbalances, while Romania needs to overcome historical infrastructure shortcomings, improve connectivity across complex terrain, and enhance administrative efficiency to accelerate the absorption of EU funds and project implementation. However, this also signifies immense cooperation potential between China and Romania. China's engineering capabilities, equipment manufacturing, and capital advantages complement Romania's geographical advantages, EU market access, and urgent need for modernization upgrades.

Currently, the global transportation industry is at a critical juncture of transitioning from traditional development models to green, low-carbon, and intelligent ones. Many countries and regions are building competitive advantages through measures such as carbon tariffs and green technology barriers. The EU's Carbon Border Adjustment Mechanism (CBAM) will tax the embedded carbon emissions of imported goods starting in 2026. The U.S. Clean Competition Act provides subsidies for domestic new energy vehicle supply chains. In international shipping, the draft IMO Global Carbon Emissions Trading System for 2025 may increase operating costs for Chinese vessels. China faces both historical opportunities for technological breakthroughs and policy support, as well as severe challenges from environmental constraints and international competition. In the future, it will be necessary to take technological innovation as the core driving force, industrial synergy as an important support, and policy innovation as an institutional guarantee to promote the transformation of transportation enterprises from scale expansion to quality leap.

References

1. https://www.fmprc.gov.cn/gjhdq_676201/gj_676203/oz_678770/1206_679426/1206x0_679428/ Accessed: 27 October 2025
2. https://www.gov.cn/lianbo/bumen/202412/content_6994691.htm Accessed: 13 October 2025
3. https://baijiahao.baidu.com/s?id=1829439763123436467&wfr=spider&for=pc Accessed: 27 October 2025
4. https://news.sohu.com/a/880808676_120468014 Accessed: 27 November 2025

Contents

Chapter 1
Development Status, Challenges, and Strategies of Road Transport Enterprises in China

Jianlin Wang and Xueqin Wen

Abstract This paper conducts an in-depth analysis of the development status of China's road transport enterprises during the period from the 12th Five-Year Plan (2011–2015) to the 13th Five-Year Plan (2016–2020). Against the backdrop of national policy support, it explores significant achievements in enterprise structural optimization, market expansion, and advancements in informatization, intelligent technologies, and new energy technologies. Research indicates that recent advancements in information technology, intelligent transportation systems, and online freight platforms have effectively enhanced transport efficiency and diversified transportation services. However, the industry continues to face challenges including imbalances between transport capacity and demand, insufficient informatization levels, inadequate safety management systems, and operational cost pressures, which constrain further development. To address these challenges, the paper proposes solutions including optimizing transport capacity allocation, enhancing information technology levels, improving safety management systems, promoting multimodal transport, and strengthening corporate collaboration. Looking ahead, the article emphasizes that technological innovation, diversification of the industrial ecosystem, and optimization of organizational management will be key directions for industry development. These measures aim to foster the sustainable development of road transport enterprises, meet the demands of national economic growth, and enhance their international competitiveness.

J. Wang · X. Wen (✉)
Anhui University of Finance and Economics, Bengbu, China
e-mail: wxq18305589742@163.com

J. Wang and M. Song (eds.), *The Green Path of Chinese Transportation Enterprises*, Resources, Climate and Sustainable Development,
https://doi.org/10.1007/978-981-95-5289-4_1

1.1 Current Development Status of China's Road Transport Enterprises

1.1.1 Policy Framework

During the period from the 12th Five-Year Plan (2011–2015) to the 13th Five-Year Plan (2016–2020), the Chinese government significantly increased policy support for the transportation sector. The goal was to cultivate it into a solid foundation for the national economy and guide the modern road freight industry toward scale and network development. (2016–2020), the Chinese government significantly intensified policy support for the transportation sector. The objective was to cultivate it into a solid foundation for the national economy and guide the modern road freight industry toward transformation characterized by scale, networking, and branding. This holds critical significance for achieving the construction of a moderately prosperous society in all respects [1]. During the 12th Five-Year Plan period (2011–2015), the Ministry of Transport formulated the "12th Five-Year Development Plan Outline for the Road Transport Industry (2011–2015)" signaling enhanced government policy support for the industry in areas such as market access, vehicle renewal, technological upgrades, and information infrastructure development. This aimed to promote the evolution of modern road freight enterprises toward scale, networking, and branding [1]. During the 13th Five-Year Plan period (2016–2020), the State Council issued the "13th Five-Year Development Plan for a Modern Comprehensive Transportation System (2016–2020)," establishing guiding principles, development objectives, and core tasks for transportation. This sector was positioned as a vital pillar of the national economy, with its development plan prioritized. During this period, transportation advancements played a significant role in promoting steady growth of the national economy, marking a critical phase in building a moderately prosperous society in all respects. For instance, ZGJJJ Co., Ltd. benefited from national policy support, continuously strengthening its capabilities in highway infrastructure construction.

1.1.2 Market Scale

The market scale of China's transportation enterprises has been growing steadily, particularly in infrastructure construction, fixed-asset investment, transportation service capacity, and the smart transportation market. From an infrastructure perspective, by the end of 2023, China's operational railway mileage reached 159,000 km, with high-speed rail accounting for 45,000 km. The double-track ratio and electrification rate reached 60.3% and 75.2% respectively, marking substantial improvements in infrastructure and transport service capacity [2]. The total highway mileage increased to 5.4368 million kilometers, an increase of 82,000 km from the end of the previous year. Navigable inland waterways reached 128,200 km, with ports operating 22,023 berths [3]. Among these, COSCO Shipping, a global leader in container transportation and integrated supply chain services, holds a significant position in the worldwide maritime market. Global maritime trade volume grew to 11.407 billion tons in 2023. Civil aviation transportation airports reached 259, with 4270 transport aircraft registered. In urban passenger transport, the total number of public bus and tram routes operated was 79,800, while urban rail transit operated 308 lines with a total operational mileage of 10,158.6 km.

Regarding fixed-asset investment, transportation fixed-asset investment has remained above 3 trillion yuan for seven consecutive years, reaching 3.9 trillion yuan in 2023. The daily average investment in transportation infrastructure hit a record high of 10.7 billion yuan. Analyzing transport capacity, commercial freight volume in 2023 reached 54.747 billion tons, an 8.1% year-on-year increase, while freight turnover amounted to 240.646 trillion ton-kilometers, up 6.3% year-on-year [3]. Road passenger traffic reached 56.556 billion trips, up 26.1%. Among these, interregional passenger traffic totaled 61.288 billion trips, growing by 30.7%. Regarding smart city transportation, preliminary statistics indicate China's smart transportation industry market size exceeded 230 billion yuan in 2023. By 2029, this market is projected to potentially expand to over 360 billion yuan. Particularly in the A3 sector, continuous innovation and development in smart transportation and modern logistics services have enhanced logistics efficiency and service quality through technologies like big data, cloud computing, and artificial intelligence. China's transportation industry has maintained steady market growth over the past five years. Infrastructure development, fixed-asset investment, transportation service capacity, and the advancement of the smart transportation market have provided a solid foundation for the sector's sustained progress.

1.1.3 Optimization of Enterprise Structure

Supported by national policies, China's road transport enterprises have undergone significant structural transformation, reflected in policy backing, market expansion, enhanced transport services, and improved operational efficiency. In terms of market scale, by the end of 2023, the national highway mileage reached 5.4368 million kilometers, an increase of 82,000 km from the previous year, with highway density rising to 56.63 km per 100 square kilometers. Regarding transportation services, the annual commercial freight volume reached 40.337 billion tons, up 8.7% year-on-year, while freight turnover reached 73.95 trillion ton-kilometers, growing by 6.9% (Fig. 1.1).

Online freight platforms play a vital role in optimizing the structure of road transport enterprises. Among them, MABAJ, as one of China's leading online freight platforms, captured over half of the market share in China's online freight platform industry in 2022. According to data from the "2023 China Logistics Platform Development Report," by the end of December 2023, the number of online freight companies nationwide reached 3069, connecting 7.989 million vehicles and 6.476 million drivers. The total number of shipping orders uploaded throughout the year reached

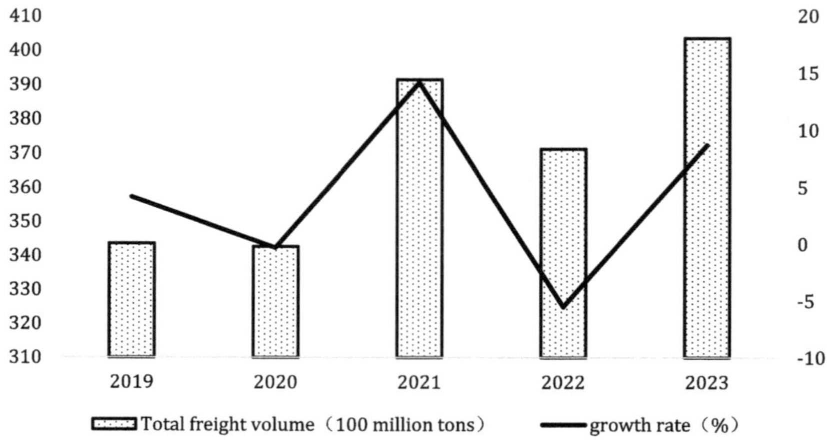

Fig. 1.1 Total commercial freight volume and year-on-year growth rate of China's road transport enterprises, 2019–2023. *Data Source* Statistical Bulletin on the Development of China's Transportation Industry, 2019, 2020, 2021, 2022, 2023

130 million, representing a 40.9% increase. Network freight platforms have played a significant role in resource integration, cost reduction, and efficiency enhancement, reflecting the industry's transition from rapid expansion to high-quality development. ZCHZH, through its intelligent multimodal transport management system, has pioneered a new model for network freight and was selected as one of the first "Typical Cases of Cost Reduction, Quality Improvement, and Efficiency Enhancement in Transportation and Logistics." Additionally, road transport enterprises have achieved notable operational gains. Representative companies like ZYSDW saw growth in both express delivery volume and revenue. In 2023, ZYSDW handled 132.07 billion express parcels, a 19.4% year-on-year increase, with express revenue reaching 1.2074 trillion yuan, up 14.3% from the previous year [3]. The total market freight volume in 2023 was approximately 700 billion yuan, accounting for about 12% of the nation's total road transportation costs, highlighting the significant role road transport enterprises play in the broader logistics industry (Fig. 1.2).

1.1.4 Technological Advancements

China's road transport enterprises have achieved epoch-making progress in technological development, primarily manifested in multiple fields including informatization and intelligentization, autonomous driving technology, vehicle–road coordination, application of the BeiDou Navigation Satellite System (BDS), new energy and materials, integrated logistics handling systems, and investment in road informatization. Technological advancement has propelled the management and development of China's transport enterprises. Through its "Internet Plus" high-efficiency

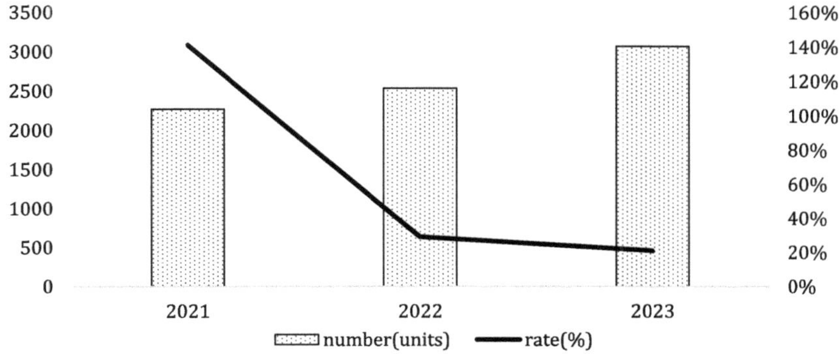

Fig. 1.2 Number of network freight platform enterprises and year-on-year growth rate over the past three years. *Data Source* http://www.chinawuliu.com.cn/lhhzq/202406/21/633767.shtml

logistics innovation model, ZCHZH has established an intelligent transport capacity trading platform and a digital-intelligent logistics platform. This enables end-to-end digital and visual management, enhancing transport efficiency while reducing logistics costs. ZCHZH also leverages its smart multimodal transport management system to deliver comprehensive customized logistics solutions, thereby boosting operational efficiency. ZYOUK specializes in the R&D, design, manufacturing, and sales of intelligent logistics systems, alongside the modification and sales of smart specialized vehicles. The company provides high-quality intelligent logistics solutions, such as the "Robotics+" indoor-outdoor integrated AI delivery solution for ZYSDW headquarters and the intelligent robotic sorting system project for WHYOU [4]. ZJTIA delivers logistics automation solutions centered on automated conveying systems, sorting, and storage systems. ZJTIA has participated in smart logistics projects across multiple industries, including a smart automated warehouse project for a major domestic chemical industrial base and a smart logistics project for a leading enterprise in the new energy sector. LAJIZ provides end-to-end smart logistics system solutions serving industries such as tobacco, large-scale retail, military, pharmaceuticals, and automotive. The company implemented projects including the Phase I MEB Final Assembly Workshop LOC Automated Storage System at FAW-Volkswagen Foshan Plant (a major manufacturing facility of the FAW-Volkswagen joint venture in Foshan, Guangdong province), demonstrating its technical prowess in logistics handling system integration. ZKWEZ delivers comprehensive solutions for intelligent conveying, sorting, and warehousing, serving industrial warehousing, e-commerce retail, aviation, and other sectors. The company has implemented projects including the integrated logistics hub for Tongda Express in Yiwu, Zhejiang, and the intelligent automated sorting system at ZTO Express's Wuxi Shuofang Airport Sorting Center, demonstrating its development capabilities in highway information investment and the smart transportation market (Table 1.1).

Table 1.1 Technological advancements and outcomes in road transport enterprises

Technological progress	Effects
Informatization & Intelligentization	Improves transportation efficiency
Autonomous Driving Technology	Enhances safety and efficiency
Vehicle–Road Collaboration	Facilitates smart transportation system development
BeiDou System Application	Improves transport accuracy and reliability
New Energy & Materials	Low-carbon and environmentally friendly
Logistics Handling System Integration	Increases efficiency, reduces costs

1.1.5 Industry Competitive Environment

China's road transport enterprises face a complex competitive landscape characterized by regional development disparities, intense inter-enterprise competition, continuous service model innovation, and gradually increasing industry concentration. In regional competition, the eastern regions of China exhibit significantly more developed road freight networks than the western regions. Taking Shandong Province as an example, its annual road freight volume exceeds 2.5 billion tons, ranking first nationally at 2.769 billion tons—a Figure illustrating regional development imbalances. At the corporate level, certain enterprises within the industry have demonstrated exceptional competitiveness. According to the 2021 Survey Report on Key Enterprises in the Road Freight Industry, among 30 major road freight companies, five achieved main business revenues exceeding 10 billion yuan. Among them, SHUFE Express, as an industry leader, delivered particularly outstanding performance. Regarding service model innovation, road freight enterprises are adapting to market demands through continuous service enhancements. For instance, SHUFE has differentiated itself from the "Three Tong and One Da" carriers in supply chain offerings by providing customized and diversified services to boost its market competitiveness [5]. Regarding industry concentration, despite the presence of numerous small and medium-sized private logistics enterprises, overall concentration remains low. Large freight companies are actively expanding into regional and niche markets, intensifying competition and signaling a shifting competitive landscape. These enterprises are leveraging scale and networked operations to increase their market share (Table 1.2).

1.1.6 International Competitive Environment

China's road transport enterprises face significant challenges and opportunities in the international competitive environment. Despite the continuous expansion of their market scale, low concentration limits their international competitiveness. In 2022,

Table 1.2 List of key enterprises in the road freight industry (2021)

Enterprise name	Main revenue (100 million yuan)	Main business
SHUFE	302.3	Less-than-Truckload, Urban Distribution
ZCHZH	276.8	Full Truckload, Less-than-Truckload
CHHZH	274.2	Full Truckload, Less-than-Truckload, Highway Terminals (Parks)
DBANK	106.8	Less-than-Truckload, Full Truckload, Urban Distribution
ANEKU	96.5	Less-than-Truckload
TIDIH	83	Less-than-Truckload, Full Truckload, Freight Brokerage, Highway Terminals (Parks)
JIHUW	75.6	Full Truckload
ZHTOK	64	Less-than-Truckload, Full Truckload, Urban Distribution, Contract Logistics, Freight Brokerage, Highway Terminals (Parks)
SAZHW	62	Less-than-Truckload

China's road freight market reached approximately RMB 5 trillion, yet its concentration level remains relatively low. The CR4 (market share of the top four enterprises) stands at only 7.78%, far below the 60% held by the top five enterprises in the United States. This low concentration reflects intense market competition and the widespread presence of small-scale operators. However, opportunities and challenges always coexist. With deepening international cooperation and the advancement of the Belt and Road Initiative, China's road transport enterprises have gained new platforms for global competition. Notably, the implementation of the TIR system is projected to boost annual import–export trade between China and major Belt and Road countries by up to $13.583 billion, indicating the rapid enhancement of China's road transport enterprises' international competitiveness [6].

1.2 Development Trends of China's Road Transport Enterprises

1.2.1 Fiscal Policy

From a fiscal policy perspective, the development trends of China's road transport enterprises exhibit a distinctly positive trajectory, reflecting strong government support and promotion. Firstly, demand in the road freight market continues to grow. By May 2023, road freight turnover reached 28,650.67 billion ton-kilometers, marking a year-on-year increase of 2.70%. It is projected that by 2029, the scale of road freight will exceed 42.8 billion tons. This trend is driven by the government's increasing fiscal support for the transportation sector. During the 13th Five-Year Plan period (2016–2020), total fiscal investment at all levels nationwide reached 7.5 trillion yuan, leveraging over 16 trillion yuan in fixed-asset investment and significantly advancing the construction and upgrading of transportation infrastructure [7]. Backed by fiscal policies, the government also implemented a series of tax and fee reduction measures, further optimizing the operating environment for road transport enterprises. For instance, the value-added tax rate for the transportation sector was lowered from 11% to 9%, and the scope of input tax credits was expanded, effectively reducing enterprises' logistics and operational cost burdens. Furthermore, national fiscal policies actively support the development of new energy technologies. To promote the adoption of new energy vehicles in road transportation, central and local governments jointly bear the subsidy funds for new energy urban buses and power battery upgrades at a 90:10 ratio. Particularly in national ecological civilization pilot zones and key air pollution prevention areas, the proportion of new energy vehicles in public transport, taxis, and logistics distribution has significantly increased. Highway service areas in 26 provinces (autonomous regions and municipalities) now offer charging services, cumulatively saving approximately 6 million tons of fuel over five years and advancing the green development of the road transport industry [8]. Fiscal policies have also played a pivotal role in driving innovation. Through the implementation of the Transportation Industry Science and Technology Innovation Talent Promotion Program, the state has selected and funded leading scientific innovators and innovation teams, providing robust technical support to road transport enterprises. With the continuous advancement of digital and intelligent technologies, the application of new technologies such as smart logistics and green transportation will become key means for road transport enterprises to enhance their competitiveness in the future. The Highway "14th Five-Year" Development Plan (2021–2025) further clarifies development goals and tasks for the next five years. The government will continue to support sustained investment in the transportation sector through multi-channel funding arrangements, ensuring the industry's steady development.

1.2.2 Enterprise Ecosystem

The corporate ecosystem of China's road transport enterprises can be analyzed across multiple dimensions, including core enterprises, supporting enterprises and service providers, upstream and downstream enterprises, innovation partners, markets and customers, as well as government and regulatory bodies. Core enterprises in road transport are typically large logistics companies or transportation firms that dominate the industry chain, determining the direction and development of the entire ecosystem. In recent years, these enterprises have progressively established comprehensive business development systems. By closely tracking market trends and customer demands, they prepare for market expansion to enhance competitiveness and capture market share. Take SHUFE, China's largest integrated express logistics service provider and the world's fourth-largest player, as an example. SHUFE

Holdings' operations span multiple sectors including time-definite express, economy express, freight express, cold chain and pharmaceutical logistics, same-city urgent delivery, supply chain services, and international logistics. Time-definite express constitutes its primary revenue source, accounting for 39.51% [9]. Moreover, SHUFE Holding's international operations span 78 countries and regions, delivering comprehensive logistics solutions including international express delivery, cross-border e-commerce, international freight forwarding, global warehousing, and consolidation/transshipment services. Driven by policy support, market demand, and technological innovation, industry leaders like SHUFE Holding continuously consolidate and expand their leadership position within the road transportation sector.

A collaborative, win–win corporate mindset has become increasingly vital. Supporting enterprises and service providers—such as parts suppliers, technical service providers, and logistics companies—offer backing to leading enterprises, forming cooperative relationships that collectively drive the operation of the industrial ecosystem. For instance, HAWEE and its partners jointly released a smart highway non-stop overweight vehicle control solution. This solution applies 5G, AI technology, and a cloud-edge-end collaborative architecture to non-stop overweight vehicle control operations on highways, advancing the modernization of the transportation industry's governance system and capabilities [10]. Concurrently, China's transportation enterprises have innovated their partnerships to include R&D institutions, technology companies, and universities. These entities contribute innovative technologies, research outcomes, or intellectual property to advance the ecosystem's technical progress and innovation capabilities. For instance, the establishment of the BeiDou Industry Collaborative Development Platform has further accelerated the large-scale application and industrialization of the BeiDou system, marking a new phase in central enterprises' BeiDou innovation and development.

Government policies, regulations, and standards form the institutional safeguards for the industrial ecosystem. The Ministry of Transport, in particular, oversees the highway construction market, formulates relevant policies, systems, and technical standards, and monitors their implementation. Additionally, regulatory efforts must be strengthened during oversight, the business environment optimized, and bidding activities standardized. The "14th Five-Year" Development Plan for a Modern Comprehensive Transportation System (2021–2025) explicitly sets targets: expanding expressway mileage to 190,000 km and increasing the share of new energy buses in urban fleets to 72%. These objectives chart the industry's developmental course. The development trend of China's road transport enterprises is characterized by core enterprises continuously enhancing market competitiveness, supporting enterprises and service providers offering technical support and logistics services, innovation partners driving technological advancement, and government and regulatory bodies ensuring the industry's healthy development through policies and regulations.

1.2.3 Technological Advancement and Innovation

China's road transport sector is undergoing a critical transition from traditional to modern models. The integration of technologies such as Intelligent Transportation Systems (ITS), GPS tracking, and electronic logistics platforms, alongside vehicle modernization and widespread IT adoption, signals a new wave of technological transformation for road transport enterprises. Digital transformation has become central to the strategic planning of numerous transportation companies. The Ministry of Transport guides the construction of smart highways nationwide, requiring the adoption of new technologies, introduction of new equipment, and creation of new travel service models. These measures not only enhance capabilities in perception, analysis, and decision support but also achieve full-process digitalization and deep integration of people, vehicles, roads, and the environment. The Ministry of Transport places high priority on highway digital transformation, advancing smart highway construction through technologies like 5G and AI to integrate physical highways with digital twin counterparts [11]. By 2027, digital transformation in the highway sector is projected to achieve comprehensive lifecycle management, establishing a unified model and unified data system covering design, construction, maintenance, and operations. Additionally, the Ministry of Transport has issued the "Technical Guide for Highway Infrastructure in Support of Automated Driving" (a national guideline/standard issued by MOT), aiming to provide policy support and technical guidance for the development of automated driving technology.

The level of intelligence in highway transportation is also significantly improving. Mobile communication technologies, such as mobile radio communication and in-vehicle navigation systems, have been widely adopted, while vehicle-to-everything (V2X) technology continues to advance. Through in-vehicle navigation systems and V2X technology, the accuracy of vehicle positioning and real-time monitoring capabilities have been enhanced, optimizing route planning and resource allocation. Autonomous driving technology is expected to play a pivotal role in future highway transportation. The Ministry of Transport has issued guiding opinions to advance the development and application of autonomous driving technology in road transportation, further implementing the Intelligent Vehicle Innovation and Development Strategy. In constructing Intelligent Transportation Systems (ITS), these systems integrate advanced information and communication technologies to enhance the safety, efficiency, and sustainability of transportation networks, reduce traffic congestion, and optimize the driving experience. Key ITS technologies primarily encompass positioning systems, communication protocols, data processing, data sharing, and artificial intelligence [12]. The application of these technologies enables more efficient and safer transportation systems. Overall, the modernization of China's road transport sector is accelerating. Through the integration of intelligent transportation systems, GPS tracking, electronic logistics platforms, and the advancement of digital transformation, the efficiency and safety of road transport have been significantly enhanced.

1.2.4 Organization and Management

Facing new market conditions, China's road transport enterprises have enhanced service quality and operational efficiency through a series of organizational and management optimizations to adapt to and serve the demands of economic and social development. Within China's road transport sector, the trend toward logistics group development has become increasingly pronounced. Through resource integration, enterprises have not only achieved economies of scale and effectively reduced operational costs but also strengthened their competitiveness in the fiercely competitive market. Taking ZGJJJ as an example, as a leading enterprise in the industry, it has enhanced its service capabilities and significantly boosted its market competitiveness by implementing a group development strategy [13]. In 2023, ZGJJJ's operating revenue reached RMB 267.49 billion, a 29.11% increase from the previous year, while net profit attributable to shareholders of the listed company amounted to RMB 6.17 billion, marking a 44.62% year-on-year growth. This demonstrates the company's sound financial health and robust growth potential. While focusing on its core business, ZGJJJ has advanced its group-based and scaled development, further solidifying its leading position in the industry. This group-based approach has not only optimized internal resource allocation and improved operational efficiency but also enabled the company to better respond to market demands and deliver more efficient and reliable logistics services. Simultaneously, this group-based approach has generated synergistic effects in capital, technology, talent, and other areas, laying a solid foundation for the company's sustained innovation and long-term development (Fig. 1.3).

Information technology development is emerging as a key driver for the growth of China's highway transportation enterprises. Through technological innovation and policy support, the industry is advancing toward greater efficiency, intelligence,

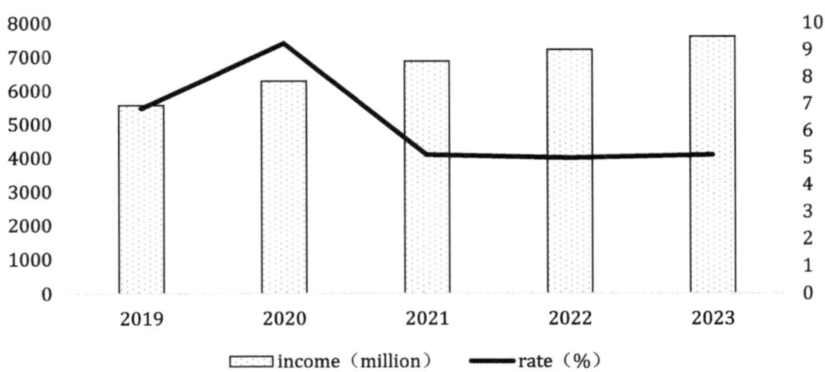

Fig. 1.3 China Communications Construction Company Limited (CCCC) revenue and year-on-year growth rate, 2019–2023. *Data Source* 2019–2021 data sourced from Shanghai Securities News; 2022–2023 data sourced from CCCC's Annual Report Summary

and collaboration. In recent years, enhanced IT infrastructure has enabled companies to optimize logistics information service platforms, effectively address information asymmetry, and improve supply chain coordination efficiency. The Ministry of Transport has issued guiding opinions on advancing the digital transformation of highways, outlining two phased objectives: By 2027, significant progress will be made in highway digital transformation, establishing a "single model, single data set" for full-lifecycle digitalization and achieving end-to-end digital transformation. By 2035, comprehensive digital transformation of highways will be fully realized. In implementing information infrastructure, highway transportation enterprises have actively adopted advanced data collection methods such as drone-based LiDAR surveying, oblique photography, high-resolution remote sensing, and BeiDou positioning. They utilize BIM + GIS technology to integrate data management and optimize surveying and mapping processes. Simultaneously, design units are encouraged to establish BIM-based forward design workflows and collaborative design platforms. This enables 3D collaborative design, automated bill of quantities generation, parametric design, and 3D simulation analysis for complex projects. Through refined, intelligent design, efficiency is enhanced and project costs are reduced. In smart highway construction, enterprises like ZGJJJ have achieved economies of scale through resource integration, boosting service capabilities and market competitiveness. ZGJJJ operates over 60 wholly-owned subsidiaries, conducts business across 157 countries and regions, and employs 168,060 staff. Policy support also provides a solid foundation for information infrastructure development. The "Long-term Development Plan Outline for Scientific and Technological Innovation in the Transportation Sector (2021–2035)" underscores the critical role of technological innovation in transportation. It proposes measures to refine innovation mechanisms, cultivate high-caliber scientific talent, and accelerate the commercialization of research outcomes.

China's road transport enterprises have made positive strides in forming alliances, sharing resources, and advancing multimodal transport, laying a solid foundation for building an efficient, green, and intelligent modern comprehensive transportation system. By forming large-scale alliances with other transport modes (such as sea, land, and air), road transport enterprises achieve resource sharing and complementary advantages, enhancing overall transport capabilities. This trend of industry alliances is explicitly reflected in the "14th Five-Year" Development Plan for a Modern Comprehensive Transportation System (2021–2025). The plan underscores the importance of multimodal transport and sets the goal of significantly enhancing its development level by 2025. Nationwide rail and waterway freight volumes are projected to increase by approximately 10% and 12% respectively compared to 2020, with container rail-water intermodal transport growing at an annual rate exceeding 15%. As a globally leading mega-scale infrastructure integrated service provider, ZGJJJ operates over 60 wholly-owned and controlled subsidiaries, maintains 289 overseas offices, and conducts business across 157 countries and regions with a workforce of 168,060 employees. This demonstrates how ZGJJJ has developed robust comprehensive transportation service capabilities worldwide through group-based, large-scale development. Concurrently, the National Railway Administration,

Ministry of Industry and Information Technology, and China State Railway Group have jointly introduced measures supporting rail transport of new energy passenger vehicles to bolster the industry's development. These policies clarify railway support for new energy vehicles and standardize rail transport services, representing a crucial step in advancing multimodal transport. At the local level, provinces like Hubei subsidize river-sea combined transport routes and refine constraints such as multimodal service regulations and technical standards. These efforts enhance connectivity between transport modes and address challenges within industry alliances.

China's road transport enterprises have achieved comprehensive and in-depth development in institutional innovation. Internally, companies have optimized processes and enhanced efficiency by adopting advanced management models like lean management and Six Sigma. Employee training and development have also been prioritized, with systematic training programs significantly improving professional skills and overall competence. Additionally, enterprises actively explore advanced management approaches such as supply chain management and customer relationship management to better adapt to market demands and elevate service quality. Technological innovation and R&D investment constitute another vital component of institutional innovation. Research and development in logistics informatization and intelligentization, in particular, have provided impetus for technological advancement and industrial upgrading. Concurrently, in response to national green development initiatives, enterprises have implemented energy-saving and emission-reduction measures in areas such as vehicle renewal and transport organization. These include promoting new energy vehicles and optimizing transport routes. Regarding safety production management, enterprises have elevated safety standards by establishing and refining safety responsibility systems and strengthening safety education and training. Finally, supportive policies from national and local governments—such as tax incentives and financial subsidies—have created favorable external conditions for enterprise development. The implementation of these institutional innovations has not only enhanced corporate core competitiveness but also laid a solid foundation for the industry's sustainable growth. Logistics group development has enabled resource integration and economies of scale; informatization initiatives have driven the industry's digital transformation; industry alliances have facilitated resource sharing and multimodal transport; while institutional innovation provides enterprises with sustained momentum and safeguards for continuous innovation. Collectively, these measures have reshaped the landscape of China's road transport enterprises, laying a solid foundation for the industry's long-term development and enhanced international competitiveness. Looking ahead, as technological advancements and market demands continue to evolve, China's road transport enterprises will persist in seeking breakthroughs through innovation and forging new paths through transformation.

1.3 Challenges Facing China's Road Transport Enterprises

1.3.1 Imbalance Between Transport Capacity and Demand

A pronounced supply–demand contradiction exists in China's road passenger and freight markets, with transport capacity growth outpacing cargo volume growth. Since 2017, China's total freight volume has shown fluctuating upward trends, yet capacity expansion has exceeded cargo growth. From 2021 to 2023, the average annual growth rate of drivers reached approximately 20%, while the corresponding period saw China's total freight volume increase by only 5% and road freight volume grow by a mere 3%. Data from RUWOJ Yiliu indicates that freight volume from January to June 2023 remained largely unchanged from the previous year but showed a significant decline compared to the year before last. The mismatch between transport capacity growth and cargo demand has created an oversupply of transport capacity, putting pressure on driver incomes and increasing the survival pressure on truck drivers. Among them, MABAJ provides a large volume of freight services through its platform. However, market feedback indicates that drivers face significant survival pressure due to the situation of many trucks but few cargoes. Moreover, the low actual loading rates on some routes have led to poor economic efficiency, affecting the overall profitability of transportation enterprises. Despite improved transportation efficiency, driver compensation has become the primary cost in road freight, accounting for 24.4% of total expenses. Combined with structural shifts in cost composition due to toll-free highway policies and reduced fuel costs, transportation enterprises are facing increased operational expenses. Shanghai Zeyi Supply Chain Management Co., Ltd. notes that while logistics companies' fulfillment costs are rising, cargo owners persistently suppress logistics fees. To compete for business volume, transport providers resort to cutthroat competition and price inversion tactics, creating disorderly market competition. These data and insights reveal that China's road transport enterprises face supply–demand imbalances in transport capacity and cargo volume, coupled with declining profit margins (Figs. 1.4 and 1.5).

1.3.2 Low Level of Informatization

China's road transport enterprises face a long journey in advancing informatization, requiring systematic improvements across infrastructure, technology adoption, information sharing, data management, talent development, and management innovation. Infrastructure for information technology remains inadequate, particularly in remote regions where development lags significantly, hindering the industry's overall digital transformation. While technologies like GPS are applied in specific areas, their adoption within transport enterprises remains limited, failing to fully leverage their potential for enhancing operational efficiency and service quality. Furthermore, information silos are prevalent. The absence of effective information-sharing mechanisms

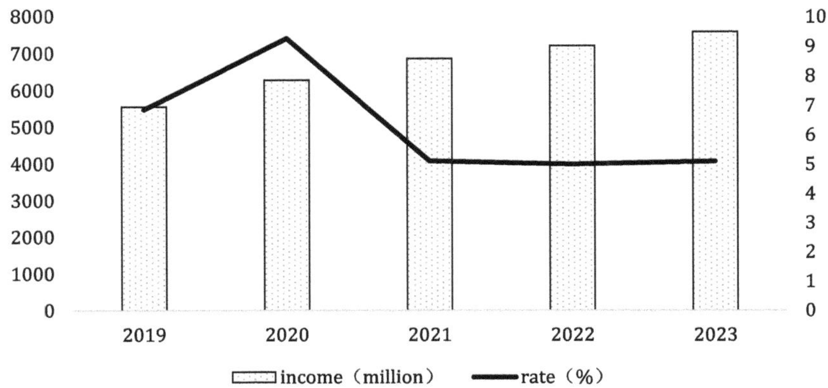

Fig. 1.4 ZGJJJ 2019–2023 operating revenue and year-on-year growth rate

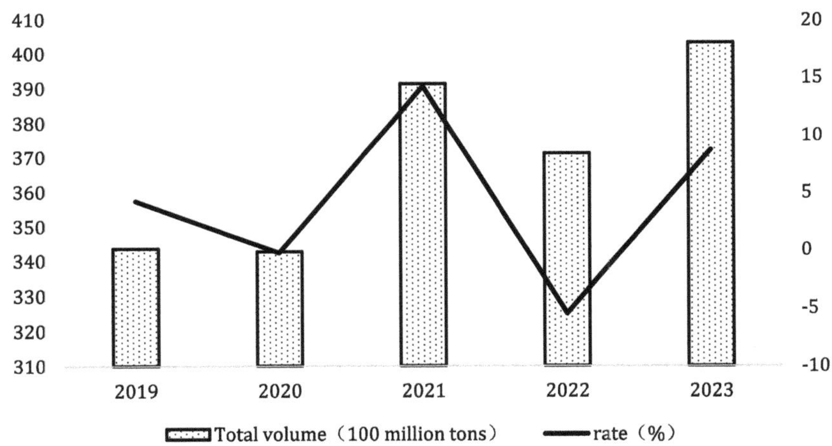

Fig. 1.5 Total road transport volume and growth rate in China, 2019–2023. *Source* Statistical Bulletin of the Ministry of Transport

hinders data flow between departments or enterprises, reducing operational efficiency and increasing costs. Concurrently, data management and utilization are inadequate. While vast amounts of transportation data hold immense value, challenges in data collection, storage, analysis, and application persist. Data quality is difficult to guarantee, and data openness requires improvement. A critical issue is the shortage of IT professionals. The development of information technology urgently requires a large number of specialized talents, but China currently lacks sufficient reserves in this area, with a particularly acute shortage of senior IT personnel. Furthermore, the management concepts and approaches of some transportation enterprises have failed to keep pace with the times and adapt to the demands of the information age.

This hinders the in-depth promotion and effective implementation of information technology initiatives.

1.3.3 Inadequate Safety Management and Regulatory Compliance

China's road transport sector still requires improvement in safety governance and regulatory compliance. These shortcomings heighten the risk of traffic accidents, necessitating urgent enhancement through concrete and effective measures. Firstly, transport personnel often lack sufficient safety awareness, partly due to inadequate professional training provided by some enterprises before employees commence work. This frequently results in violations of regulations and relevant laws and regulations. Additionally, key personnel in some enterprises do not sufficiently prioritize production safety, leading to suboptimal safety management standards. Second, deficiencies exist in the safety management system. Current safety governance materials, such as enterprise safety production regulations, remain incomplete, and many companies urgently need to enrich their safety management documentation. As the industry advances, it is essential to update concepts and incorporate new elements to achieve effective oversight of traffic safety regulations. Furthermore, enterprises often achieve poor results in safety production education and training. Some companies conduct training activities in a perfunctory manner, failing to integrate practical production processes, specific workflows, and employee circumstances to deliver effective, targeted training. Hazard identification and remediation activities also exhibit irregularities. Some enterprises fail to conduct regular inspections as stipulated by hazard management systems, neglect to accurately document identified hazards, and do not formulate practical corrective measures.

1.3.4 Significant Cost and Operational Pressures

Road transport enterprises face dual pressures from rising labor costs and declining transport demand, intensifying operational difficulties. Among the cost structure of road freight companies, tolls, fuel expenses, driver wages, and vehicle depreciation constitute the four primary costs. Fuel costs, in particular, significantly impact corporate expenses amid rising oil prices. With the rapid development of rail and civil aviation, coupled with the government's accelerated promotion of the "road-to-rail" freight shift, both freight and passenger transport volumes on highways have declined to varying degrees. This directly impacts toll revenue from expressways. Affected by the supply–demand structure of transport capacity, freight rates in the road freight market remain generally depressed. The oversupply situation is unlikely to improve in the short term, limiting the potential for freight rate increases and further squeezing corporate profit margins. Rising costs and shifts in the supply–demand structure have resulted in overall low operational efficiency for enterprises. While network-based less-than-truckload (LTL) transport and contract logistics companies demonstrate relatively stronger profitability, the overall efficiency of the sector still needs improvement.

1.4 Recommendations for the Development of China's Road Transport Enterprises

1.4.1 Optimize Transport Capacity Allocation

Conduct rational assessments and adjustments of existing transport capacity to optimize its allocation. Road transport enterprises can leverage data analysis to identify regions and routes with excess or insufficient capacity, thereby adjusting resource distribution to ensure alignment between capacity and cargo volume. Additionally, encourage flexible transport scheduling mechanisms utilizing internet platforms for real-time matching of road transport resources to optimize capacity allocation. First, establish a comprehensive freight matching platform based on big data, cloud platforms, and blockchain technologies. This platform should integrate functions such as information matching, transport capacity optimization, and route planning to create a holistic highway freight service ecosystem. Second, adopt cloud-edge collaboration technology. By coordinating cloud and edge computing, real-time data collection, processing, and storage can be achieved to meet users' real-time requirements. This reduces network transmission latency, enhances information response speed, and optimizes vehicle loading rates and route costs. Furthermore, implementing a bilateral matching decision method allows for the construction of a freight-vehicle matching indicator system tailored to real-world demands. This system considers cargo attributes, truck attributes, shipper preferences, and driver preferences, utilizing a dual-objective optimization model to improve matching efficiency. Subsequently, a comprehensive credit evaluation system is established. Using the Analytic Hierarchy Process (AHP) and Fuzzy Comprehensive Evaluation Method, credit rating models for vehicle owners and shippers are developed to screen high-quality operators and cargo sources. Collaborating with these vetted parties, a virtual fleet is formed to achieve controllable transport capacity. Finally, applying real-time multi-task logistics route optimization for individual vehicles on road networks—combining real-time Dijkstra algorithms with genetic algorithms—can provide customers with the shortest delivery time routes while accounting for the importance of time-varying road network characteristics in last-mile logistics route planning.

1.4.2 Enhancing Information Technology Levels and Strengthening Data Management and Utilization

To elevate the overall competitiveness and operational efficiency of China's road transport enterprises, a comprehensive development strategy must be implemented, encompassing infrastructure construction and IT talent cultivation. First, strengthening infrastructure is fundamental. Investment in IT infrastructure in remote areas must be increased. By improving network coverage and enhancing data processing capabilities, the digital divide between urban and rural areas can be narrowed, ensuring all regions benefit from IT advancements. Second, promoting IT application is crucial. Road transport enterprises should be encouraged and supported to adopt advanced technologies such as GPS, the Internet of Things (IoT), and big data analytics to enhance transportation efficiency and service quality. Concurrently, training and education programs should elevate employees' understanding and proficiency in these technologies to ensure their effective implementation. Furthermore, establishing information-sharing mechanisms is paramount. By constructing cross-departmental and cross-enterprise information-sharing platforms, we can resolve data silos, promote information flow, and optimize resource allocation. This not only boosts operational efficiency but also reduces costs. Moreover, strengthening data management and utilization is central to enhancing competitiveness. The quality and efficiency of data collection, storage, analysis, and application must be improved to ensure data accuracy and security. Increasing data openness promotes resource sharing and reuse, providing robust data support for corporate decision-making. Ultimately, cultivating IT talent serves as a critical foundation. Collaboration with universities and research institutions must be strengthened to cultivate a cohort of versatile professionals proficient in both technology and management. Concurrently, internal training programs should enhance existing employees' IT skills, building a specialized team capable of adapting to digital advancements. By bolstering infrastructure, promoting IT applications, establishing information-sharing mechanisms, enhancing data management and utilization, and developing IT talent, China's road transport enterprises can effectively advance their digital transformation. This will elevate service quality, reduce operational costs, and achieve sustainable development.

1.4.3 Improving the Safety Management System and Strengthening Supervision Mechanisms

To comprehensively elevate the safety management standards of China's road transport enterprises, the first step is to refine the safety management system at the institutional level. Enterprises must establish and improve safety production regulations, covering critical aspects such as vehicle maintenance, driver management, and accident reporting and handling procedures. Simultaneously, a comprehensive safety management system should be implemented, spanning all levels from senior management to frontline staff, ensuring every employee clearly understands their safety responsibilities. This creates an impenetrable safety assurance network. Next, strengthening the enforcement of regulations is crucial for ensuring the effective implementation of these rules. Enterprises must strictly adhere to all national laws and regulations governing road transportation, covering not only driver qualifications and vehicle technical conditions but also aspects like route planning. For individuals or departments violating these regulations, enterprises should implement strict disciplinary measures to enhance overall compliance awareness and execution. Furthermore, to further improve the objectivity and effectiveness of safety management, enterprises may consider introducing third-party assessment and supervision mechanisms. By engaging third-party intermediaries, such as safety assessment agencies or professional consulting firms, to conduct regular audits and evaluations of the enterprise's safety management system, potential safety hazards can be identified. This also promotes continuous improvement and optimization of the enterprise's safety management. Through this series of coherent measures, China's road transport enterprises will be able to establish a more robust safety defense, effectively prevent and reduce traffic accidents, and provide a solid foundation for sustainable enterprise development.

1.4.4 Optimizing Cost Structures and Strengthening Industry Collaboration

The cost and operational pressures faced by China's road transport enterprises necessitate effective measures to optimize cost structures. Enterprises must achieve cost reductions through technological upgrades and management innovation. This includes adopting high-efficiency vehicles and equipment, implementing energy management systems to reduce fuel consumption, and leveraging centralized procurement and economies of scale to decrease fixed costs such as toll fees and driver wages. The government's role remains equally crucial in this context. It should continue providing policy support such as tax incentives and financial assistance to help road transport enterprises navigate challenges. Furthermore, intra-industry collaboration must be strengthened by encouraging partnerships with other transport modes like railways and civil aviation. Through resource sharing and leveraging complementary strengths, more efficient transport networks can be formed. The collective implementation of these measures aims to enhance the overall competitiveness and market adaptability of road transport enterprises, thereby achieving the industry's healthy and sustainable development.

References

1. Compilation of the 12th Five-Year Plan for Transportation Development. Ministry of Transport of the People's Republic of China (Ed.) [M]. Beijing: People's Transportation Press, 2012.
2. Feng Haiyu. Research on the Impact of High-Speed Rail on the Quality of Urbanization in Small and Medium-Sized Cities in China [D]. Master's Thesis, Beijing Jiaotong University, 2021.
3. Jing Linbo, Wang Xuefeng. Current Situation, Problems, and Countermeasures in the Development of China's Circulation Industry [J]. China Business Economics, 2012.
4. Zhang Yun. Empirical Study on Infrastructure Construction and Economic Growth in Gansu Province [D]. Master's Thesis, Northwest Normal University, 2013.
5. Zhang Tong. Mechanism and Path Selection of Logistics Services Promoting Manufacturing Upgrading under the New Normal [J]. Commercial Economic Research, 2015.
6. Zheng Han. Risk Analysis and Development Strategies of Personal Wealth Management Services in Commercial Banks [D]. Master's Thesis, Zhejiang University, 2014.
7. Yang Shu. Progress Made in the Belt and Road Initiative [N]. International Business Daily, 2017.
8. Fulfilling Duties According to Law to Protect the "Money Bag" and Safeguard State-Owned Assets – National People's Congress of China website.
9. Gong Min. Evolution of the Green Technology Innovation Ecosystem from a Multi-Agent Perspective [D]. Doctoral Dissertation, Nanchang University, 2022.
10. Zhao Boyu. The Impact of Resource Integration on Strategic Performance: A Case Study from the Perspective of Diversification Strategy [D]. Master's Thesis, Inner Mongolia University, 2022.
11. International Express Press Conference.
12. Huawei and Partners Jointly Launch the Smart Highway Overload Detection 1.0 Solution – Huawei China Government & Enterprise Business.
13. Wang Tao (Ed.). Proceedings of Economics and Management 2015 [M]. Beijing: University of International Business and Economics Press, 2015.

Chapter 2
Railway Transport Enterprises: Operational Realities, Strategic Challenges, and Future Pathways

Jianlin Wang and Xin Lu

Abstract As a vital pillar of the national economy, railway transportation holds a core position in China's modern transportation system. This paper first explores the development background of the railway transportation industry, then provides an in-depth analysis of the current status, challenges, response strategies, and future development directions of railway transportation enterprises. Through case studies of companies such as ZGTIE, DQITE, and JHGTE, it illustrates their specific development situations. Finally, future prospects are proposed. Railway transportation enterprises will continue to advance in technological innovation, green development, and service optimization to consolidate their central role. Only by fully understanding the development and existing problems of railway transportation and carefully studying relevant policies can China's railway transportation industry achieve sustainable development.

2.1 Introduction

2.1.1 Background of Railway Transportation Development

As a crucial part of China's transportation infrastructure, railways play a significant role in driving economic development, promoting industrial upgrading, and reducing logistics costs. With advantages such as long-distance capability, large capacity, low cost, high efficiency, and environmental friendliness, railways have established a backbone position in China's comprehensive transportation network and have profoundly impacted sustainable economic and social development. From a management perspective, China's railway system can be divided into four major categories, all under the supervision of the National Railway Administration.

J. Wang · X. Lu (✉)
Anhui University of Finance and Economics, Bengbu, China
e-mail: luxinn1224@163.com

J. Wang and M. Song (eds.), *The Green Path of Chinese Transportation Enterprises*, Resources, Climate and Sustainable Development,
https://doi.org/10.1007/978-981-95-5289-4_2

In 1949, the Ministry of Railways of China was officially established. In 1953, the Chinese government proposed and implemented the First Five-Year Plan, initiating the construction of major railway lines in the southwest and northwest. By 1980, the skeleton of China's railway network was largely formed. Against the backdrop of reform and opening-up, China's railway industry became increasingly market-oriented, significantly accelerating the modernization process. The development of railways can be summarized into four periods: the Founding Period, the Slow Development Period, the Reform and Opening-Up New Development Period, and the High-Quality Development Period [1], as detailed in Table 2.1.

As a key component of the modern transportation system, the development of railway transportation has always been closely monitored and highly valued by relevant national government departments. In recent years, the state has successively formulated and implemented a series of supportive policies and regulatory measures, creating a favorable development environment for the railway transportation industry. For example, the *14th Five-Year Plan for the Development of a Modern Comprehensive Transportation System* issued by the State Council in 2021 proposed building a high-quality comprehensive three-dimensional transportation network, deepening supply-side structural reforms, and scientifically advancing railway planning and construction. Some specific policies are shown in Table 2.2.

Table 2.1 Periods of Chinese railway development

Time	Events
Founding Period (1876–1893)	China's first self-constructed railway—the Beijing-Zhangjiakou Railway
Slow Development Period (1894–1978)	After the Qing government's defeat in the First Sino-Japanese War in 1894, the Eight-Nation Alliance designed and built a number of railways The Nanjing Nationalist government built railways through "joint ventures" between bureaucratic comprador capital and imperialist monopoly capital
Reform & Opening-Up Period (1979–2013)	The State Council executive meeting discussed and approved the Medium and Long-Term Railway Network Plan, clarifying long-term construction goals for China's railway network
High-Quality Development (2013–present)	The State Council institutional reform plan split the railway system, separating government and enterprise functions. The National Railway Administration was established under the Ministry of Transport, and China Railway Corporation was formed to assume the enterprise duties of the former Ministry of Railways

Table 2.2 Selected recent policies related to the railway transportation industry

Release Year	Issuing Authority	Document Name	Key Interpretation
2024	National Railway Administration, etc.	《Implementation Plan for Promoting Low-Carbon Development in the Railway Industry》	Proposes 12 specific tasks across three major areas to accelerate the low-carbon transition of the railway industry
2023	Ministry of Transport	《Five-Year Action Plan for Accelerating the Building of a Leading Transport Nation (2023–2027)》	Defines the thoughts, goals, and action tasks for building a leading transport nation over the next five years and constructs a framework of ten major actions
2022	State Council	《Outline of the Strategy for Expanding Domestic Demand (2022–2035)》	Optimizes the construction of five major coal supply guarantee bases and improves coal railway transportation capacity
2022	Ministry of Transport	《Comprehensive Transportation Standard System (2022)》	Aims to establish a comprehensive, well-structured, and cohesive comprehensive transportation standard system by 2025
2021	State Council	《14th Five-Year Plan for the Development of a Modern Comprehensive Transportation System》	Build a high-quality comprehensive three-dimensional transport network, deepen supply-side structural reforms, and scientifically advance railway planning and construction
2021	State Council	《Guiding Opinions on Further Improving Railway Planning and Construction Work》	Refines the strategic goals for railway construction, clarifying the general principle of seeking progress while maintaining stability
2020	State Council	《Outline of the New Practice for Building a Leading Transport Nation with Railways Pioneering》	Clarifies the “two-step” strategic goal for railway construction over the next 30 years and defines key focuses from 9 aspects
2019	State Council	《Outline for Building a Leading Transport Nation》	Emphasizes the importance of optimizing the transportation structure and enhancing overall efficiency and service quality

2.1.2 *Current Development Status of Railway Transportation*

2.1.2.1 Development of the Railway Transportation Industry

1. Transportation Volume

In recent years, with the rapid development of high-speed rail in China, its comfort and speed have made it the preferred mode of travel for many, promoting a continuous increase in railway passenger traffic. In terms of passenger transport, due to the impact of the COVID-19 pandemic in 2020, China's railway passenger volume dropped sharply. However, as the pandemic was brought under control, passenger volume gradually recovered. By 2023, national railway passenger traffic reached 3.855 billion person-trips, an increase of 2.182 billion person-trips or 130.40% compared to the previous year. Passenger turnover showed a similar trend, declining significantly in 2020 due to the pandemic before starting to recover. In 2023, national railway passenger turnover reached 1,472.936 billion passenger-kilometers, an increase of 815.183 billion passenger-kilometers or 123.90% year-on-year. Meanwhile, freight volume also continued to grow. Specifically, in freight transport, the total national railway freight volume reached 5.035 billion tons in 2023, an increase of 51 million tons or 1.00% compared to the previous year. Freight turnover also sustained growth, reaching 3,646.039 billion ton-kilometers, an increase of 51.47 billion ton-kilometers or 1.40% year-on-year (Table 2.3; Figs. 2.1 and 2.2).

2. Operating Mileage

Due to China's continuous increased investment in the railway transportation industry, especially in high-speed rail construction, the total operating mileage and network density of railways have been steadily increasing, laying a solid foundation for the industry's development. According to the 2024 National Railway Supervision and Management Conference, by the end of 2023, the national railway operating mileage reached 159,000 km, an increase of 4000 km year-on-year. High-speed rail reached 45,000 km, an increase of 3000 km year-on-year. Local railway operating mileage reached 24,000 km, of which self-operated mileage was 11,700 km. There

Table 2.3 National railway transportation volume

	2018	2019	2020	2021	2022	2023
Passenger traffic (100 mn persons)	33.75	36.60	22.04	26.12	16.73	38.55
Passenger turnover (100 mn pkm)	14,146.58	14,706.64	8266.19	9567.81	6577.53	14,729.36
Total freight volume (100 mn tons)	40.93	44.12	45.52	47.74	49.84	50.35
Total freight turnover (100 mn tkm)	28,927.84	30,217.36	30,514.46	33,238.00	35,945.69	36,460.39

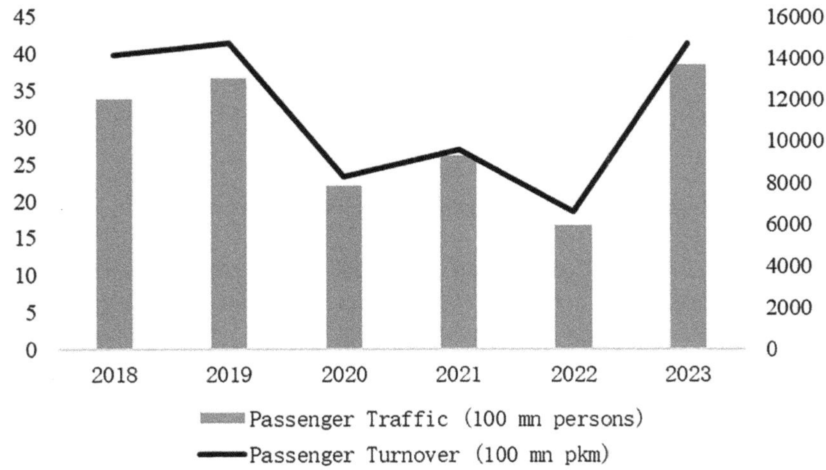

Fig. 2.1 National railway passenger transport situation

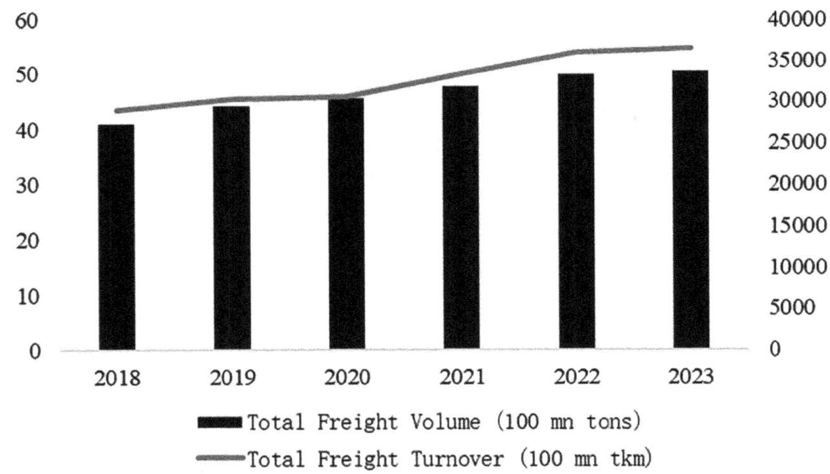

Fig. 2.2 National railway freight transport situation. *Data source* 2023 Railway Statistical Bulletin

were over 9000 dedicated railways and railway lines. Both the scale and quality of the network have significantly improved. By the end of 2023, there were 79 licensed railway transportation enterprises [2].

3. Transportation Equipment

By the end of 2023, according to data on railway transportation equipment, China had 22,400 railway locomotives, an increase of 200 from the previous year-end, including 7800 diesel locomotives and 14,600 electric locomotives. The number of railway passenger coaches nationwide was 78,000, an increase of 1000 from the previous year-end, including 4427 standard sets (35,416 vehicles) of EMUs, an increase of 209 standard sets (1674 vehicles). Additionally, the number of railway freight cars nationwide was 1.007 million, an increase of 11,000.

2.1.2.2 Development of Railway Transportation Enterprises

Supported by national policies and driven by the market economy, railway transportation enterprises have achieved significant accomplishments in various aspects. Economically, many enterprises have seen steady revenue growth, benefiting from the continuous improvement of the railway network and increased transportation demand. For example, ZGTIE achieved total operating revenue of 579.4 billion yuan in the first half of 2024, with a net profit of 1.7 billion yuan, demonstrating sound operational performance. In terms of transportation, railway enterprises have also achieved remarkable results. In the first eight months of 2024, national railways completed a freight volume of 3.378 billion tons, a year-on-year increase of 1.7%, and transported 2.996 billion passengers, a year-on-year increase of 14.9%. With continuous technological progress and optimized transportation organization, the efficiency of railway transportation enterprises has been constantly improving. ZGTIE, for instance, has effectively enhanced transportation efficiency by optimizing train schedules and increasing operating speeds. In terms of service, railway enterprises are placing greater emphasis on improving service quality. Measures such as optimizing ticket purchasing processes and improving onboard environments have increased satisfaction among passengers and shippers. Additionally, strengthened cooperation with local governments and enterprises has promoted the construction and improvement of infrastructure like railway logistics parks and freight yards.

2.2 Representative Railway Transportation Enterprises

The structure of the railway industry chain can be divided into upstream, midstream, and downstream sectors, assuming different responsibilities each other. The upstream sector focuses on the construction of railway infrastructure and equipment manufacturing. This can be further subdivided into project engineering and equipment manufacturing. Project engineering includes preliminary work for railway construction, specifically survey, design, and construction. Equipment manufacturing focuses on the production of railway vehicles and specialized equipment, including locomotives, carriages, and related technical devices. The midstream sector is core to the railway transportation link, subdivided into railway freight transport and passenger transport. Freight transport involves the railway transportation of various goods, connecting the whole country through an efficient logistics network. Enterprises like DQITE and ZTITH hold important positions in this field. Passenger transport is dedicated to providing safe and convenient railway passenger services. For example, companies like JHGTE and GSTEU have become the preferred choice for many travelers due to their extensive railway network and high-quality service. The downstream sector focuses more on the operation and maintenance of railways, involving the daily operation, upkeep, and emergency repair of railway lines to ensure the safe and stable operation of the railway system [3].

2.2.1 Representative Enterprise 1

The leader in China's railway sector—ZGTIE—is a state-owned sole proprietorship enterprise established in accordance with the *Company Law of the People's Republic of China* with the approval of the State Council and directly managed by the central government. Its core business covers railway passenger and freight transport, and it implements a diversified operation strategy. Additionally, ZGTIE is responsible for the unified dispatch and command of railway transport, the overall allocation of national transport capacity resources, undertaking state-designated public welfare transport tasks, and managing the settlement of transport revenue and funds within the railway industry.

2.2.1.1 Transportation Aspect

1. Passenger Transport. ZGTIE's passenger traffic reached 3.685 billion person-trips, an increase of 2.075 billion person-trips compared to the previous year, representing significant growth of 128.8%. Passenger turnover also reached 1,471.712 billion passenger-kilometers, an increase of 814.536 billion passenger-kilometers year-on-year, a sharp increase of 123.9%.
2. Freight Transport. ZGTIE's total freight volume reached 3.911 billion tons, a slight increase of 8 million tons compared to the previous year, representing a minor growth of 0.2%. Among this, container shipment volume achieved an annual growth of 7.3%. The total freight turnover remained at 3,263.850 billion ton-kilometers, roughly the same as the previous year (Table 2.4; Figs. 2.3 and 2.4).

2.2.1.2 Revenue Situation

In 2023, ZGTIE achieved operating revenue of 1245.4 billion yuan, a year-on-year increase of 10.5%, and a net profit of 3.3 billion yuan. National railway fixed asset investment reached 764.5 billion yuan, with 3637 km of new lines added, including 2776 km of high-speed railway. The total national railway operating mileage was 159,000 km, of which high-speed rail accounted for 45,000 km. The national railway network density was 165.0 km per 10,000 square kilometers, the double-tracking rate was 60.3%, and the electrification rate was 75.2%. The operating mileage of railways

Table 2.4 ZGTIE transportation situation

	2018	2019	2020	2021	2022	2023
Passenger traffic (100 mn persons)	33.17	35.79	21.67	25.33	16.10	36.85
Passenger turnover (100 mn pkm)	14,063.99	14,529.55	8258.10	9599.09	6571.76	14,717.12
Total Freight volume (100 mn tons)	31.91	34.40	35.81	37.26	39.03	39.11
Total freight Turnover (100 mn tkm)	25,800.96	27,009.56	27,397.83	29,950.01	32,668.36	32,638.50

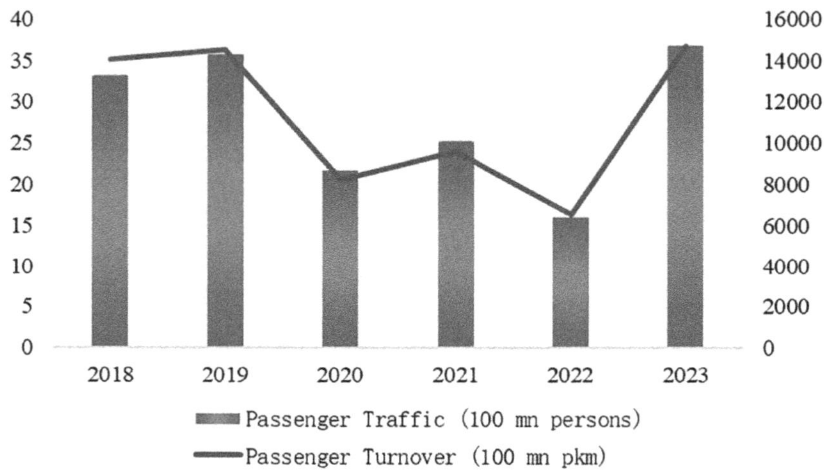

Fig. 2.3 ZGTIE passenger transport situation

in the western region was 64,000 km. The operating mileage of national railways was 133,000 km, with a double-tracking rate of 62.5% and an electrification rate of 76.7%.

2.2.1.3 Development Advantages

1. Strong Railway Construction Capability

According to publicly released information, ZGTIE implements the Party Central Committee's decisions on building a modern infrastructure system and promotes railway construction with high quality and efficiency. By the end of 2023, the national railway operating mileage reached 159,000 km, including 45,000 km of high-speed rail. These figures fully demonstrate ZGTIE's strong capability and significant achievements in railway construction.

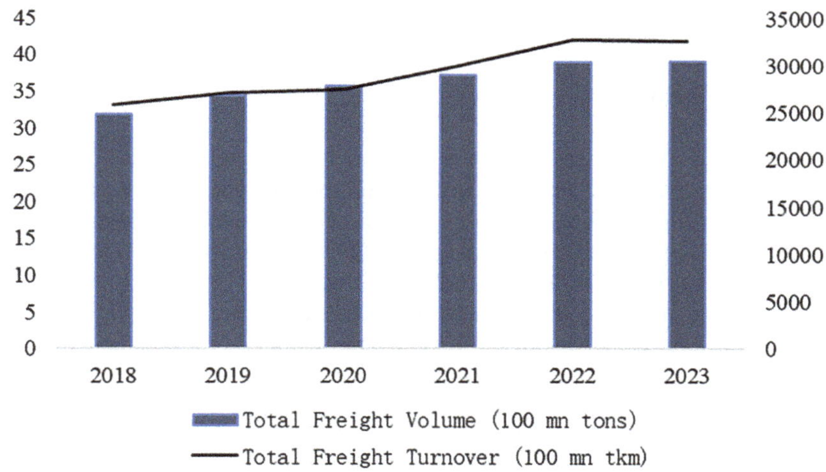

Fig. 2.4 ZGTIE freight transport situation. *Data source* Company annual reports, 2023 Railway Statistical Bulletin

2. Significant Transport Capacity

In passenger services, ZGTIE fully utilizes the operational advantages of the high-speed rail network, optimizes train schedules, and enhances peak-hour passenger capacity. In 2023, ZGTIE successfully transported 3.68 billion passengers, with the single-day peak passenger traffic exceeding 20 million and the average daily passenger traffic surpassing 10 million, both setting new historical records. In freight transport, ZGTIE ensures the smooth transportation of key materials such as thermal coal, grain, and fertilizers, increases port collection and distribution and "highway-to-railway" transport efforts, and actively promotes the development of rail-water intermodal and total logistics contract services. The annual freight volume completed in 2023 reached 3.91 billion tons, a record high. These data indicate that ZGTIE has achieved remarkable results in enhancing transport capacity, providing strong support for economic and social development.

3. Technological Innovation

ZGTIE took the lead in establishing the Railway Technology Innovation Alliance, promoting deep integration of industry, academia, research, and application. It compiled and issued the *Digital Railway Plan* and released and implemented 122 important technical standards, including those for the Fuxing Hao CR400 EMU series. Furthermore, ZGTIE has achieved impressive innovations in smart high-speed rail technology. For example, the CR450 technology innovation project made critical progress, with high-speed trains reaching 453 km/h on open track and 891 km/h during meeting maneuvers. These technological innovations not only enhance the safety and efficiency of railway operations but also earn China's railways more influence and competitiveness in the international market.

2.2.2 *Representative Enterprise 2*

2.2.2.1 Development Overview

DQITE is a regional, comprehensive railway transportation enterprise focused on the transport of goods such as coal, coke, and non-metallic ores, as well as passenger transport [4]. It is directly controlled by ZGTIE and serves as an important vehicle for ZGTIE's coal and other freight transport. DQITE is characterized by heavy-haul transport, and its freight volume has long been among the highest in the country, ensuring national energy security. The railway is 653 km long, connecting 37 stations, and manages several railway lines including the Daqin and Beitongpu lines. Its network runs north–south through Shanxi and spans two provinces and two municipalities: Shanxi, Hebei, Beijing, and Tianjin, with an operating mileage of 3149 km. Its operational area covers provinces and cities like Shanxi, Shaanxi, and Inner Mongolia, playing a crucial role in China's "Coal Transport from West to East" and "Coal Transport from North to South" energy transportation systems. As the world's largest producer and consumer of coal, ensuring stable coal transportation is vital for economic operation. Therefore, DQITE holds an irreplaceable position in the national energy strategy and economic and social development.

2.2.2.2 Revenue Situation

In terms of operating revenue, DQITE achieved revenue of 81.02 billion yuan in 2023, a year-on-year increase of 6.95%. Its freight volume reached 726.13 million tons, accounting for 14.5% of the national railway freight volume. Coal constitutes the majority of its freight, with 620 million tons transported, accounting for 85% of the enterprise's total freight volume and 22.5% of the national railway coal transport volume. In 2023, DQITE's transport revenue totaled 79.442 billion yuan, accounting for 98.05% of its operating revenue. Freight business was the largest source of income, with revenue of 61.238 billion yuan in 2023, accounting for 75.58% of operating revenue; passenger transport revenue was 8.919 billion yuan, accounting for 11.01% of operating revenue [4] (Fig. 2.5).

2.2.2.3 Development Advantages

1. Driving Economic Development Along the Railway Line

According to statistics, with the continuous growth of DQITE's transport volume, the circulation efficiency of raw materials and finished products in areas along its line has significantly improved, providing solid logistics support for industrial development. This logistical convenience not only significantly reduces corporate transport costs but also speeds up cargo turnover, thereby enhancing overall economic benefits. Compared to other railway transport enterprises, DQITE's contribution to promoting

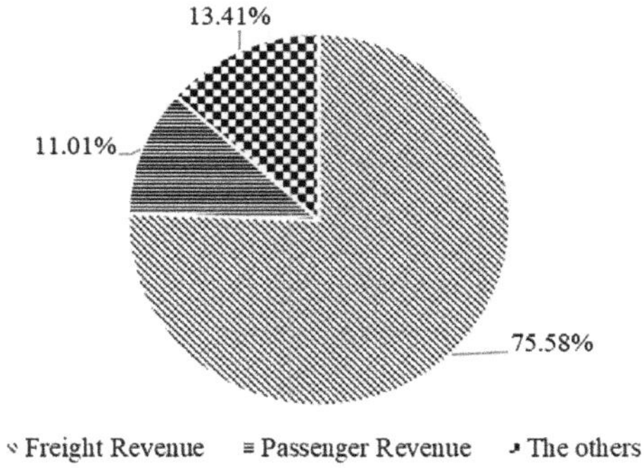

Fig. 2.5 DQITE revenue structure. *Data source* Qianzhan Network

local economic development is more prominent, with the economic vitality and growth rate of areas along its line leading the way.

2. Freight Efficiency and Cost Control

Compared to other forms of transport, DQITE's unit volume is not only larger but also more environmentally friendly. Against the backdrop of growing global awareness of green environmental protection, DQITE's green transport characteristics are particularly outstanding, becoming another significant advantage compared to other railway transport enterprises [5]. In 2024, its daily transport volume broke the 1 million ton mark. This record not only demonstrates China's railway transport efficiency and technological level but also signifies that China's technological accumulation and innovation capability in the field of heavy-haul railway transport have reached internationally leading levels.

2.2.3 Representative Enterprise 3

JHGTE is also an important part of ZGTIE's passenger transport and logistics network, renowned for its efficient and convenient passenger transport services. JHGTE not only has the longest line and the highest technical standards but is also the construction project with the largest single investment scale since the founding of the People's Republic of China. The total length of the JHGTE line is 1318 km, including 1060 km of mainline bridges, 16 km of tunnels, and 242 km of roadbed. The route passes through four provinces and three municipalities: Beijing, Tianjin, Hebei, Shandong, Anhui, Jiangsu, and Shanghai, with 24 stations along the way, including major transport hubs such as Beijing South, Tianjin West, Nanjing South, and Shanghai Hongqiao stations. JHGTE not only connects the two major economic zones of the Bohai Rim and the Yangtze River Delta but also links Northeast China, North China, and East China, becoming a vital artery for promoting regional economic development [6].

2.2.3.1 Transportation Situation

JHGTE's core business covers two main aspects: providing passenger transport services for its own line trains and providing necessary network support for cross-line trains. In 2023, the railway industry overall showed significant recovery, specifically seen in the dual substantial growth of passenger traffic and turnover. JHGTE's passenger traffic growth rate far exceeded the industry average. In 2023, national railways completed passenger traffic of 3.68 billion person-trips, achieving significant growth compared to the previous year. During the same period, JHGTE's own line trains transported 53.252 million passengers, a growth rate of 209.1%; cross-line train running kilometers accumulated to 92.044 million train-kilometers, a year-on-year increase of 66.2%, effectively promoting regional connectivity and economic development.

2.2.3.2 Revenue Situation

The two main sources of JHGTE's revenue are passenger transport business and network services. Passenger transport business involves providing high-speed rail travel services for passengers on JHGTE-operated trains and charging ticket fees accordingly. Network services refer to charging other railway companies' trains for using infrastructure such as track access and catenary usage when they operate on the high-speed rail lines managed by JHGTE and its subsidiaries. Since 2016, JHGTE's performance has continued to grow, reaching a peak in 2019 with operating revenue of 32.942 billion yuan and a net profit of 11.94 billion yuan. Affected by the COVID-19 pandemic, which reduced railway passenger flow, JHGTE's performance declined for the first time in the first quarter of 2020. In 2022, it was severely hit, even recording its first net loss of 1.552 billion yuan. However, in 2023, it achieved operating revenue of 40.683 billion yuan, a year-on-year increase of 110.40%, marking the first time JHGTE's revenue exceeded 40 billion yuan. In terms of net profit, JHGTE profited 11.546 billion yuan in 2023, a dramatic increase of 2103.63% compared to the net loss of 576 million yuan in the same period last year, achieving a turnaround to profitability year-on-year. JHGTE's main railway transport business revenue was 40.305 billion yuan, a year-on-year increase of 112.45%; the gross profit margin was 45.29%, a significant increase of 33.41 percentage points year-on-year [7].

2.2.3.3 Situation After Acquiring JFUAH

In 2019, JHGTE used funds raised from its listing to acquire a 65% equity stake in JFUAH. JFUAH owns four operational lines, all high-speed passenger dedicated lines with a design speed of 350 km/h, consistent with JHGTE's technical standards. However, at the time of acquisition, JFUAH was still operating at a loss. In the first three quarters of 2018 and 2019, JFUAH reported net losses of 1.2 billion yuan and 884 million yuan, respectively. Due to the impact of the COVID-19 pandemic in

2022, the actual loss reached 2.795 billion yuan. However, after 2023, passenger flow gradually recovered. JFUAH achieved revenue of 5.468 billion yuan, with a net loss of 968 million yuan, a reduction of 65.35% year-on-year. Train running kilometers completed 35.595 million train-kilometers, a year-on-year increase of 53.7%. Although the acquisition of JFUAH did not immediately bring sufficient profits to JHGTE and even faced significant loss pressure, in the long run, JFUAH's business model is similar to JHGTE's. Line expansion is conducive to expanding the company's development space, and its operational condition is expected to further improve, bringing more considerable benefits to JHGTE. Meanwhile, JHGTE will continue to explore and optimize its operational model to enhance overall profitability [8].

2.2.3.4 Development Advantages

1. Geographical Location and Network Advantages

JHGTE connects the two major economic zones of Beijing-Tianjin-Hebei and the Yangtze River Delta, where the level of economic and social development along the line is relatively high. According to data from the National Bureau of Statistics, the administrative area of the provinces and cities along the JHGTE line accounts for about 6.5% of the national total, holds 27% of the national permanent population, and contributes 33% of the national GDP. Furthermore, JHGTE is a key part of the "Eight Vertical and Eight Horizontal" high-speed rail network, interconnecting with multiple high-speed lines such as Beijing-Harbin, Taiyuan-Qingdao, and Xuzhou-Lanzhou, and linking a series of critical railway transport hubs like Beijing South Station, Tianjin West Station, Jinan West Station, Nanjing South Station, and Shanghai Hongqiao Station. These location and network advantages bring stable passenger flow and continuously growing transport demand to JHGTE.

2. Efficient Operational Model and Profitability

JHGTE adopts a "separation of network and operation" operational model. Specifically, the company entrusts passenger transport management to the Beijing, Jinan, and Shanghai Railway Bureaus, and entrusts power supply and electrical facility maintenance to ZHTDJ. This entrusted transport management model helps reduce operational costs, improve transport efficiency, and avoid duplicate construction and resource waste. On the one hand, JHGTE has a relatively small number of employees on its payroll but owns the lines, stations, and related supporting assets. On the other hand, in terms of profitability, in 2023, the company achieved annual net profit attributable to shareholders of 11.9546 billion yuan and maintained a net profit margin level of over 25%, performing excellently. Its gross profit margin is also high, reaching 45.29% in 2023. These data indicate that JHGTE has strong profitability and operational efficiency.

2.3 Difficulties and Solutions for Railway Transportation Enterprises

2.3.1 Problems Encountered

2.3.1.1 Uneven Development

As a major artery of China's national economy, the development of railway transportation varies significantly across different regions. Specifically, due to historical accumulation, economic foundation, and characteristics of coastal openness, the railway transport system in the eastern region is relatively developed, giving it a developmental advantage in logistics, passenger movement, and economic development. Taking ZTSHA as an example, the railway transport network under its jurisdiction is dense and widely covered, connecting many important domestic cities and achieving connectivity with multiple countries and regions overseas. In contrast, the central and western regions, especially the western region, have relatively backward railway transportation construction due to complex geographical environments and a relatively weak economic base, among other factors. This not only limits the effective development and utilization of resources but also hinders the process of regional economic integration. For example, although ZTCDJ has achieved certain accomplishments in railway construction in recent years, influenced by complex geographical environments and a weak economic base, its railway transport construction remains relatively backward by comparison.

2.3.1.2 Lagging Infrastructure Construction

First, aging railway lines and outdated infrastructure. Due to insufficient capital investment and environmental factors, some railway transport lines and supporting facilities are in disrepair and struggle to handle the growing transportation demand. This not only reduces transport efficiency but also increases operational risks, posing a certain threat to the safety of railway transportation. Second, insufficient railway network coverage and a lack of overall planning in layout. Taking WLUUM in the western region as an example, although railway construction has accelerated in recent years, in some remote mountainous and ethnic minority areas, the extension of the railway network is still lagging, with low line density and poor connectivity, unable to provide strong support and guarantee for railway transport work. Third, due to the lack of unified scientific guidance, the layout of the existing railway network also appears fragmented and dispersed, making it difficult to form an efficient and coordinated transport system [9].

2.3.1.3 Low Level of Informatization

In today's rapidly evolving information technology landscape, railway transport management still relies on traditional methods and lacks advanced information technology support. This not only leads to low efficiency in transport management but also affects the level of intelligence and personalization of transport services, making it difficult to meet the diverse needs of the market and customers. GUZHJ, as a representative enterprise in the railway transportation industry, faced challenges related to lagging informatization levels during its digital transformation process. This resulted in insufficient digital transformation in areas such as product innovation, production operations, and user service for GUZHJ, making it unable to provide intelligent digital products and services that meet passenger demands.

2.3.2 *Solutions*

2.3.2.1 Narrowing Regional Development Gaps

Increase policy support and capital investment for railway construction in underdeveloped central and western regions. The government should formulate and implement targeted policies to provide more financial subsidies and incentives for the central and western regions. At the same time, priority should be given to railway construction projects passing through these regions, especially those of great significance to regional economic development. Secondly, strengthen railway planning and layout in the central and western regions, optimizing the railway network structure. It is necessary to comprehensively consider local geographical conditions, economic foundations, and future development needs, scientifically plan railway lines and stations, and improve the coverage and connectivity of the railway transport network. Strengthen cooperation and connection with other transportation modes such as highways and aviation to build a comprehensive transportation system and enhance the competitiveness of local railway transport. Furthermore, introduce advanced railway transport management concepts and technical means, strengthen railway transport organization and management, and improve transport efficiency and service quality. Strengthen talent cultivation and introduction, enhance the professional skills and service awareness of railway industry personnel, and provide higher quality and more efficient transport services for passengers and shippers. Establish and improve regional railway cooperation mechanisms, strengthen communication and collaboration between different regions, and jointly promote the implementation and operation of railway construction projects.

2.3.2.2 Enhancing Infrastructure Construction

It is necessary not only to increase capital investment for comprehensive overhaul and upgrading of existing railway lines but also to actively expand the construction of the railway network. This includes adding new railway lines, especially trunk lines connecting major economic regions and transport hubs, and improving railway networks at all levels, such as high-speed rail and conventional-speed railways. While expanding the railway network, attention must also be paid to the coverage and layout planning of the railway network. For the western region, as well as remote mountainous and ethnic minority areas, the extension of the railway network should be accelerated to increase line density and connectivity. Additionally, improving the level of facilities and equipment is an important part of strengthening infrastructure construction. Actively update key equipment such as trains and communication systems, adopt advanced technical means, and enhance the intelligence level of railway transportation. This can not only improve transport efficiency but also provide passengers with a more comfortable and convenient travel experience. At the same time, attention must be paid to strengthening the management and maintenance of facilities and equipment to ensure the safe operation of railway transportation.

2.3.2.3 Promoting the Application of Information Technology

Increase investment in information technology, adopt cutting-edge IT architectures such as cloud computing platforms, big data analytics, and artificial intelligence technologies as digital infrastructure to enhance the intelligence level of transport management. Secondly, promote product innovation by developing intelligent digital products and services that meet passenger needs. For example, utilize big data to analyze travel demands and provide customized travel suggestions and services; through intelligent systems, achieve real-time tracking and intelligent adjustment of goods, enhancing the quality and efficiency of the logistics process. Utilize Internet of Things (IoT) technology to achieve real-time monitoring and intelligent management of key elements such as vehicles and equipment, ensuring the safety and stability of the transport process [10]. Finally, build a comprehensive digital service system to enhance the intelligence and convenience of user services. Through digital means, provide more convenient and efficient services such as ticket purchasing and consultation, increasing user satisfaction.

2.4 Development Trends and Future Prospects

2.4.1 Development Trends

2.4.1.1 Rapid Development Supported by Policies

According to the strategic layout of the 《14th Five-Year Plan for the Development of a Modern Comprehensive Transportation System》, China's railway operating mileage is expected to increase to 165,000 km by 2025, with high-speed rail mileage reaching 50,000 km, and urban rail transit operating mileage expanding to 10,000 km. The core framework of this plan is to build a high-speed rail network centered on the "Eight Vertical and Eight Horizontal" main lines, supplemented by regional connecting lines and intercity railways. The aim is to ensure that high-speed rail services with speeds of 250 km/h and above cover over 95% of cities with a population of more than 500,000 and effectively address the current bottlenecks faced by conventional-speed railways. Furthermore, the *New Era Outline for Building a Leading Transport Nation with Railways Pioneering* issued by ZGTIE further outlines a grand blueprint for 2035, striving to expand the total national railway network length to over 200,000 km, including approximately 70,000 km of high-speed rail. This signifies that Chinese railways will comprehensively strive for and reach global top levels in multiple dimensions, including infrastructure construction, application of high-end technical equipment, technology innovation drive, service quality improvement, safety guarantee capability, management system, and governance capacity, laying a solid foundation for building a leading transport nation.

2.4.1.2 Digitalization Trend

With the rapid development of big data technology, data has become an important resource for enterprise development. The effective utilization and analysis of data resources enhance the professional management and operational level of railway transportation, reduce costs, and improve corporate competitiveness. Utilizing advanced computer science and technology to achieve the leap from manual management to IT-based management and control can become a new strategic high point for enterprises [11]. The railway transportation system of GALIU optimized its overall digital planning with the help of automated intelligent systems, investing over 30 million RMB in digital software and hardware resources for this purpose. In the future, railway transportation enterprises will rely on a series of digital-intelligent systems to achieve comprehensive information collection and seamless connection, while promoting the automation and intelligence of operations, ensuring that information can flow smoothly through various stages of the production process and be monitored in real-time. In the next 5 years, China's railway transportation will enter a new stage of digital construction, completing all information interconnection online, promoting the innovation of "unmanned" and "fewer people" intelligent construction in railway operation areas, and improving the level of informationalized, automated, and intelligent operations [12].

The informatization and digital construction of railway transportation are crucial for railway management. The application of information technology can overturn traditional management models [13]. Furthermore, this also opens up new development space for enterprises. For example, ZGTIE has established the Railway Technology Innovation Alliance, promoting the deep integration of industry, academia, research, and application. It also compiled and issued the *Digital Railway Plan* and released and implemented 122 important technical standards, including those for the Fuxing Hao CR400 EMU series. These measures not only enhance the safety and efficiency of railway operations but also earn China's railways more influence and competitiveness in the international market.

2.4.1.3 Green Development Trend

The green development of railway transportation enterprises is an inevitable requirement for achieving sustainable development. First, railway transportation enterprises should actively promote the application of clean energy, such as using solar and wind energy to power railway facilities and equipment, reducing dependence on traditional energy sources and lowering carbon emissions. Second, improve energy efficiency. By adopting advanced energy-saving technologies and equipment, such as energy-saving locomotives and intelligent lighting systems, reduce energy consumption and improve energy utilization efficiency. Additionally, strengthen environmental protection measures. During railway construction and operation, strictly comply with environmental protection regulations, strengthen environmental monitoring and assessment, and ensure that railway construction and operation have minimal impact on the environment. For example, the *Implementation Plan for Promoting Low-Carbon Development in the Railway Industry* jointly issued by the National Railway Administration and six other departments in 2024 proposed 12 specific tasks across three major areas to accelerate the low-carbon transition of the railway industry. These tasks include promoting the use of clean energy, improving energy efficiency, and strengthening environmental protection measures. These measures will help railway transportation enterprises achieve green development and contribute to building a beautiful China.

2.4.2 Future Prospects

2.4.2.1 Technological Innovation

Technological innovation is a crucial driving force for the development of railway transportation enterprises. In the future, railway transportation enterprises should increase investment in technological research and development, promote the application of intelligent and digital technologies, and enhance the intelligence level of railway transportation. For example, the CR450 technology innovation project has made critical progress, with high-speed trains reaching 453 km/h on open track and 891 km/h during meeting maneuvers. These technological innovations not only enhance the safety and efficiency of railway operations but also earn China's railways more influence and competitiveness in the international market. Additionally, railway transportation enterprises should strengthen cooperation with research institutions and universities to jointly promote technological innovation and application. By establishing technology innovation alliances and platforms, promote the deep integration of industry, academia, research, and application, and accelerate the transformation and application of technological achievements.

2.4.2.2 Service Optimization

Service optimization is an important way for railway transportation enterprises to enhance their competitiveness. In the future, railway transportation enterprises should focus on improving service quality, optimizing service processes, and enhancing user satisfaction. For example, optimize ticket purchasing processes, improve onboard environments, and provide more convenient and efficient services such as ticket purchasing and consultation. Additionally, railway transportation enterprises should strengthen cooperation with local governments and enterprises to promote the construction and improvement of infrastructure such as railway logistics parks and freight yards. By providing comprehensive logistics services, meet the diverse needs of customers and enhance the added value of services.

2.4.2.3 International Cooperation

International cooperation is an important path for the development of railway transportation enterprises. In the future, railway transportation enterprises should actively participate in international cooperation, promote the "going global" of Chinese railway technology and equipment, and enhance their international influence and competitiveness. For example, strengthen cooperation with countries along the "Belt and Road," promote the construction and operation of railway projects, and achieve mutual benefit and win–win outcomes. Additionally, railway transportation enterprises should learn from advanced international management experience and technical standards, enhance their own management level and technical capabilities, and lay a solid foundation for international competition.

2.5 Conclusion

As a vital pillar of the national economy, railway transportation holds a core position in China's modern transportation system. This paper first explores the development background of the railway transportation industry, then provides an in-depth analysis of the current status, challenges, response strategies, and future development directions of railway transportation enterprises. Through case studies of companies such as ZGTIE, DQITE, and JHGTE, it illustrates their specific development situations. Finally, future prospects are proposed. Railway transportation enterprises will continue to advance in technological innovation, green development, and service optimization to consolidate their central role. Only by fully understanding the development and existing problems of railway transportation and carefully studying relevant policies can China's railway transportation industry achieve sustainable development.

References

1. Qianzhan Industry Research Institute. Foreseeing 2023: An in-depth analysis of the market status, competitive landscape, and development trends of China's railway industry. 2023. https://bg.qianzhan.com/trends/detail/506/230104-a4dccabc.html
2. Ministry of Transport of the People's Republic of China. Statistical bulletin on the development of the transport industry in 2023. 2024. https://xxgk.mot.gov.cn/2020/jigou/zhghs/202406/t20240614_4142419.html
3. Qianzhan Industry Research Institute. 2024 panoramic analysis of the railway industry chain and regional heat map. 2024. https://www.qianzhan.com/analyst/detail/220/240703-8fb756c2.html
4. Qianzhan Industry Research Institute. 2024 analysis of leading enterprises in China's railway industry: Daqin Railway holds a leading position.2024. https://www.qianzhan.com/analyst/detail/220/240529-c0756361.html
5. Yang, B.On improving the efficiency of railway transportation organization [J]. Urban Construction Theory Research, 2024(4):178–180.
6. Cai Shi. Research report on Beijing-Shanghai High-Speed Railway: A mid-to-long-term perspective. 2022. https://baijiahao.baidu.com/s?id=1739379424573527055
7. Ding, J. J. The "cash cow" Beijing-Shanghai High-Speed Railway returns: Revenue exceeds 40 billion yuan for the first time in 2023, with record-high revenue and net profit in Q1 this year. 2024. https://www.163.com/dy/article/J1149RO80534A4SC.html
8. Ren Shi. Research report on Beijing-Shanghai High-Speed Railway: Passenger transport performance gradually recovers, and price increase elasticity becomes apparent. 2022. https://zhuanlan.zhihu.com/p/462912943
9. Hou, Y. A study on countermeasures for improving the transportation organization efficiency of Taiyuan railway hub [J]. Railway Freight Transport, 2021, 39(3):20–25.
10. Liu, Y. A discussion on the development strategies of railway transportation in integrated transport systems [J]. China Transportation Weekly, 2024(4):57–59.
11. Shi, Y. C., & Chen, C. Research on the design of an integrated information platform for railway multimodal transport [J]. Railway Freight Transport, 2020, 38(7):6–10.
12. Liu, W. Z. Applying information and digital technology to enhance enterprise railway transportation management [J]. Automation Application, 2024, 65(10):251–253.
13. Song, J. Y. Railway shunting operation planning [M]. China Railway Publishing House. 2000.

Chapter 3
Urban Rail Transit Enterprises: Financial Performance, Operational Barriers, and Strategic Evolution

Jianlin Wang and Jialuo Xia

Abstract This book focuses on urban rail transit companies, taking C1 as an example to explore its business scope, current status, and core competitiveness. Urban rail transit companies utilize wheel-rail trains to provide high- and medium-capacity rapid public transportation services. In China, these companies are primarily subway companies and form a critical component of the urban public transportation system. Based on ownership and operational nature, they can be categorized as state-owned and state-operated, state-privately owned, and privately owned and privately operated. However, due to their public nature and safety, state-owned and state-operated companies typically dominate the market and often receive government financial support and investment. The analysis of the companies' financial performance begins with a discussion of overall operating performance, finding that most companies have seen increasing revenues over their development, though a few remain loss-making due to cost control and technical issues. Secondly, an analysis of the operating revenue data of selected companies from 2019 to 2023 reveals that C1's net profit performance is particularly impressive, which is why this company is specifically highlighted in this article. Finally, combining corporate development reports with financial analysis, this paper summarizes the company's financial, operational, technological, and external factors. Future strategies to address these issues are proposed, aiming to promote the healthy and sustainable development of urban rail transit companies and enhance the competitiveness of China's urban rail enterprises.

3.1 Basic Information of the Enterprise

Urban rail transit is a vehicle transportation system that uses rail structures for transport and guidance. It transports large passenger volumes via trains or single vehicles along routes defined by the urban transportation network, making it a crucial form

J. Wang · J. Xia (✉)
Anhui University of Finance and Economics, Bengbu, China
e-mail: 15255381893@163.com

J. Wang and M. Song (eds.), *The Green Path of Chinese Transportation Enterprises*, Resources, Climate and Sustainable Development,
https://doi.org/10.1007/978-981-95-5289-4_3

of public transportation. Urban rail transit companies are those that operate large and medium-capacity rapid public transportation within urban areas using wheel-rail systems. They primarily profit from subways, with a small number utilizing other modes such as trams and light rail. These companies typically undertake the construction and operation of urban rail transit networks and are a vital component of urban public transportation systems.

Based on corporate ownership and operating rights, urban rail transit companies can be categorized into three types: state-owned and state-operated, state-privately owned, and privately owned and privately owned. State-owned and state-operated companies typically account for the majority of the market, with the government constructing rail transit facilities and designating government-affiliated agencies and state- owned enterprises to oversee their operations and management. Under this model, the government typically subsidizes losses incurred from operations and management. In the state-privately owned model, the government provides funding for the construction of rail transit facilities and transfers operating rights to private enterprises. Operators are subject to relevant government regulations, but the government does not interfere with the operations and management of the companies, nor does it subsidize operating losses. Privately owned and privately owned companies involve private capital funding the construction and operation of rail transit facilities. The government imposes contractual constraints on rail transit investment and construction, the equity structure of operating companies, and the range of fare fluctuations, but does not interfere with the companies' operations and management.

Regardless of the operating model, government involvement is essential. This is because urban rail transit, as a vital component of urban public transportation, offers numerous advantages, including energy efficiency, high capacity, low pollution, and safety. It is the primary mode of public transportation in large and medium-sized cities and carries significant public and safety importance. Therefore, the government typically regulates companies, primarily through strict oversight of pricing, quality, and safety, to ensure that urban rail transit companies fulfill their social responsibilities and obligations.

3.2 Enterprise Competition Landscape

3.2.1 Urban Rail Transit Construction Enterprises

In recent years, with the acceleration of China's urbanization, the urban rail transit industry has faced significant development opportunities and challenges, resulting in an increasingly fierce and diversified competitive landscape. Currently, the urban rail transit market is dominated by state-owned enterprises, with active participation from numerous private and foreign-funded enterprises [1]. Representative construction companies include large enterprises such as C2, C3, C4, and C5. These companies, leveraging their deep expertise and technological advantages in infrastructure construction, have established a dominant position in the market. Private enterprises, on the other hand, have demonstrated innovative vitality. For example, BYD has successfully entered the rail transit sector through the cross-border application of its new energy vehicle technology, injecting new momentum into the market. Furthermore, some foreign-funded enterprises have joined projects through technical cooperation, promoting the high-quality development of the rail transit industry.

3.2.2 Urban Rail Transit Operating Enterprises

Overall, my country's urban rail transit operating companies can be roughly divided into three echelons. The first echelon companies are subway operating companies in first-tier cities such as Beijing, Shanghai, Guangzhou and Shenzhen. These companies have large operating scales, cover multiple lines, have high average daily passenger volumes, have relatively rich business models, and have revenue scales that are ahead of most companies; the second echelon is represented by subway operating companies in Xi'an, Chengdu, Wuhan and Nanjing. This echelon of companies is mainly distributed in other first- and second-tier cities. The number of their operating lines is similar to that of the first echelon companies, but their passenger volume is slightly less than the former, with an average daily passenger volume of 6 million to 8 million; the third echelon is mainly composed of subway operating companies in other small and medium-sized cities. These companies have limited service scope and small operating scale, and their passenger volume is significantly lower than that of the first two echelons, but they still play an important role in meeting local travel needs [2] (Fig. 3.1).

According to information from the enterprise database Qichamao, there are more than 60,000 urban rail transit companies in China, mainly distributed in the Pearl River Delta region. From the registration trend, 2020 was a registration boom in the

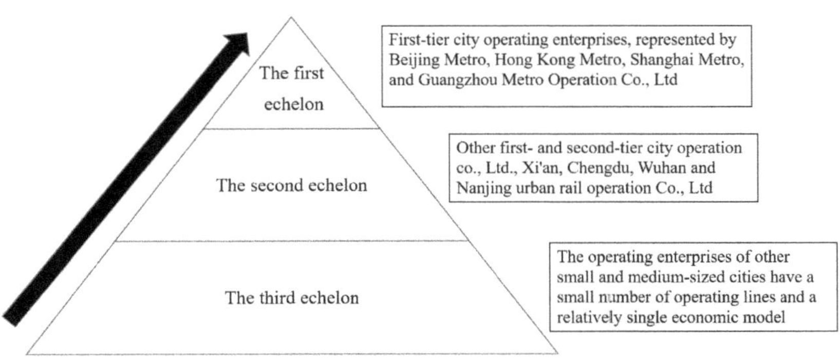

Fig. 3.1 Competitive tier of urban rail transit operators

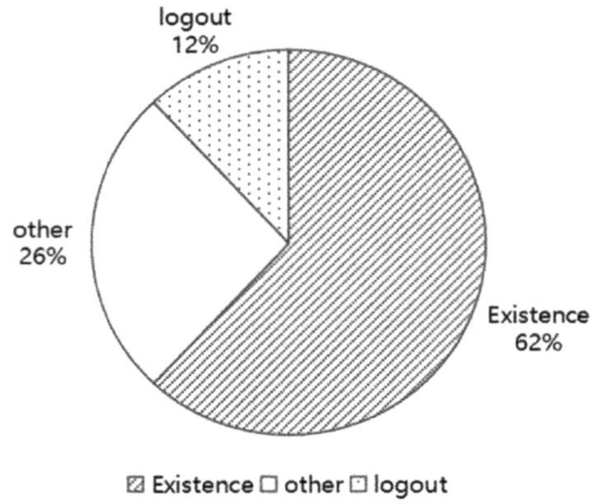

Fig. 3.2 Distribution of business operations in 2022 (unit: %). *Data source* China Enterprise Database Qichamao

industry, with 12,644 new companies added that year, and this number decreased to 8780 in 2021. As of now, there are 38,863 existing companies, accounting for 61.7% of the total number of companies. In terms of corporate form, most urban rail transit industry companies are limited liability companies and sole proprietorships, with 52,209 and 20,695 respectively. At the same time, 12.5% of companies, or 7871 companies, have chosen to cancel their registrations. In terms of registered capital scale, the registered capital of China's urban rail transit companies is currently mainly distributed in the range of 10 million to 50 million, corresponding to 18,764 companies. The number of companies with registered capital of 5 million yuan or more is close to 70%, and the number of companies with registered capital of more than 10 million is more than 50% (Fig. 3.2).

3.3 Enterprise Financial Status

3.3.1 Overall Financial Situation

With the continuous expansion of urban rail transit networks and the increase in passenger traffic, the operating revenue of most companies has shown an upward trend. Subway companies in major cities such as Beijing, Shanghai, and Shenzhen have seen rapid revenue growth in recent years. Overall, with the gradual maturity of operating lines and the continuous increase in passenger traffic, many projects have achieved break-even or even profitability. Some companies are actively exploring diversified profit models, such as advertising revenue and property development, which have effectively alleviated their financial pressures. Companies themselves are continuously improving operational efficiency and service quality through refined management, cost control, and service innovation, further enhancing profitability. As a public transportation service industry, the government often provides subsidies to loss-making companies through direct investment or guarantees, filling the operating funding gap of some loss-making projects and ensuring the sustainable operation of urban public transportation. However, profitability in this industry is not always smooth sailing, and companies still face numerous challenges.

First, companies face generally high operating costs, which constitute a major constraint on profitability. These primarily include labor costs, equipment maintenance costs, and energy costs. As urban rail transit networks expand and their operating life increases, equipment maintenance and energy costs will continue to rise. For example, the operating costs of the C6 subway system have increased by as much as 417% in recent years. Secondly, urban rail transit projects are often characterized by large construction investments, long construction cycles, and low profit-to-return ratios [3]. This results in significant financial pressure for most companies in the early stages, and it often takes a long time to recoup any profit shortfalls. Furthermore, adjustments to urban economic policies and price fluctuations directly impact corporate profits, causing fluctuations in earnings and preventing sustainable profitability.

3.3.2 Financial Status from 2019 to 2023

3.3.2.1 Corporate Revenue Showed a Downward Trend in 2019

In 2019, the outbreak of the COVID-19 pandemic prompted strict travel measures across the country. The implementation of "curfew" orders led to a significant decrease in passenger flow, which directly impacted the profitability of rail transit companies. As a result, operating revenue and expenditure for urban rail transit companies nationwide were generally tight, with most companies hovering near break-even and even relying on government subsidies to maintain operations. Statistics show that the national average operating revenue-to-expenditure ratio is 72.7%, with only a few cities experiencing ratios exceeding 100%. For example, first-tier cities such as Shenzhen, Beijing, and Guangzhou, due to their large passenger base and high operational efficiency, have maintained operating revenue-to-expenditure ratios exceeding 100%, even despite the adverse impact of the pandemic. Their profit declines have been relatively minimal, demonstrating their strong resilience to risk.

With the most impressive annual operating revenue was C1, which reached 20.99 billion yuan, an increase of 84.9% year-on-year. Its net profit reached 11.69 billion yuan, an increase of 61.2% year-on-year (see Table 3.1), far exceeding the second-highest revenue, C7 Metro, by nearly 6 billion yuan. In contrast, the profitability of urban rail transit companies in other cities was generally mediocre. Despite this, the operating mileage of most subway companies in that year increased compared to 2018, indicating that the companies had previously focused on expanding subway lines with the goal of improving transportation network coverage and achieving revenue growth by increasing passenger volume. However, the outbreak of the COVID-19 pandemic disrupted this original expectation, resulting in a decrease in passenger flow and seriously affecting the profitability of the companies.

Table 3.1 Operating mileage and revenue of some subway companies in 2019

Subway company	Operating mileage (km)	Operating income (100 million yuan)	Net profit (100 million yuan)
C1	315.1	209.9	116.9
C10	515	122.34	10.44
C11	131	66.99	12.65
C12	166	52.75	0.03
C13	341	40.59	0.27
C14	176.8	28.63	2.75
C8	329	26.00	0.01
C15	233	20.17	5.11
C16	161.11	18.04	3.35
C9	187.5	4.81	2.04

Data source 2019 Urban Rail Transit Statistics and Analysis Report, 2019 Financial Reports of Subways or Rail Transit Groups

While some companies have temporarily alleviated their financial pressures through government subsidies, the industry as a whole remains plagued by losses. For example, in Chongqing, C8 has an operating mileage of 329 km and carries over 3 million passengers daily. However, its 2019 annual report showed an operating loss of approximately 32.41 million yuan, including approximately 39.76 million yuan in non-operating income from government subsidies, barely achieving a positive net profit. The situation is equally bleak for other large companies. For example, C9's operating income is dismal. According to its financial report, total revenue for the year was 481 million yuan, with a net profit attributable to the parent company of 204 million yuan. However, its total liabilities reached 92.684 billion yuan, and its debt-to-asset ratio has been rising year by year. A review of the company's financial reports for the previous four years reveals that its debt-to-asset ratio reached 65.77%, 67.22%, 67.92%, and 68.54%, respectively. This indicates that the company's debt has been increasing year by year, and its debt ratio has been climbing steadily, maintaining a persistently loss-making state.

3.3.2.2 Corporate Income Varied Widely in 2020

In 2020, the revenue of urban rail transit companies showed significant disparity and complexity. Due to the ongoing impact of the COVID-19 pandemic, passenger traffic decreased significantly, while operating costs rose and ticket revenue decreased, putting widespread financial pressure on companies. These unfavorable factors had a significant impact on their profits. To cope with the difficulties, many companies actively adjusted their operating strategies and methods to offset losses, explored diversified revenue sources, and alleviated financial pressure through refined management, demonstrating strong resilience in this unique economic context.

Despite facing numerous unfavorable factors, some companies have managed to achieve profitability through diversified operations and effective cost control. As shown in Table 3.2, C1 maintained its leading position in 2020, achieving operating revenue of 20.928 billion yuan and a net profit of 11.102 billion yuan. While other companies achieved strong revenue performance, their relatively small revenue scale did not offset rising operating costs, and their net profits were generally low. For example, C17 Metro achieved revenue of 13.665 billion yuan in 2020, but a net profit of only 3.156 billion yuan; C10 Metro had revenue of 12.891 billion yuan, but a net profit of only 230 million yuan. Some city subway companies even incurred losses, such as C11 Metro, which lost 1.454 billion yuan in 2020.

It is worth noting that most urban rail transit companies can only achieve profitability with government subsidies. For example, C18 Company received government subsidies of 10.463 billion yuan, which became an important source of its profits. This may be related to the company's strong economic foundation, clear policy orientation, urgent strategic needs, the particularity of infrastructure construction, and the effectiveness of specific subsidy measures. Thanks to sufficient financial

Table 3.2 Net profit and actual profit and loss of 20 companies in 2020 (unit: 100 million yuan)

Enterprise	Net profit	Government subsidies	Actual profit and loss
C1	111.02	0.36	110.66
C18	31.56	104.63	−73.07
C10	2.3	5.1	−2.8
C13	14.76	91.44	−76.68
C7	16.78	1	15.78
C12	0.04	0.23	−0.19
C16	0.69	11.88	−11.19
C8	0.25	16.17	−15.92
C11	−11.62	2.92	−14.54
C14	3	37.46	−34.46
C19	5.07	5.04	0.03
C15	5.17	22.37	−17.2
C20	0.86	0.03	0.83
C21	5.67	0.12	5.55
C22	1.07	7.01	−5.94
C23	7.05	5.45	1.6
C9	0.7	5.01	−4.31
C24	7.16	36.95	−29.79
C25	1.22	0.01	1.21
C26	7.96	29.1	−21.14

Data source 2020 financial reports of various subway companies

support, Beijing can continue to invest a large amount of funds in rail transit infrastructure construction, which not only ensures the operation of enterprises, but also promotes the high-quality development of the city's economy and society.

3.3.2.3 Corporate Profits in 2021 Are Affected by Multiple Factors

In 2021, the financial performance of urban rail transit companies continued to be impacted by a variety of factors. The normalization of epidemic prevention and control measures led to a slow recovery in passenger traffic, continued increases in operating costs, and significant fluctuations in ticket revenue. However, government subsidies alleviated some of the financial pressure. These factors combined to create significant variations in the financial performance of different subway companies.

C1 Metro's revenue and profits remained among the highest in China in 2021. The company's financial report shows that total revenue for the year reached nearly 16.4 billion yuan. While slightly lower than C18's 17 billion yuan, C1 Metro's profitability was outstanding, with net profit attributable to parent company reaching nearly 3 billion yuan, ranking first nationwide. Unlike in the past, a significant portion of C1 Metro's revenue comes from non-ticketing revenue, such as integrated station-city development (i.e., real estate development at rail stations) and property management. This diversified business model enables Shenzhen Metro to maintain strong profitability despite fluctuations in ticketing revenue.

However, the financial performance of rail transit companies in other cities varied widely. Some cities' subway companies achieved profitability or reduced losses through diversified operations and cost control, while others incurred losses due to rising operating costs and reduced ticket revenue. Similar to previous years, with the help of government subsidies, although only a small number of subway companies achieved profitability, the number increased compared to 2020. This phenomenon demonstrates the flexibility and adaptability of companies to policies and measures during this special period of the epidemic, reflecting the strong resilience and adaptability of my country's urban rail transit companies (Table 3.3).

3.3.2.4 Corporate Revenue Will Gradually Recover in 2022

As the epidemic prevention and control situation gradually stabilizes and economic activity gradually resumes, some regions are relaxing restrictions on movement, and tourist attractions are gradually reopening. The increased foot traffic has brought economic benefits to businesses, resulting in an overall increase in revenue for urban rail transit companies. However, there is significant differentiation between cities. As shown in Table 3.4, C1 Metro's revenue surpassed C18 companies, reaching 23.976 billion yuan, an increase of 7.579 billion yuan compared to the previous year. C10 Metro ranked third with revenue of 12.285 billion yuan. Furthermore, the number of companies with revenue exceeding 10 billion yuan increased to six, with the addition

Table 3.3 Profits of 20 urban rail transit companies in 2021 (unit: 100 million yuan)

Subway company	Government subsidies	Net profit attributable to parent company	Profit after deducting subsidies
C1	1.37	29.97	28.6
C7	0.87	17.31	16.44
C19	14.33	16.99	2.56
C20	0.12	0.78	0.66
C10	2.04	2.65	0.61
C27	3.37	2.34	−1.03
C25	2.84	1.14	−1.7
C28	2.84	0.73	−2.11
C29	3.34	0.18	−3.16
C9	5	0.91	−4.09
C30	9.87	3.66	−6.21
C11	3.15	−5.46	−8.61
C22	10.6	0.96	−9.64
C23	20.23	6.07	−14.16
C8	23.61	0.53	−23.08
C15	30.43	6.41	−24.02
C31	16.1	−12.35	−28.45
C16	35.99	2.55	−33.44
C14	38.05	1.24	−36.81
C18	214.66	24.54	−190.12

Data source 2021 financial reports of urban rail transit companies

of C13 Rail Transit and C23 Rail Transit. However, over 70% of these companies had revenue below 5 billion yuan, and nine of them had revenue below 1 billion yuan.

In terms of revenue structure, ticketing remains a key source of income for subway companies, but non-ticketing revenue (such as property development and advertising) has seen even more significant growth. In 2022, non-ticketing revenue for urban rail transit companies reached 64.13 billion yuan, an increase of 20.14 billion yuan from the previous year, a year-on-year increase of 45.8%. This shift in operating models has expanded beyond daily ticket sales, which primarily rely on passenger flow, to include the widespread addition of revenue from real estate development. For example, revenue from the C13 subway's integrated station development business increased to 6.247 billion yuan, becoming a significant source of revenue. Revenue from the C1 subway's integrated station-city development also accounts for a significant portion of total revenue.

Despite overall revenue growth, the number of profitable companies remains relatively small. Only urban subway companies in C1, C7, and C20 achieved profitability after deducting subsidies. Most urban rail transit companies continue to face losses,

Table 3.4 Revenue of some companies in 2022

Serial number	Company	Total operating revenue (100 million yuan)	Net profit margin attributable to parent company
1	C1	239.76	10.42
2	C18	150.85	21.82
3	C10	122.85	8.38
4	C13	117.99	6.48
5	C23	107.91	3.44
6	C7	101.50	15.70
7	C12	90.76	0.08
8	C20	87.99	2.14
9	C26	48.07	1.95
10	C32	46.78	4.61
11	C16	43.05	3.05
12	C11	33.09	7.14
13	C22	31.89	1.20
14	C30	31.79	3.68
15	C19	29.94	1.20
16	C25	27.00	−4.81
17	C14	24.50	2.46
18	C9	6.29	0.79

Data source 2022 financial reports of various subway companies

primarily due to factors such as lower-than-expected passenger traffic, generally rising operating costs, and increased reliance on government subsidies. For example, C9 Company achieved total revenue of 629 million yuan in 2022, but operating costs reached 1.559 billion yuan. Ultimately, it only achieved a profit of nearly 80 million yuan thanks to a substantial 975 million yuan government operating subsidy.

3.3.2.5 Corporate Profits Will Grow Rapidly in 2023

In 2023, with the full lifting of epidemic prevention and control measures, urban vitality will rapidly recover. Tourism and catering industries will begin to recover, passenger flow will begin to increase significantly, and the operating income of most subway companies will also increase accordingly. According to data from the China Urban Rail Transit Association, cities with urban rail transit operations will complete 29.466 billion passenger trips in 2023, a 52.66% increase over 2022. During this period, as companies diversify, they will increase revenue through the "railway + property development" (TOD) model. In 2023, total revenue from urban rail transit resource operations will be 61.1 billion yuan, a decrease of 3.03 billion yuan from the previous year, a year-on-year decrease of 4.7%. However, non-ticketing revenue will continue to grow significantly, including property development revenue of 43.13 billion yuan, accounting for 71% of total revenue, and Guangtong Commercial operating revenue of 10.64 billion yuan, accounting for 17% of total revenue.

However, despite revenue growth, subway companies' profits have generally declined, primarily due to high operating and construction costs, as well as interest expenses. Statistics show that the average operating cost per vehicle-kilometer for urban rail transit nationwide in 2023 was 34.43 yuan, a year-on-year increase of 3.36 yuan. Some companies, such as C18, saw total operating costs exceeding 10 billion yuan, with total operating costs exceeding 49.094 billion yuan. Furthermore, as revenue has not covered costs in recent years, subway companies generally face high debt levels. By 2023, the total debt of subway companies in 29 cities nationwide had reached 4.3 trillion yuan, with some companies boasting debt-to-asset ratios exceeding 60% and even exceeding 80%. Consequently, subway companies generally rely on government subsidies to maintain operations, with most subsidies amounting to tens of billions or even hundreds of billions of yuan, and some even exceeding 100 billion yuan. 2023 marks a fresh start, and despite recent efforts to boost profitability, urban rail transit companies remain in deficit. Government subsidies are, therefore, the primary source of profit for most companies.

Specifically, C1 Metro achieved operating revenue of 25.15 billion yuan in 2023, continuing to rank first among national subway companies in terms of revenue. C18, the parent company of C17 Metro, achieved operating revenue of 21.8 billion yuan, ranking second. C13 and C10 Metro achieved operating revenues of 15.8 billion yuan and 14.12 billion yuan, respectively, ranking third and fourth. However, in terms of profitability, C17 Metro was the most profitable subway company in China, with a net profit attributable to parent company of 2.4 billion yuan. C13 followed closely behind with a net profit attributable to parent company of 830 million yuan. Despite its leading revenue, C1 Metro experienced a significant decline in profits, with a net profit of only 550 million yuan in 2023. Companies like C25 and C21 Metro are still operating at a loss after deducting government subsidies (Table 3.5).

3.4 Introduction to Representative Enterprises

Based on the five years of financial reports, we can see that C1 Co., Ltd. has achieved outstanding profit performance, far exceeding the profits of other companies. Therefore, we highlight this company (the following information is referenced from the company's official website: C1 (szmc.net)).

Table 3.5 Revenue and profit of some metro companies in 2023

Serial number	Company	Total operating revenue (100 million yuan)	Net profit attributable to parent company
1	C1	251.5	5.5
2	C18	218.0	24.0
3	C13	158.0	1.8
4	C10	141.2	0.2
5	C12	113.7	0.2
6	C26	105.1	3.9
7	C20	97.9	3.2
8	C7	66.9	0.4
9	C32	46.3	4.8
10	C16	44.3	1.3
11	C30	39.5	1.0
12	C14	35.0	3.9
13	C31	33.8	0.2
14	C28	33.8	3.7
15	C22	32.7	1.20
16	C8	30.4	0.3
17	C19	29.3	2.0
18	C15	22.6	7.2
19	C33	16.4	0.7
20	C25	15.7	−6.5

Data source Financial reports of urban rail transit companies in 2023

3.4.1 Company Overview

C1 Co., Ltd. (hereinafter referred to as C1), established on July 31, 1998, is a wholly state-owned enterprise directly under the Shenzhen State-owned Assets Supervision and Administration Commission (SASAC). Over the years, C1 has established a comprehensive industrial layout encompassing national railways, intercity railways, and urban rail transit, and a core value chain encompassing rail construction, rail operations, station-city development, and resource management. C1 strives to build an open, innovative, and inclusive "Railway+" ecosystem. In 2021, the Group had registered capital of RMB 45.943 billion, total assets of RMB 507.006 billion, net assets of RMB 296.184 billion, and approximately 28,000 employees. By the end of 2023, the Group's total assets reached RMB 716.589 billion.

Currently, C1 is actively implementing a "transit-oriented" TOD development model, leveraging the appreciation in value of subway overpasses and properties along the lines to support rail transit construction and operations. This "railway + property" model is increasingly integrated with the city, and "station-city integration" projects, exemplified by hubs, are gradually becoming core products. The group has established a diverse product portfolio encompassing office buildings, retail, hotels, and advertising media. The group has received numerous honors, including the National Science and Technology Progress Award, the Zhan Tianyou Award for Civil Engineering in China, the Lu Ban Award for China Construction Engineering, the National Quality Engineering Gold Award, the Guangdong Provincial Science and Technology Progress Award, the Shenzhen Municipal Science and Technology Progress Award, and the International Tunnel Association (ITA) Underground Space Innovation Contribution Award.

3.4.2 Business Scope

The company is responsible for over 95% of the rail transit construction and operation work in Shenzhen. Its business scope involves the "three railways", including its investment and construction, operation and maintenance, station and city development, and resource operation. It also covers productive service areas such as engineering survey and design, building materials, property services, and education and training related to the rail industry chain. This has shaped the "three railways in one" industrial structure and the four-in-one core value chain of "railway construction, operation, station and city development, and resource management", and is fully committed to creating an open, innovative and inclusive 'railway+" ecological system.

According to official website data, C1 Transportation operates 567.1 km (including the Longhua Tram), of which 536.4 km are operated by the company (30.7 km are operated by the MTR). The city's urban rail operating mileage ranks fifth nationwide, and its total passenger flow ranks fourth. Its network density leads mainland China, and its technical indicators, such as passenger traffic intensity, train punctuality, operational reliability, and equipment reliability, are among the best in the industry. The city's urban rail construction scale, coverage, service level, operational efficiency, and technical capabilities are among the best in the industry.

3.4.3 Subway Construction and Operation

C1 boasts strong technical expertise and extensive experience in subway construction, having completed multiple phases of subway projects. These projects have not only improved Shenzhen's urban transportation but also provided citizens with more convenient and efficient travel options. The company is also actively promoting national railway intercity projects, such as the Shenzhen-Shantou Railway and the Guangzhou-Dongguan-Shenzhen Intercity Railway from Shenzhen Airport to Qianhai, further expanding Shenzhen's rail transit network. According to official website data, Shenzhen's urban rail transit operating mileage is 567.1 km, consisting

Table 3.6 Subway-related indicators

Project (Formally operating project)	2023	2022	2021	2020
Annual passenger volume (billion passengers)	21.80	15.79	16.99	13.38
Average daily passenger volume (10,000 people)	597.14	432.63	465.51	365.53
Average ticket price (yuan/person)	2.19	2.30	2.31	2.63
Vehicle-kilometer cost (yuan/vehicle-kilometer)	16.61	16.46	16.66	17.48

Data source C1 company financial analysis report

of 16 subway lines and one tram line. C1 operates 15 of the subway lines and one tram line, while MTR C1 operates one subway line. Table 3.6 shows the key operating indicators for C1-operated subway lines. The officially operational lines include Lines 1, 2, 3, 5, 6, 7, 8, 9, 10, 11, and 20.

In addition to subway construction and operation, C1 is actively expanding its integrated development business. The company fully utilizes land resources along subway lines to promote the deep integration of urban rail transit and the city. By developing projects such as subway properties and commercial complexes, the company not only provides citizens with more convenient living facilities but also injects new momentum into its sustainable development.

3.4.4 Main Economic Indicators of the Company

At the end of the most recent 2023 reporting period, the company's total assets reached 716.005 billion yuan, an 8% year-on-year increase. Total liabilities reached 396.213 billion yuan, a 13% year-on-year increase. Total net assets reached 319.792 billion yuan, a 3% year-on-year increase. The debt-to-asset ratio reached 55.34%, a 2.43 percentage point increase year-on-year. However, compared to previous years, both the company's asset-liability ratio and its debt-to-asset ratio have shown an upward trend. This is primarily due to the company's vigorous promotion of the "Three Railways" initiative, which has led to an increase in rail transit assets. To promote financial balance throughout the rail transit cycle, the company has increased its land resource reserves and equity investments, resulting in a corresponding increase in assets. To secure funding for large-scale infrastructure investment and daily operations, the company has increased its financing efforts, resulting in a slight increase in debt. However, the debt-to-asset ratio remains within a reasonable range (Table 3.7).

Table 3.7 Important economic indicators of the company (unit: 10,000 yuan)

Project	2023	2022	2021	2020
Operating income	2,515,432	2,397,578	1,639,666	2,082,810
Operating costs	2,191,172	2,105,939	1,333,913	1,347,239
Business taxes and surcharges	131,723	322,415	227,259	410,758
Investment income	479,884	678,796	640,875	1,165,414
Total profit	163,863	232,912	296,590	1,115,029
Net profit	78,929	86,590	288,988	1,110,229

Data source C1 Group's four-year financial reports

3.4.5 Core Competitiveness Analysis

C1 Enterprise boasts a complete rail transit industry system, with a high level of industrial chain capabilities and supporting facilities. Currently, the company leads the industry in key economic development indicators, holds an AAA credit rating, and possesses comprehensive financing capabilities exceeding 100 billion yuan. Shenzhen Metro implements a platform and ecosystem strategy, leveraging the driving and guiding role of rail transit construction across the entire industrial chain. Adhering to the principle of "either first or only" in cooperation, it has driven the coordinated development of over 2000 companies along the entire industrial chain. The company is actively cultivating and developing strategic emerging industries, improving the modern rail transit industry system, and promoting the clustered development of Shenzhen's rail transit industry.

C1 has achieved world-class standards in rail transit products and services, leading the nation in the TOD development index. Shenzhen's metro construction scale, coverage, service levels, operational efficiency, and technological capabilities are among the best in China and internationally. The company has created a number of world-class products that embody Shenzhen's modern urban character, including the Gangxia North Hub ("Eye of Shenzhen"), the Huangmugang Hub, the Dayun Hub, and the Shenzhen Bay Ruiyun Center ("Sky Realm"). The company also innovatively promotes "railway + TOD integrated development," ranking among the top in Shenzhen in terms of both development volume and sales performance. The "railway + property" model is increasingly integrated into the city, with Shenzhen ranking first nationwide in terms of underground space access per kilometer, underground space development volume, and the urban rail TOD index.

The company has demonstrated strong technological innovation capabilities in the rail transit sector, boasting a high level of green development. Adhering to an innovation-driven development strategy, the company has comprehensively planned and advanced its "Technology C1" strategy. Currently, the company boasts 11 innovation platforms at or above the municipal level, including one at the national level, three at the provincial level, and seven at the municipal level. It also boasts two national-level high-tech subsidiaries. In 2023, the company received numerous rail transit technology awards from provincial, municipal, and industry associations. This initiative is driving digital transformation across the board, including the construction of a "two-site, three-center" data center and a core group management network. The NOCC Phase II project has been selected as a "New Urban Construction" exhibition and display center by the Ministry of Housing and Urban–Rural Development and a pilot demonstration project for Shenzhen's "New Urban Construction" initiative.

C1 Group has continuously strengthened its governance system and capabilities, effectively promoting the full integration of Party building with modern corporate governance to drive high-quality corporate development. It has established a "strategy + operations" management and control model, building a "function + business" management framework to meet the needs of cross-disciplinary, cross-industry, cross-regional, and large-scale business development. Regarding business management and control, a "standardized, normalized, refined, and information-based" management system has been established, resulting in capabilities in construction management, operational safety assurance, passenger service assurance, maintenance quality assurance, internal control management, and human resources that exceed industry standards.

In terms of talent, the company boasts a highly qualified and professional human resources team. As of the end of 2023, the company had 26,649 employees. Among them, 12,680 hold bachelor's degrees or higher, 1185 hold master's degrees or higher, including 48 doctoral students. 91 hold senior professional titles and 1029 hold associate senior professional titles, forming a professional, high-quality, and innovative talent team across all aspects of urban rail transit. At the same time, the company continues to deepen reforms in its job positions, compensation, and assessment systems, establishing a market-based selection mechanism that allows for promotion and demotion of cadres, increase and decrease of compensation, and employee recruitment and exit. This has fostered a new normal in talent selection and employment that prioritizes career development, matches people to positions, and matches personnel to jobs, and fosters a performance-focused talent evaluation and incentive mechanism.

In summary, C1 Group surpassed the competitiveness of other companies at the time in many aspects. In the future, the company will fully implement the concept of "leading by example", firmly shoulder the development mission of "transportation pioneer", accelerate the construction of the "Greater Bay Area on Rails", and strive to build a world-leading rail transit company with first-class facilities, top technology, excellent management, high-quality services, significant benefits and sustainable development, providing strong support for the construction of the Guangdong-Hong Kong-Macao Greater Bay Area and the Shenzhen Metropolitan Area.

3.5 Common Obstacles to Enterprise Development

3.5.1 *Financial Issues*

The construction and operation of urban rail transit systems require substantial capital investment and high operating costs. The capital investment required to construct subways and other rail transit systems is substantial, with construction costs per kilometer reaching hundreds of millions of yuan, and in some cities, even exceeding 2 billion yuan for specific projects. Once a rail transit system enters operation, operators face not only the expenses of daily operations but also the costs of periodic equipment maintenance and upgrades, as well as the financial costs of project financing. Data and industry experience indicate that over a 30-year operating lifecycle, total operating costs range from 3.3 to 3.7 times the construction cost. Project construction cycles are also long, typically lasting three to five years or even longer. This prolonged period results in slow payback, with some projects potentially taking decades to recoup their investment, placing significant financial pressure on companies. Furthermore, as public infrastructure, high costs and controlled ticket prices make subways and other rail transit systems major capital drains, relying on government subsidies to sustain their operations. Taking Beijing as an example, the single ticket price of 4.5 yuan per person can only cover about one-third of the subway operating costs, and the remaining two-thirds rely on government financial subsidies.

Judging from the operating revenue data of the aforementioned companies over the past five years, most face the severe challenge of insufficient financial sustainability. Most subway companies nationwide rely on explicit or implicit government subsidies to support their operations. Once these subsidies are removed, many companies will face immense operational pressure. Some subway companies have extremely high debt ratios, such as C9, which has a debt ratio of 82%. They face billions of yuan in annual interest payments, which undoubtedly represents a significant financial burden. If subsidies continue to support business operations, it will become a financial burden on the government, and the companies themselves will lose competitiveness and struggle to gain a foothold in the market.

3.5.2 *Operational Issues*

Many urban rail transit systems have failed to meet expected passenger traffic levels. This may be because, in today's urban transportation landscape, subways and other rail transit systems are still not the preferred mode of transportation for residents, resulting in low operational efficiency and insufficient revenue to cover costs. However, there is still room for improvement in service quality and operational efficiency, including in vehicle scheduling, passenger service, and safety management. Specifically, lines experience insufficient capacity during peak hours, meaning they are unable to accommodate all passengers during peak hours. Safety management needs to be strengthened to ensure the safety of passengers and employees, as well as the safety of the vehicles themselves.

Regarding safety, the company's rail transit systems suffer from inadequate daily safety management and lack a comprehensive safety management system. As operations increase, some facilities and equipment experience aging and wear. Failure to promptly maintain and replace these facilities could lead to safety incidents. If the company's safety strategies and technologies are not promptly updated and improved during routine inspections, unexpected operations could be disrupted.

However, passengers' primary concern remains the experience. As customer expectations rise, some urban rail transit companies are currently facing shortcomings in passenger service. These include factors such as station cleanliness, service attitudes, the timeliness and accuracy of information announcements, and onboard crowding. Failure to effectively and promptly improve the overall passenger experience will lead to a decline in public trust in rail transit, further impacting passenger traffic and operator revenue. Therefore, improving service quality and operational efficiency while ensuring safety is a critical issue facing urban rail transit companies.

3.5.3 Technical Issues

Urban rail transit companies lag behind in technological innovation, impacting their competitiveness and sustainable development. Specifically, some companies lack independent R&D capabilities, rely on imported technology and equipment, and invest insufficiently in technological innovation, making it difficult to adapt to the demands of industry development. While companies have experimented with digital transformation, adopting advanced technologies like cloud computing, big data, and the Internet of Things, these efforts often lack thorough consideration and planning, preventing these technologies from fully realizing their potential and delivering substantial benefits. Furthermore, the rapid development of these companies has led to significant duplication of efforts. For example, the CBTC system (Communication-based Train Automatic Control), a Western technology from the 1980s, has been replicated across multiple companies in my country, resulting in fragmented R&D capabilities and a struggle to develop competitive products. This duplication not only wastes R&D funds but also hinders technological iteration and advancement.

3.5.4 External Factors

As a fundamental form of public transportation, government policies have a profound impact on the revenue and operations of companies. However, in reality, policies often fluctuate with the socioeconomic landscape, and their instability and uncertainty can lead to significant risks for companies in investment decisions and project implementation. For example, the efficiency of government approvals for urban rail transit construction, price controls, and the amount of operating subsidies all directly impact a company's operating costs and profitability. Furthermore, my country's urban rail transit industry's legal and regulatory framework is still underdeveloped, potentially exposing companies to legal disputes and compliance risks during operations.

The current market competition is becoming increasingly fierce, and many industries have entered a period of saturation. Enterprises need to continuously innovate and improve in technology, services, management, etc. to cope with the competitive pressure from their peers, and even need to take into account social environmental factors. Due to the development of social economy and the diversification of people's travel methods, passengers' demand for urban rail transit is constantly increasing. Enterprises need to adjust their operating strategies and service content in a timely manner according to market demand to meet the travel needs of passengers. However, this adjustment requires a lot of manpower, material and financial resources, which will increase the operating costs of the enterprise.

3.6 Development Strategies

3.6.1 Strengthen Planning and Management

Since my country's urban rail transit is not yet fully mature, there are still many problems in the actual operation process. To achieve stable development, enterprises need to pay attention to their own construction model and introduce corresponding rail transit construction plans. However, the company's development plan needs to consider many aspects. First, it must meet the city's economic status and road needs, so that the city's rail transit system can achieve sustainable and stable development and ensure that it will not put too much economic pressure on the financial department during the development process. Secondly, enterprises need to establish and improve the urban rail transit management system, clarify the responsibilities and powers of management departments at all levels, strengthen coordination and cooperation with other relevant departments, form a working synergy, introduce advanced management concepts and technical means, improve management efficiency and service levels, and finally be able to flexibly adjust in actual implementation and find appropriate methods according to specific problems.

Enterprises need to consider a wide range of factors when considering comprehensive indicators for urban construction. Feasibility analysis and simulated test routes can help improve the layout and structure of urban rail transit networks. Based on urban development and population mobility trends, scientific planning of urban rail transit networks ensures a rational layout and extensive coverage, fully considering future urban development needs, reserving development space, and avoiding duplication and waste of resources. This ensures accurate implementation and ultimately promotes stable enterprise development.

3.6.2 Improving the Level of Equipment and Facilities

Companies need to regularly inspect and maintain rail transit equipment and facilities to ensure they remain in proper working order. Updating and renovating outdated equipment, without compromising safe transportation, is a cost-saving approach. However, for significantly outdated equipment or other facilities, timely replacement is essential. Companies can establish robust equipment and facility maintenance management systems, clarifying the safety responsibilities of managers and operators at all levels. Improving the safety and reliability of the entire rail transit system and fostering greater passenger trust requires continuous investment and research.

Technological advancement is a key factor in business development. Increasing investment in technological upgrades and innovation can improve operational efficiency and reduce failure rates. Focusing on key core factors requires increased R&D investment to promote technological innovation and industrial upgrading. Collaboration with universities and research institutions is encouraged to jointly conduct technological R&D and talent development [4]. For example, the application of the Internet of Things (IoT) can enable real-time monitoring and maintenance of equipment, improve the feasibility of operational decisions through the use of big data and artificial intelligence, and streamline ticket processing through electronic payment systems. Businesses also need to upgrade their environmental protection efforts, promoting the use of energy-saving and environmentally friendly technologies. With the goal of reducing energy consumption and pollution emissions, they should adopt green lighting, energy-efficient air conditioners, and other equipment to improve energy efficiency and reduce environmental pollution.

3.6.3 Innovative Business Models

In addition to traditional government investment and bank loans, companies continue to explore diversified financing channels, such as public–private partnerships (PPPs), REITs (real estate investment trusts), and bond issuances, to attract private capital to rail transit construction. Innovative financing models, such as establishing dedicated funds and introducing strategic investors, are also being developed to enhance financing efficiency and flexibility, tailored to project characteristics. In practice, they are leveraging land resources along rail transit lines for commercial development, such as shopping malls, office buildings, and residential buildings, to increase revenue streams. They are also offering a variety of value-added services, such as advertising, Wi-Fi, and mobile payments, to meet diverse passenger needs and increase non-ticketing revenue. Furthermore, they are collaborating with other companies to develop joint projects and fully leverage resources.

3.6.4 Improving Operational Management

Optimize operational management models and improve operational efficiency and service quality. Enterprises can use intelligent and information-based means to achieve operational scheduling, passenger flow analysis, and safety assurance, establish a scientific operational management system, and achieve refined and standardized operational management [5]. Strengthen cost control and budget management, improve resource utilization efficiency and management level, and reduce operating costs through energy conservation and consumption reduction, resource sharing, and other measures. Specifically, enterprises should concretize and clarify management responsibilities, requiring every manager to be in place and perform their duties diligently, conduct daily inspections of the day's situation, and promptly correct and address any problems found.

As a public transportation service, companies need to improve their customer service. First, they should establish a comprehensive customer relationship management system to strengthen communication and interaction with passengers to increase passenger satisfaction and loyalty. Furthermore, they can establish a passenger feedback mechanism to collect passenger opinions and suggestions through questionnaires and suggestion boxes, thereby promptly understanding passenger needs. To enhance the passenger experience, they can improve the sanitation of stations and carriages, provide a more comfortable riding environment, and ensure smooth and accurate information transmission. Furthermore, they can introduce mobile applications, self-service kiosks, and more personalized customer service to provide passengers with more convenient and personalized services. Finally, they can strengthen employee training to enhance professional skills and service awareness.

3.6.5 Obtaining Government Support

The government can formulate and issue a series of policy documents to support the development of urban rail transit, such as the "Guiding Opinions on Comprehensively Promoting the Construction of an Urban Integrated Transportation System," to clarify development goals, tasks, and policy measures, and provide policy guarantees for urban rail transit companies. Urban rail transit construction should be incorporated into urban master plans and comprehensive transportation plans to ensure the coordination of rail transit with urban development. At the same time, through planning guidance, the layout of the rail transit network should be optimized to improve rail transit coverage and accessibility. Regarding safety supervision and emergency response, the government should strengthen safety supervision of urban rail transit companies, establish and improve safety management systems and emergency response plans, ensure the safety of rail transit operations, support urban rail transit companies in strengthening emergency drills and training, improve emergency response speed and handling capabilities, and ensure that they can respond quickly and effectively to emergencies.

The government should promptly provide appropriate financial subsidies to urban rail transit companies for policy-driven losses resulting from low fares, fare reductions, and operating off-the-beaten-path routes, ensuring their normal operations. Urban rail transit companies should also be given preferential tax policies, such as reductions and exemptions for corporate income tax and value-added tax, to reduce their tax burden and improve their profitability. On the technical side, the government can support urban rail transit companies in establishing partnerships with universities and research institutions to jointly cultivate rail transit professionals and improve the quality of industry talent. Furthermore, the government can formulate preferential policies to attract outstanding domestic and international rail transit professionals to work in my country, providing a strong talent base for urban rail transit companies.

Urban rail transit companies are poised to maintain rapid growth in the future. With the acceleration of urbanization and increasing travel demand, urban rail transit companies will further expand their operations and enhance their technology and service quality. At the same time, they will also face additional challenges, such as intensified market competition and rising operating costs. Against this backdrop, urban rail transit companies must continuously innovate and enhance their competitiveness to achieve sustainable development.

The letters in this article that represent the companies are:

C1-Shenzhen Metro, C2-China Railway Construction, C3-China Railway Group, C4-Tunnel Group, C5-Guangdong Hydropower, C6-Jinhua Metro, C7-Wuhan Metro, C8-Chongqing Rail Transit, C9-Kunming Rail Transit, C10-Guangzhou Metro, C11-Hangzhou Metro, C12-Suzhou Rail Transit, C13-Chengdu Rail Transit, C14-Nanjing Metro, C15-Tianjin Rail Transit, C16-Changsha Rail Transit, C17-Beijing Metro, C18-Beijing Infrastructure Investment Co., Ltd., C19-Nanchang Rail Transit, C20-Jinan Rail Transit, C21-Shenyang Metro, C22-Tianjin Metro, C23-Xiamen Rail Transit, C24-Changchun Rail Transit, C25-Lanzhou Rail Transit, C26-Ningbo Rail Transit, C27-Wenzhou Metro, C28-Fuzhou Metro, C29-Changzhou Metro, C30-Nanning Rail Transit, C31-Xi'an Rail Transit, C32-Qingdao Metro, C33-Zhengzhou Metro.

References

1. Yao Shifeng, Lei Tingting, Bai Li, et al. 2018–2023 Urban Rail Transit Enterprise Digital Transformation Report (II): Ten Major Development Trends [J]. Urban Rail Transit, 2024, (06): 35–43.
2. Yao Shifeng, Lei Tingting, Bai Li, et al. 2018–2023 Urban Rail Transit Enterprise Digital Transformation Report (I): Research Background and Case Insight Analysis [J]. Urban Rail Transit, 2024, (05): 16–23.
3. Wang Yu. Research on the management issues and optimization measures of operating costs of urban rail transit enterprises [J]. Modern Business Research, 2024, (06): 77–79.

4. Li Ying. Paths for urban rail transit enterprises to build a team of highly skilled talents [J]. Modern Corporate Culture, 2023, (12): 125–128.
5. Huang Tianyin, Xu Wenjie, Guo Peng, et al. Strategies and applications for improving the efficiency of urban rail transit enterprises [J]. Urban Rail Transit Research, 2023, 26(02): 6–10+15.

Chapter 4
Environmental Challenges, Risk Governance, and Strategic Adaptation in China's Water Transport Enterprises

Jianlin Wang and Xue Xing

Abstract China's water transport industry occupies an important position in the national economy. As a vital component of domestic logistics and international trade, its development is of great significance for promoting China's economic and social development and the prosperity of international trade. This article first outlines the current development status of China's water transport industry. It then analyzes the impact of factors such as the policy environment and innovation environment on the development of China's water transport enterprises, further revealing the practices and innovations of these enterprises in improving operational efficiency, promoting green and low-carbon transformation, and utilizing big data and intelligent technologies to enhance service quality. Finally, this paper proposes future development trends and corresponding strategic suggestions for Chinese water transport enterprises, aiming to provide reference and guidance for them.

4.1 Introduction

With the continuous deepening of global economic integration, water transport has become a key link connecting the world due to its significant advantages such as large carrying capacity and low cost. On the one hand, progress in the water transport sector supports the stability of China's economy and promotes the upgrading of its industrial structure. On the other hand, as the world's second-largest economy and a key player in international trade, China's advancements in water transport also enhance its international competitiveness. In recent years, the implementation of the "Belt and Road" Initiative and the strategic deployment of building a maritime power have opened up broad development space for China's water transport industry. However, the sector also has to cope with multiple challenges such as uncertainties in the international trade environment, fluctuations in the shipping market, and increasingly

J. Wang · X. Xing (✉)
Anhui University of Finance and Economics, Bengbu City, Anhui Province, China
e-mail: 19956325055@163.com

J. Wang and M. Song (eds.), *The Green Path of Chinese Transportation Enterprises*, Resources, Climate and Sustainable Development,
https://doi.org/10.1007/978-981-95-5289-4_4

stringent environmental regulations. In the new era, Chinese water transport enterprises urgently need to closely track industry dynamics and adopt proactive strategies to address various challenges to ensure high-quality and sustainable development.

4.2 Overall Development Situation of Chinese Water Transport Enterprises

Water transport relies on ships. Compared to other modes of transport, ship transport has the following advantages: I. Strong carrying capacity. Water transport uses ships as carriers, allowing for massive cargo volumes. II. Low transport cost. The cargo capacity of a standard freighter can reach tens of thousands of tons, equivalent to hundreds of trucks of similar capacity. III. Low energy consumption. Water transport operates on natural waterways such as rivers, lakes, and oceans, reducing the occupation and destruction of land resources, and it can effectively utilize natural energy sources like water currents, wind, and gravity, further reducing energy consumption. IV. Suitable for large-scale long-distance transport. Due to its characteristics, ships are suitable for large-scale and long-distance maritime transport, capable of continuous sailing on the ocean unrestricted by land traffic constraints, achieving seamless connections between major ports and ensuring the efficiency of global cargo flow.

4.2.1 Overview of China's Water Transport Industry

4.2.1.1 The Market Size of China's Water Transport is Increasingly Expanding

Currently, the market size of China's water transport industry continues to grow. Particularly in terms of freight volume, it repeatedly hits new highs. According to data from the Ministry of Transport, from 2019 to 2023, China's water transport freight volume showed an overall increasing trend. In 2023, China's water transport freight volume reached 9.367 billion tons, an increase of 9.5% compared to 2022. In the first half of 2024, China's water transport freight volume reached 4.606 billion tons, a year-on-year increase of 6.28%. These figures indicate strong growth momentum in China's water transport market.

Data shows that from 2020 to 2022, China's water transport passenger volume experienced a certain decline. Especially in 2022, due to the pandemic outbreak and strict control measures, water transport passenger volume dropped to a low of around 150 million person-times. With the end of the pandemic, passenger volume began to recover in 2023. According to official data, the national waterway passenger volume completed in 2023 was 258 million person-times, a significant increase of 121.6% compared to 2022. As of the first half of 2024, passenger volume increased

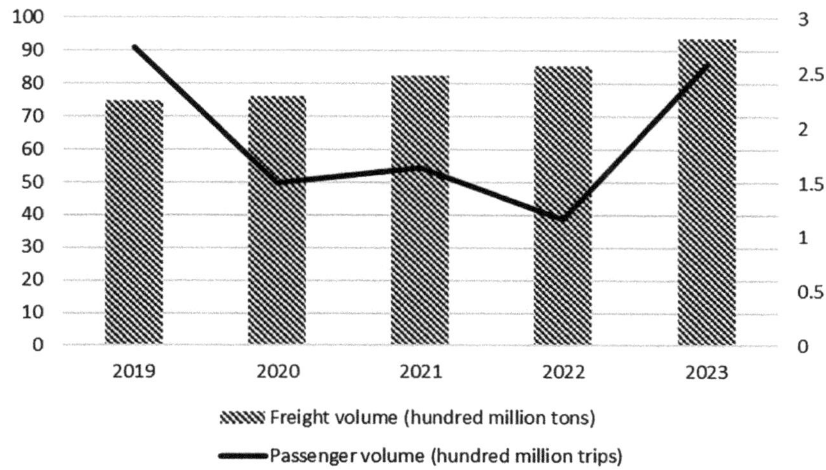

Fig. 4.1 Freight volume and passenger volume of water transport enterprises from 2019–2023. *Data source* Ministry of transport

to 124 million person-times, a year-on-year increase of 5.23%, further proving the continuous recovery and growth of the water transport passenger market (Fig. 4.1).

4.2.1.2 Sustained Growth in Fixed Asset Investment in China's Water Transport Industry

Between 2019 and 2023, China's waterway investment grew significantly, with fixed asset investment in water transport rising from 113.7 billion RMB to 201.6 billion RMB. During this period, fixed asset investment in the water transport industry demonstrated strong growth momentum and continuous development potential, reflecting the state's emphasis and investment in the construction of water transport infrastructure. Waterway fixed asset investment reached a peak growth in 2023, climbing to 201.6 billion RMB, a year-on-year increase of 20.1%. This growth benefits not only from national policy support but also relies on the enhanced capabilities and innovation of water transport enterprises themselves (Fig. 4.2).

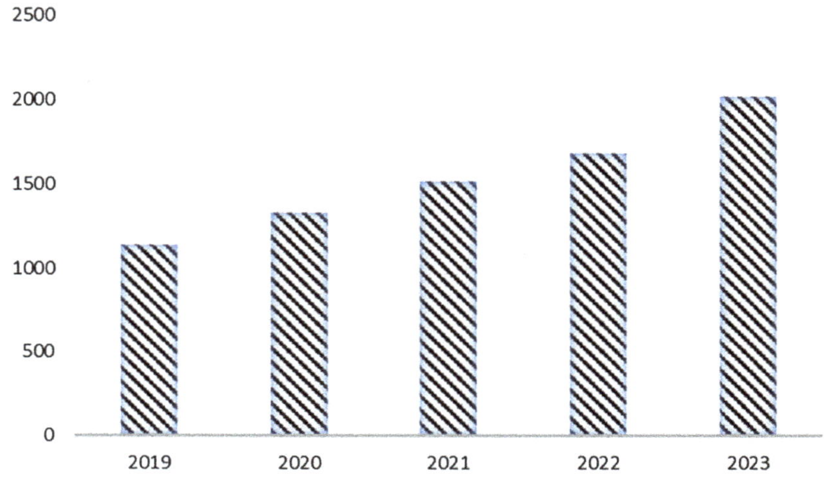

Fig. 4.2 Fixed investment in the water transport industry from 2020–2023. *Data source* Transportation Network

4.2.2 *Analysis of the Current Status of Chinese Water Transport Enterprises*

4.2.2.1 Current Development Status of Chinese Water Transport Enterprises

Large-scale water transport enterprises in China possess a vast fleet and a wide network of routes, playing an important role in the water transportation industry both domestically and globally. In terms of water transport business revenue, ZHKGGF leads significantly with 168.126 billion yuan in 2023, reinforcing its industry-leading position while showcasing its strong foundation and extensive influence in the global shipping market. Similarly, ZSNYGF and ZYHYYS have also demonstrated impressive results, with both exceeding 20 billion yuan in water transport business revenue. In terms of fleet size, ZHKGGF maintains its leading position with its large fleet, while ZSNYGF and ZYHYYS also have fleets of over 200 vessels each. These vessels cover ocean, coastal, and inland river areas, allowing them to provide more efficient and convenient logistics services to domestic and international customers (Fig. 4.3).

4.2.2.2 Introduction to Key Enterprises

(1) ZHKGGF was established in 2005 and is a global digital supply chain operation and investment platform, which boasts strong competitiveness in the global shipping industry. The company focuses on container shipping and is committed to

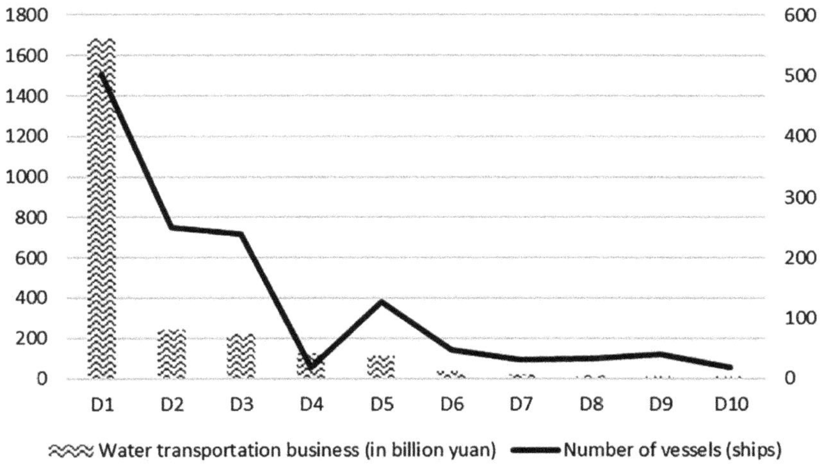

Fig. 4.3 Water transport revenue and number of ships of Chinese water transport enterprises in 2023. *Data source* Annual reports of respective Chinese water transport enterprises

creating an efficient, intelligent, and sustainable global logistics service system. As one of the leading shipping companies globally, it possesses several major advantages: First, financial advantage. ZHKGGF has ample monetary funds and strong financial strength, providing a solid financial foundation for the company's daily operations and strategic investments. According to data, in 2023, ZHKGGF's waterborne transportation business revenue reached 168.126 billion RMB. Since 2024, ZHKGGF has focused on developing productivity, accelerating the "digital intelligence" and "green low-carbon" transformation, and in the first quarter, the net profit attributable to the parent company reached 6.755 billion RMB, a quarter-on-quarter increase of 277.6%, achieving a "good start" for the first quarter of 2024. Second, the advantages of fleet size and route network. ZHKGGF has a large fleet size and a wide range of routes. In 2024, the company's capacity steadily increased to about 3.1 million TEUs, with a self-operated fleet of 510 container ships, operating 436 routes, docking at 611 ports in approximately 144 countries and regions globally, with a significant increase in route coverage compared to the end of last year (source: company website). This scale advantage gives the company stronger bargaining power and market competitiveness in the global shipping market. Meanwhile, according to data, the revenue from the container shipping business and the terminal business reached 46.533 billion RMB and 2.384 billion RMB respectively, with year-on-year growth of 1.82 and 4.35%. At the same time, the cargo volume of container shipping services climbed to 6.0273 million TEUs, while the total throughput of terminal operations reached 33.2622 million TEUs, with these two indicators showing year-on-year growths of 10.53 and 9.76%. In the future, ZHKGGF will undergo full-scale transformation and upgrading, as well as high-quality

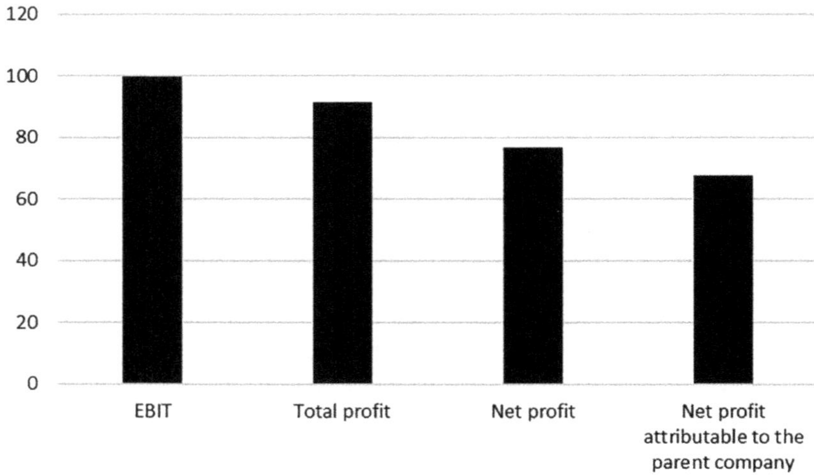

Fig. 4.4 ZHKGGF operating performance in Q1 2024 (Unit: billion RMB). *Data source* Company annual report

development to effectively resist external challenges and provide better services to customers (Fig. 4.4).

(2) ZSNYGF was established in 2004 and owns a world-class fleet of supertankers and very large ore carriers. Its fleet size, transport capacity, and service quality are all at the leading level in the industry, occupying an important position in the global shipping field. The company's business is diversified, with its core business focusing on ocean-going tanker transportation, while also providing comprehensive transportation services for various types of ships including bulk carriers, container ships, and roll-on/roll-off ships. (Source: Company website) Compared to other water transport enterprises, ZSNYGF has the most historically long-standing and experienced ocean-going tanker fleet in China. Through years of development, the fleet has accumulated rich experience and professional knowledge in the transportation field. From route planning to ship maintenance management, from handling complex maritime situations to the safe delivery of cargo, ZSNYGF's crew and management team demonstrate excellent professional skills. This expertise and technology not only ensure the safe use of ships but also provide customers with efficient and reliable shipping services. Financially, ZSNYGF shows a steady growth trend. According to the latest financial report, the company's operating revenue continues to grow. In the first half of 2024, the company achieved an operating revenue of 13.235 billion RMB, an increase of 1.88 percentage points compared to the same period last year. Net profit reached 2.497 billion RMB. Since its A-share listing, the company has cumulatively distributed 8.385 billion RMB in dividends. Although the overall economic environment is not optimistic, the company still maintains a high

level of profitability relying on its strong market competitiveness and excellent operational management capabilities. (Data source: Company annual report)

(3) SZGWL was established in 2003 and is one of the pioneers in China's domestic container shipping industry, always committed to informatization, digitalization, and scientific development. Its main businesses include liner shipping, integrated logistics, ship management, and entity investment. SZGWL integrates highway, railway, and waterway resources is committed to building a diversified intermodal transport system with containers as the carrier and water transport as the leading mode. This system covers multiple ports across the country, has multiple routes, and integrates numerous trailer companies, container trucks, and barges, aiming to provide a comprehensive logistics service solution to ensure the smooth completion of the entire transportation process for goods. At the same time, as a pioneer in domestic shipping big data and e-commerce platforms, SZGWL has obvious advantages in informatization and customized services. The company provides services such as booking, cargo tracking, and sailing schedule inquiries through its information platform, focusing on customer needs and aiming to provide customers with one-stop, comprehensive logistics services including warehousing, distribution, and information processing. (Source: Company website) In recent years, SZGWL's operating results have been significant. According to the latest financial data released, the company's total revenue reached 12.439 billion RMB in 2023, net profit attributable to the parent company was 1.717 billion RMB, and net profit was 1.722 billion RMB. A specific analysis of the revenue composition shows that water transport, as the company's core business segment, plays a pivotal role. In 2023, the company's water transport revenue reached 10.112 billion RMB, accounting for 81% of the company's total revenue. This figure highlights the strong strength of the company's water transport sector and consolidates its position as the main source of company revenue. (Data source: Company annual report)

4.3 Analysis of the Development Environment for Chinese Water Transport Enterprises

4.3.1 Internal and External Environment Facing Chinese Water Transport Enterprises

4.3.1.1 Analysis of Domestic and International Policy Environment

In 2022, China's transportation department released the "14th Five-Year Plan for Water Transport," an initiative marking the entry of the water transport field into a critical development period. This plan mainly focuses on improving the quality of multimodal transport and enhancing the overall efficiency of the water transport system. Meanwhile, a joint notice issued by the Ministry of Finance and the Ministry of Transport clarified the development direction of water transport, emphasizing the active promotion of the digital transformation of water transport infrastructure. This includes digital transformation projects, the construction of perception networks, intelligent management systems, and the improvement of network service systems. Furthermore, the 20th National Congress of the Communist Party of China further emphasized the core role of water transport in the comprehensive transportation network, opening up broad development space for enterprises related to water transport.

Globally, the increasing prominence of international trade protectionism has led to intensified trade barriers. This directly affects the access of Chinese export goods to international markets, indirectly reducing the global demand for Chinese water transport services and dealing a huge blow to the business volume of domestic water transport enterprises. Additionally, maritime regulations formulated by institutions such as the International Maritime Organization (IMO) have a significant impact on the global water transport industry. For Chinese water transport enterprises, compliance with these international standards is not only the fulfillment of international responsibilities but also an important measure to enhance international competitiveness and improve the national image.

4.3.1.2 Analysis of the Innovation Environment for Water Transport Enterprises

Technological progress drives the water transport industry to quickly adopt innovative technologies, effectively promoting the leap in operational efficiency of Chinese water transport enterprises. Taking JSGJ as an example, according to the group's performance in the first half of 2024, not only did operating revenue and profit both exceed half-year targets, but cargo and container handling volumes also achieved year-on-year growth of 6.09 and 13.6%, respectively. In terms of ship technology innovation, the group actively promotes the introduction of automated and intelligent ships, effectively enhancing the efficiency of loading and unloading operations and the safety and stability of ship navigation. In terms of route strategy planning, enterprises use big data and intelligent algorithms to accurately predict sailing time, fuel consumption, and potential waterway congestion. In terms of optimizing route design, enterprises use smart waterway technology to effectively reduce sailing cycles, lower fuel consumption and carbon emissions, thereby achieving a win–win situation for economic benefits and environmental protection [1]. (Data source: Company annual report)

4.3.2 Challenges Currently Facing Chinese Water Transport Enterprises

4.3.2.1 Increasing Pressure for Environmental Protection and Sustainable Development

Globally, environmental protection and sustainable development issues have received much attention in recent years, bringing unprecedented challenges to the water transport industry. Ships generate a large amount of sewage and waste oil during transport navigation. These wastes include water discharged from ship cooling systems containing heavy metals and various chemical components, untreated domestic sewage, and oily liquids released by mechanical parts such as ship engines and lubrication systems during operation. If this sewage and waste oil are discharged into water bodies without proper treatment, they will seriously affect the water quality of the local waters, subsequently damaging the ecosystem structure. Furthermore, the fuel used in ship navigation produces air pollutants such as sulfur dioxide, which are emitted into the atmosphere through exhaust gases, further exacerbating global climate warming and becoming a key factor contributing to frequent haze events and declining air quality in cities and surrounding areas. This poses a serious threat to human respiratory health, especially for children, the elderly, and people with respiratory diseases. In the maritime sector, ships also face the potential threat of maritime accidents, including spills of harmful substances such as fuel and lubricating oil caused by ship collisions or grounding on reefs. These spilled substances can quickly spread to surrounding waters, posing a serious threat to the marine ecosystem. The ship collision incident that occurred in the waters of Jiaxing, Zhejiang, in 2018 resulted in the spill of about 800 tons of base oil, causing severe pollution to the nearby sea area. This oil spill not only caused serious damage to the marine ecology of the nearby sea area but also brought a huge economic burden. Oil pollution can not only directly damage the seabed ecological structure but may also threaten human food safety and health through enrichment in the food chain. Moreover, ecological restoration of polluted areas often takes a long time, ranging from several years to decades. This continuous ecological intrusion seriously affects local economic development, tourism, and residents' daily lives.

4.3.2.2 Increasingly Fierce Market Competition

With the continuous acceleration of globalization, market competition in the water transport industry is becoming increasingly fierce, a change that has severely affected various companies within the industry. Large water transport companies strive to expand their market share seeking a more advantageous position, while small enterprises face the severe challenge of being acquired or forced out of the market. According to relevant research reports, China's water transport industry has shown a clear market competition hierarchy, with COSCO SH as the leading enterprise, its water transport business revenue exceeding 100 billion RMB, far surpassing other competitors. Such market competition has led to increasing market concentration within the industry, with competition increasingly focused among a few large companies, and the resulting competitive pressure is also intensifying. To maintain a competitive advantage in such a complex market environment, various water transport companies have implemented various strategies. These strategies have significantly improved the operational efficiency of enterprises, but they have also inevitably led to further intensification of market competition. Furthermore, with the deepening of international trade, competition in the water transport market has expanded from local to global. These international water transport giants, leveraging their advantages in scale, technology, and management, continue to capture market share, putting enormous pressure on domestic water transport enterprises. At the same time, the constant evolution of international trade rules, the rise of global trade protectionism, and geopolitical instabilities such as the Red Sea crisis, the Russia-Ukraine conflict, and the Israel-Palestine conflict have further increased the risks and uncertainties for Chinese water transport enterprises. In this context, under the increasingly complex environment of the global water transport market competition landscape, Chinese water transport enterprises need to adhere to more prudent principles when formulating future development strategies and actively meet the challenges [2].

4.3.2.3 High Risks in Dangerous Goods Transport

Hazardous chemicals, due to their characteristics such as flammability, explosiveness, and toxicity, pose extremely high dangers and uncertainties in the logistics and transportation link. In the water transport of dangerous goods, if accidents such as leaks, fires, or explosions occur, they will directly threaten the nearby water environment, ship safety, and the life and health of the crew and surrounding personnel. Specifically, during the transportation of dangerous goods, if the packaging materials do not meet standards or the ship's sealing performance is insufficient, it may lead to leakage of dangerous goods, causing serious water pollution, and even more seriously, triggering disasters such as fires or explosions. Moreover, during navigation, ships face challenges from extreme weather, such as typhoons, strong storms, and huge waves [3]. These factors may cause ship damage or even sinking, further increasing the risk of dangerous goods leakage. Furthermore, if the crew fails to strictly follow relevant operating procedures during the loading, unloading, and transportation of dangerous goods, and lacks necessary understanding of the properties of dangerous goods and emergency handling methods, they may not be able to quickly and effectively take countermeasures when sudden accidents occur, thereby exacerbating the severity of the accident. Take the accident that occurred on August 9, 2024, as an example. A container ship in the Beilun Port Area of Ningbo-Zhoushan Port experienced a combustion and explosion during transportation. The transit dangerous goods loaded included Class 5.2 oxidizers and organic peroxides. According to relevant investigation results, it was precisely these types of dangerous goods that caused the combustion and explosion incident. Class 5.2 oxidizers and organic peroxides have high oxidation activity, are prone to react with other substances, and can cause combustion or explosion under specific conditions. This incident reminds us once again that the transportation safety of dangerous goods cannot be ignored, and all parties must strengthen the management and supervision of the transportation process to ensure transport safety [4].

4.4 Future Development Trends and Strategic Recommendations

4.4.1 Adhere to Environmental Protection and Sustainable Development Strategies

4.4.1.1 Optimization of Transport Modes

Water transport enterprises need to continuously promote the innovation and upgrading of transport modes and are committed to building a technology-led, intelligent, efficient, and environmentally friendly green logistics network system. Enterprises should use advanced technological tools such as big data and cloud computing to accurately predict cargo flow trends and demand directions, achieve intelligent planning of transport routes and strategies to effectively reduce empty loads and traffic congestion, thereby achieving the goal of energy saving and emission reduction. In this process, enterprises should fully exploit the unique advantages of water transport (low cost, large capacity, environmental protection) [5], strengthen coordination and cooperation with other transport modes such as railway and highway, promote efficient and environmentally friendly door-to-door integrated transport services, and build a set of multimodal transport mechanisms to enhance overall transport efficiency [6].

4.4.1.2 Cargo Packaging Materials

When selecting cargo packaging materials, water transport enterprises need to comprehensively consider the entire life cycle of the packaging, covering all key links from raw material procurement, use, to waste disposal, and ensure that environmental factors are always integrated throughout. Water transport enterprises should establish close collaboration with suppliers to jointly develop new environmentally friendly packaging materials, such as bio-based materials and nanocomposite materials with excellent physical properties. This is conducive to promoting their rapid degradation in the natural environment to reduce reliance on landfill and incineration treatment. Furthermore, water transport enterprises should also proceed to establish packaging recycling mechanisms and actively advocate customer participation in recycling programs to promote the construction of a circular economy system, striving to minimize the environmental impact of packaging waste.

4.4.2 Strengthen Risk Management and Control for Dangerous Goods Transport

Since China proposed and promoted the "Belt and Road" Initiative, the importance of water transport has become increasingly significant. However, the safety hazards of dangerous chemicals in water transport cannot be ignored. Due to their highly flammable and explosive dangerous attributes, if an accident occurs, it will inevitably cause serious casualties and property losses, and have a profound adverse impact on the ecological environment. In view of this, regarding the transportation management of dangerous chemicals, this article puts forward relevant suggestions from three aspects: personnel, transport vehicles and facilities, and cargo packaging and loading management [7].

4.4.2.1 Personnel Aspects

Chinese water transport enterprises must provide comprehensive training for key personnel involved in the transportation of dangerous goods, such as drivers and escorts. This mainly includes the physical and chemical properties of dangerous goods, emergency handling methods, etc., to enhance the safety vigilance, risk identification ability, and handling capacity of practitioners. Water transport enterprises should establish a strict qualification certification system, requiring relevant personnel to pass professional exams and obtain corresponding qualification certificates before engaging in the transportation of dangerous goods. This ensures that every staff member involved in the transportation of dangerous goods possesses the necessary professional knowledge and skills, providing a solid human resource guarantee for safe transportation work.

4.4.2.2 Transport Vehicles and Facilities Aspects

Water transport enterprises should give priority to selecting dangerous goods transport ships that comply with international and national safety standards to ensure that the ship's structural strength, equipment configuration, and fire and explosion protection performance meet the special requirements of dangerous goods transportation. Furthermore, shipping companies need to establish a reliable ship maintenance and inspection system, conducting regular comprehensive inspections and maintenance of ships to ensure they are always in good operating condition. Water transport enterprises should pay special attention to the management of key safety elements such as fire prevention, explosion prevention, and leakage prevention on ships, strengthening inspections to ensure the ship's safety performance is always in the best condition. At the same time, ships must be equipped with sufficient emergency facilities and personal protective equipment, including but not limited to fire-fighting equipment, sealing materials, and protective gear, to ensure that countermeasures can be taken quickly and effectively in case of sudden emergencies.

4.4.2.3 Cargo Packaging and Loading Management Aspects

During cargo transportation, packaging materials that meet safety requirements must be selected. These materials must be able to withstand various stresses generated during transportation, such as vibration, impact, and temperature changes, and have good sealing to prevent cargo leakage. In terms of packaging methods, enterprises should follow the principles of scientificity and sufficiency, ensuring stable packaging structure and protecting goods from external damage. For fragile goods, additional reinforcement measures should be taken, such as adding shock-absorbing layers or using impact-resistant packaging. Furthermore, information such as the product name, type, hazard level, and emergency measures must be clearly displayed on the packaging to facilitate quick identification and appropriate action in case of an emergency. During loading and unloading of goods, operating procedures must be strictly adhered to, appropriate loading methods and tools must be selected, and operations should be carried out according to the characteristics of the goods. When stacking goods, follow the stacking method for the goods and ensure sufficient spacing between goods. This is conducive to ventilation and heat dissipation, avoiding dangers caused by heat accumulation.

4.4.3 Scale Operation and Strategic Mergers and Acquisitions

4.4.3.1 Scale Operation

Water transport enterprises can achieve scale operation by adopting strategies such as increasing the number of ships and expanding transport capacity, thereby enhancing their competitiveness in the global shipping market. Enterprises can increase the number of ships in a timely manner based on the operational efficiency of existing ships to improve transport capacity and speed. In terms of expanding transport capacity scale, enterprises should not be limited to simply increasing hardware facilities but also undertake comprehensive upgrades and modernization of ship performance. Adopting more energy-saving and environmentally friendly ship design concepts and integrating automation and intelligent technologies can further enhance ship transport efficiency. Furthermore, enterprises should deeply analyze multi-dimensional factors such as market demand dynamics, cargo flow trends, and waterway conditions to design more reasonable and efficient route networks, shorten ship transport cycles, reduce enterprise operating costs, more accurately meet the diverse needs of customers, and thereby further enhance customer satisfaction.

4.4.3.2 Strategic Mergers and Acquisitions and Alliances

Water transport enterprises can significantly expand their market share and markedly enhance their comprehensive competitiveness in a relatively short period by implementing strategic mergers and acquisitions (M&A) or forming close strategic alliances. This strategy not only enables some enterprises to rapidly expand their scale but also promotes the effective allocation of resources and the effective integration of businesses, opening up broader growth paths for water transport enterprises and strengthening their market influence. Taking shipping giants like COSCO SH as an example, these groups have successfully achieved rapid expansion of their enterprise scale by means of various methods such as M&A and alliances. This move has not only significantly increased their market share but also further consolidated their leadership position in the industry. Through M&A, enterprises can quickly absorb high-quality assets, broaden business areas, and enhance overall strength; through strategic alliances, they can achieve mutually beneficial cooperation in areas such as resource sharing and risk sharing, jointly strengthening their resilience to market risks and their ability to seize market opportunities.

4.4.4 Strengthen Cooperation and Exchange

With the accelerated advancement of globalization and the prosperity of international trade, the water transport industry is facing unprecedented development opportunities. Strengthening cooperation and exchange with other enterprises and other modes of transport is particularly crucial for the sustainable development of the water transport industry.

4.4.4.1 Deepen Regional Cooperation to Build a Development Foundation

Water transport enterprises should strengthen close cooperation with local governments, ports, and shipping companies to jointly discuss cooperation strategies for information exchange and resource sharing. At the same time, water transport enterprises need to strengthen collaboration in areas such as capacity allocation, information interoperability, and risk prevention and management. This will help optimize ship resource allocation, improve transport efficiency, and thereby enhance the comprehensive competitiveness and risk management capabilities of water transport enterprises. Furthermore, water transport enterprises should actively participate in various regional cooperation forums, such as regional organizations like the Yangtze River Delta Shipping Cooperation Alliance. Participation in regional cooperation forums not only helps deepen understanding and trust with other enterprises in the region but will also promote information sharing and coordinated development among regional enterprises [8].

4.4.4.2 Promote Multimodal Transport to Enhance Efficiency

Water transport enterprises need to deepen close cooperation with other transportation fields such as highway and railway. Enterprises can achieve seamless connection between water transport and other transportation modes by promoting innovative transport models such as multimodal transport and river-sea combined transport, thereby enhancing the overall efficiency of cargo transportation. Multimodal transport is an efficient and flexible transportation method that achieves optimization and seamless connection of the entire freight transportation process through the organic combination of different transportation modes. With the development of world trade, the application of multimodal transport in international trade and logistics is increasing. In the future, it will develop towards informatization, intelligence, and greening.

4.4.4.3 Actively Strengthen International Cooperation

In the context of globalization, China's water transport enterprises should actively seek new avenues for international cooperation. Strengthen communication and collaboration with overseas ports and leading shipping enterprises, and actively participate in the activities of international water transport organizations and keenly capture the dynamics of the international business environment. Based on this, flexibly adjust market strategies and trade network layouts to enhance the competitiveness of China's water transport enterprises on the international stage. Furthermore, China's water transport enterprises need to deepen technical exchanges and cooperation with internationally top-tier enterprises, absorbing and learning from foreign advanced technologies and management experience to enhance their own technical strength and management efficiency. Such international cooperation will help further expand China's water transport market and develop new international routes, enabling China's water transport enterprises to effectively cope with various challenges in international trade and occupy a more favorable position in the international market.

4.5 Conclusion

Water transport has always played an important role in the development of China's national economy. It is a vital bridge connecting domestic and international markets, concerning the smooth flow of the logistics chain, and influencing the efficiency of resource allocation, the stability of the industrial and supply chains, and the prosperity of foreign trade. With the deepening of global economic integration, Chinese water transport enterprises face many opportunities and challenges. Therefore, enterprises must respond actively and always uphold the concept of environmental protection and sustainable development, integrating this concept into the lifeblood of their own development. Enterprises can promote the use of clean energy ships and implement ship energy efficiency management to reduce carbon emissions; strengthen waste treatment and recycling measures to reduce enterprise operating costs, thereby enhancing the comprehensive competitiveness of Chinese water transport enterprises.

References

1. Yongshen Ling. Analyzing the Impact of River Transport Planning on the Safety of Water Transportation [J]. Automobile Weekly, 2024, (08): 13–15.
2. Yuzhen Zhang, Lu Dong. Discussion on Opportunities and Challenges Faced by Water Transportation under the New Economic Situation [J]. China Navigation Weekly, 2024, (08): 45–47.
3. Dong Xiao. Analysis of Causes of Marine Vessel Accidents and Preventive Measures [J]. Pearl River Water Transport, 2024, (02): 116–118.
4. Yanhua Hu, Linlin Lu, Caiyun Xue. Innovation Leads to Guaranteeing the Safety of Dangerous Goods Waterway Transportation [N]. China Transportation News, 2024-07-19(004).
5. Haifeng Liu. Reflections on Economic Development of Water Transportation under a Low-Carbon Economy [J]. Cooperative Economy and Technology, 2024, (14): 38–39.
6. Xiaoli Ai. The Waterway Transportation Demand in the Hinterland North of the Yellow River along the Beijing-Hangzhou Grand Canal [J]. China Ports, 2024, (06): 20–23.
7. Li Ding. Prospects for the Development of China's Port Industry [J]. Port Technology, 2017, (01): 1–3.
8. Sheng Zhang, Yi Lu, Hui Wu, et al. Accelerating Water Transport Development to Promote the Adjustment of Transportation Structure in Hunan [J]. China Water Transport, 2024, (02): 15–17.

Chapter 5
The Evolution and Economic Landscape of Civil Aviation Enterprises in China

Jianlin Wang and Mengjie Zhu

Abstract This paper provides an in-depth analysis of the current development status, market environment, economic conditions, and challenges and opportunities facing China's civil aviation enterprises. It highlights that after years of development, China's civil aviation industry has become one of the nation's vital strategic sectors, playing an irreplaceable role in driving economic growth and facilitating international exchanges. However, aviation enterprises also confront challenges such as oil price volatility and safety incidents. The article particularly emphasises the application of digital transformation and artificial intelligence technologies within the civil aviation sector, alongside the development potential in areas such as green transition and general aviation. Moving forward, China's civil aviation enterprises must seize opportunities arising from technological innovation, growing market demand, and policy support to achieve high-quality development, thereby establishing China as a leading aviation power.

5.1 Current Development Status of Civil Aviation Enterprises in China

5.1.1 Historical Development of Civil Aviation Enterprises

The development journey of China's civil aviation enterprises is a history of starting from scratch, growing from small beginnings, and evolving from weakness to strength. Under the guidance of the Party and the government, China's civil aviation has traversed multiple phases including establishment, development, and reform, gradually transforming into a vital pillar industry supporting national economic and social progress. Looking ahead, civil aviation enterprises will continue to deepen

J. Wang · M. Zhu (✉)
Anhui University of Finance and Economics, Bengbu City, Anhui Province, China
e-mail: zikw2859091185@163.com

J. Wang and M. Song (eds.), *The Green Path of Chinese Transportation Enterprises*, Resources, Climate and Sustainable Development,
https://doi.org/10.1007/978-981-95-5289-4_5

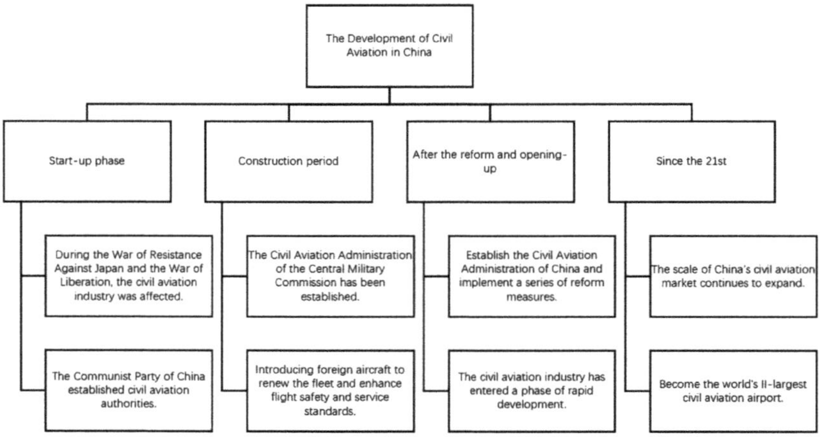

Fig. 5.1 Development History of Civil Aviation Enterprises

reforms across all fronts, foster high-quality growth, and strive towards the goal of establishing China as a leading aviation power (Fig. 5.1).

5.1.2 The Significant Position of Civil Aviation Enterprises Within China's Transport Sector

The civil aviation industry is regarded as a vital strategic sector for the nation and constitutes a key component in building China into a transport powerhouse. It plays a pivotal role in advancing Chinese-style modernisation, representing a transport mode within the integrated transport system that possesses significant latecomer advantages and growth potential. Not only does it facilitate internal and external circulation, but it also propels the transformation and upgrading of the economic structure. In terms of major national strategies, the civil aviation sector aligns with initiatives such as the coordinated development of the Beijing-Tianjin-Hebei region and the integrated development of the Yangtze River Delta, fostering synergy with world-class urban clusters while fully supporting the Belt and Road Initiative. By the end of 2022, China had concluded bilateral air transport agreements with 100 countries participating in the Belt and Road Initiative, maintaining regular passenger and cargo services with 64 of these nations. This initiative has further optimised overseas market and expanded global coverage. The sector's development level not only reflects national modernisation and economic competitiveness but also plays an irreplaceable role in facilitating international exchanges, providing public transport services, and supporting emergency rescue operations. Through air-rail intermodal transport and integrated air-ground operations, the civil aviation sector has achieved seamless connectivity between air transport and ground transportation, enhancing the synergistic effects of integrated air-ground transport networks. Although the value added by civil aviation and its related industries constitutes a relatively small proportion of GDP, its development is significantly influenced by macroeconomic factors and is closely linked to the nation's overall economic performance. Research indicates that passenger traffic on domestic civil aviation routes exhibits an elasticity of 1.9 times relative to GDP, while cargo and mail transport demonstrates an elasticity of 1.6 times. This underscores the sector's substantial economic contribution. Furthermore, characterised by high technological content, the civil aviation sector possesses an extended industrial chain and integrates advanced technologies, thereby providing extensive scope for technological innovation in related fields. Notably, the upstream aviation manufacturing sector may drive innovative development in industries such as materials science, metallurgy, chemical engineering, and mechanical manufacturing.

The civil aviation sector also makes substantial contributions to national infrastructure development. In 2023, total fixed-asset investment reached RMB 193.326 billion, with civil aviation infrastructure construction and technological upgrading accounting for RMB 124.13 billion. This has significantly advanced the modernisation and high-quality development of China's transport sector. Despite continuous growth in passenger volumes and flight frequencies, China's civil aviation sector achieved a 14.3% reduction in fuel consumption per tonne-kilometre in 2023 compared to 2005. Furthermore, the average energy consumption per passenger and carbon dioxide emissions per passenger at airports decreased by 38.4 and 60.5% respectively from their baseline levels. This demonstrates the sector's commitment to environmental stewardship, actively reducing pollution and advancing green, sustainable development.

5.2 Overview of Private Aviation Enterprises

5.2.1 Current Development Status of Civil Aviation Enterprises

5.2.1.1 Current State of Global Civil Aviation Development

China's civil aviation sector encompasses numerous aviation-related enterprises, particularly air transport companies. These entities primarily rely on aircraft for commercial transportation, generating the sector's principal economic returns. Beyond air carriers, ancillary businesses such as fuel suppliers, aviation materials providers, and sales enterprises operate in support of transport companies. Airlines, specifically, utilise various aircraft as transport vehicles, specialising in the aerial carriage of passengers or cargo. Furthermore, these companies may operate independently, form partnerships with other airlines, or establish alliances, with their service scope spanning intercontinental, intra-continental, or domestic routes.

Following the pandemic's impact, the global civil aviation sector has entered a new developmental phase. According to projections by the International Air Transport Association (IATA), global air passenger traffic is anticipated to exceed 4.7 billion passengers in 2024, with revenues expected to reach US$964 billion, while cargo revenues are forecast to amount to US$111 billion. Specific data indicates that global aviation demand maintained growth momentum throughout 2023. By May 2023, global revenue passenger kilometres (RPKs) had risen by 39.1% year-on-year, recovering to 96.1% of 2019 levels [1]. In 2023, the global aviation industry achieved total transport turnover of 1,188.3 billion tonne-kilometres, carried 620 million passengers, and transported 7.354 million tonnes of cargo and mail. Compared with the same period last year, these figures increased by 98.3, 146.1, and 21% respectively, recovering to 91.9, 96.1, and 98.3% of 2019 levels [2].

Key operational metrics for the entire civil aviation sector in 2024, including total transport turnover, passenger traffic, and cargo/mail volume, are projected to comprehensively surpass pre-pandemic levels. Passenger traffic is projected to achieve double-digit year-on-year growth, while cargo and mail volume is anticipated to see modest expansion [3]. This trajectory signals an optimistic outlook for China's civil aviation sector. Financially, the International Air Transport Association (IATA) forecasts global aviation net profits to reach US$30.5 billion in 2024, with a net profit margin of 3.1%—both figures surpassing 2023's performance. These figures reflect airlines' robust post-pandemic recovery and enhanced profitability. The global civil aviation sector is projected to maintain rapid and healthy development in the coming years, making significant contributions to global economic growth and societal progress.

5.2.1.2 The Current State of Development in China's Civil Aviation Industry

Since the founding of the People's Republic of China, and particularly following the implementation of reform and opening-up policies, the progressive deepening of market-oriented reforms has driven sustained rapid development across multiple facets of China's civil aviation sector. These include air transport, general aviation, fleet expansion, route design, regulatory framework development, and transport support systems [4]. In 2023, the total transport turnover of China's civil aviation industry reached 118.83 billion tonne-kilometres, representing a 98.3% increase compared to the previous year. Domestic routes achieved a total transport turnover of 86.733 billion tonne-kilometres, representing a year-on-year increase of 123.6%. Within this, routes to Hong Kong, Macao and Taiwan accounted for 1.00 billion tonne-kilometres, a 334.2% year-on-year rise. International routes recorded a total transport turnover of 32.101 billion tonne-kilometres, reflecting a 51.8% year-on-year increase (Fig. 5.2).

China's civil aviation industry transported 620 million passengers throughout the year, representing a year-on-year increase of 146.1%. Domestic routes carried 600 million passengers, up 136.3% year-on-year, while routes to Hong Kong, Macao and

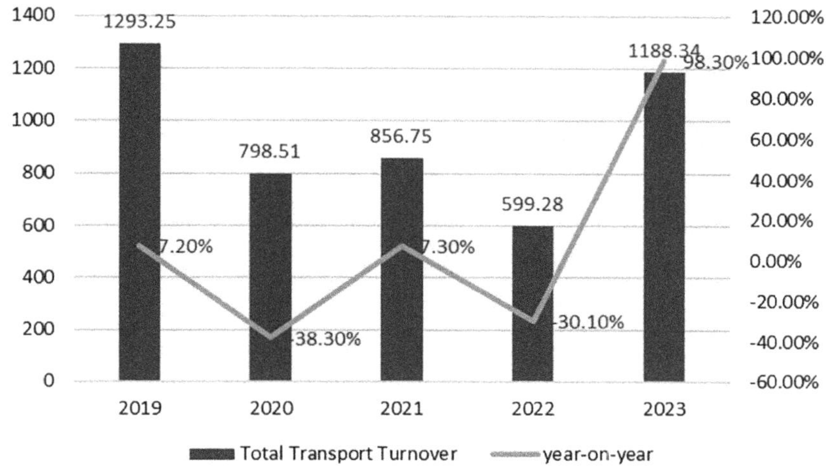

Fig. 5.2 Total civil aviation traffic volume, 2019–2023. *Data Source* 2023 Statistical bulletin on the development of the civil aviation industry

Taiwan handled 6.6845 million passengers, a 1324.7% year-on-year rise. International routes transported 29.05 million passengers, marking a 1461.7% year-on-year increase (Fig. 5.3).

China's civil aviation sector transported 7.3538 million tonnes of cargo and mail throughout the year, marking a 21% increase compared to the previous year and nearly returning to 2019 levels. Domestic routes handled 4.5639 million tonnes of

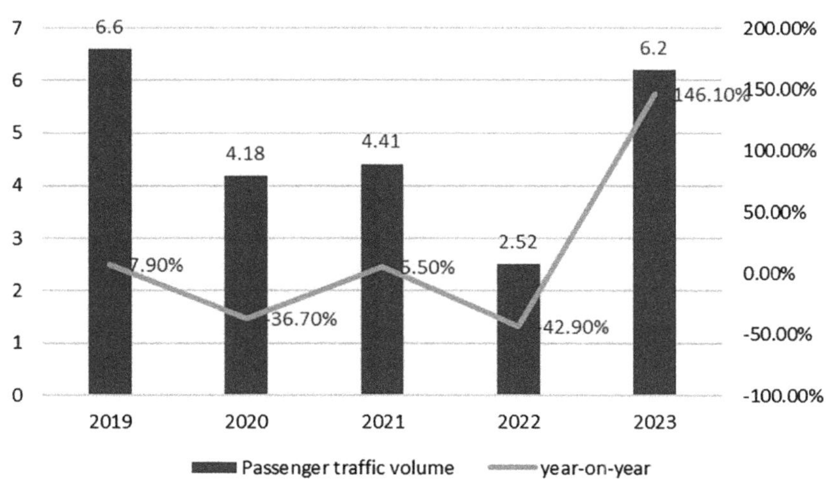

Fig. 5.3 Civil aviation passenger transport volume, 2019–2023. *Data source* 2023 Statistical bulletin on the development of the civil aviation industry

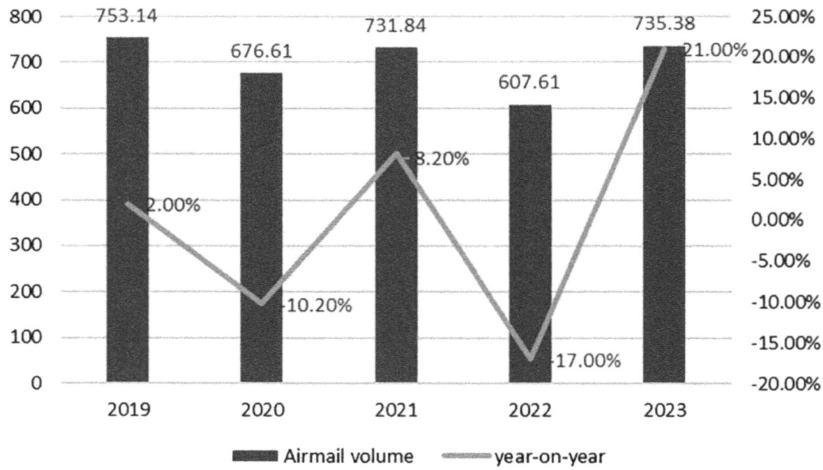

Fig. 5.4 Civil aviation cargo and mail transport volume, 2019–2023. *Data Source* 2023 Statistical bulletin on the development of the civil aviation industry

cargo and mail, representing a 32.8% year-on-year rise. Within this, routes to Hong Kong, Macao and Taiwan transported 151,200 tonnes, a 2.5% increase. International routes handled 2.7899 million tonnes of cargo and mail, a 5.8% increase year-on-year [2]. By the end of 2023, China had 66 transport airlines operating a fleet of 4,270 civil transport aircraft, an increase of 105 aircraft compared to the previous year-end. Scheduled flight routes reached 5,206, with international scheduled services connecting to 57 countries or regions (Fig. 5.4).

China's civil aviation sector is simultaneously advancing its digital transformation to enhance international competitiveness while addressing challenges in achieving high-quality development. Objectives include establishing an integrated data platform to efficiently consolidate flight data, surveillance information, traffic flows, meteorological data, and operational intelligence. The practices of DOFAK in digital transformation offer valuable insights for the sector. By 2035, China aims to become a leading aviation digital transport nation, achieving three-hour coverage for air travel between major domestic cities, securing fundamental self-reliance in core technologies and equipment, and enabling large-scale operations of domestically produced aircraft [5].

Despite opportunities arising from domestic demand growth and international market recovery, civil aviation enterprises must still overcome challenges including competitive pressures, cost control, and recurring pandemic disruptions. By 2023, passenger traffic on China's domestic routes had recovered and surpassed pre-pandemic levels, increasing by 1.5% compared to 2019; while international scheduled passenger flights have resumed to 4,782 weekly departures, representing approximately 62.8% of pre-pandemic levels [6]. Moving forward, China's civil aviation sector will advance towards a phase of high-quality, high-efficiency development, entering a new era of industrial convergence—particularly the low-altitude economy underpinned by traditional general aviation and emerging drone industries. Concurrently, the sector's growth has unlocked consumer market potential, enhanced transport efficiency, driven economic diversification, and elevated overall consumption levels. Through policy support, digital transformation, market recovery, and forward planning, China's civil aviation industry has achieved significant progress, demonstrating robust momentum and promising prospects.

5.2.2 The Contribution of Civil Aviation Enterprises to the Global Economy

China's civil aviation enterprises have assumed a leading position within the global aviation sector through sustained profitability, digital transformation initiatives, and significant contributions to the domestic economy, thereby making substantial contributions to the global economy. These enterprises have not only enhanced their own competitiveness but also provided impetus and exemplary models for the development of the global aviation industry.

As one of China's largest airlines, GUJIK has played a vital role during the global economic recovery. Despite challenges posed by sluggish global economic recovery and slowing domestic growth, it has maintained stable operations and high profitability through measures such as optimising its hub network, refining its fleet structure, and strengthening collaborations with external partners. The company has also actively engaged in social welfare initiatives, fulfilling its social responsibilities and contributing to the creation of both social and economic value. The launch of the C919 large passenger aircraft by COMAC also represents a significant milestone for the global aviation industry. The C919's success not only demonstrates the technological prowess and innovation capabilities of China's aviation sector but also elevates China's influence and standing in the global aviation market. It has broken the long-standing monopoly of Western nations in mainline passenger aircraft manufacturing, signalling a shift in the competitive landscape of global aviation manufacturing towards a tripartite structure of 'Airbus + Boeing + COMAC' [7]. Data indicates that the overall performance of China's civil aviation sector has exerted a significant impact on the global economy, contributing over 25% to worldwide civil aviation growth [8]. Since 1978, the average annual growth rate of China's total air transport turnover has been approximately 1.87 times that of its concurrent GDP growth, while its growth rate has reached 3.2 times the global civil aviation average. Beyond this, the rapid expansion of China's civil aviation sector has catalysed advancements across related industrial chains. Over the past three decades, aviation equipment conglomerates have procured over 1,800 aircraft for China's civil aviation industry while introducing substantial quantities of specialised equipment for airport and air traffic control systems, thereby enhancing flight safety and air traffic management capabilities.

5.3 The Market Environment for Civil Aviation Enterprises

5.3.1 The Relationship Between Airlines

5.3.1.1 Competitive Relationship

Competition among airlines primarily manifests in market share, route networks, fares, and service quality. CHQUK, as a low-cost carrier, has established a novel operational system through its low-cost philosophy and modern aviation enterprise framework, becoming China's first state-owned low-cost airline. The competition between RUANN and EAJIY in the Irish market also exemplifies this dynamic: EAJIY announced in 2004 the launch of services between Cork and London Gatwick, prompting RUANN's withdrawal from that market.

5.3.1.2 Partnership

Despite fierce competition within the aviation industry, extensive cooperation exists between airlines. Primary forms of collaboration include airline alliances, code-sharing, and frequent flyer programmes. Through strategic partnerships, Star Alliance, Oneworld, and SkyTeam achieve complementary resource sharing. This not only assists airlines in circumventing barriers to entering international markets but also effectively enhances cooperative benefits, expands global route networks, and reduces operational costs. Furthermore, airlines establish partnerships to navigate market competition, such as CHQUK's equity participation in E7. Such collaborations not only bolster Spring Airlines' market share but also channel low-cost carrier passenger traffic to NAFAK. Cooperation proved particularly vital during the pandemic, exemplified by WEAOH's alliance with MLLLL, which significantly reshaped the competitive landscape of the trans-Pacific aviation market.

5.3.1.3 Competitive-Cooperative Relationship

Within the contemporary aviation sector, the competitive yet collaborative relationship between airlines and online travel agencies (OTAs) has become a defining characteristic of the industry. Amidst the trend of 'direct sales promotion and commission reduction', airlines and OTAs engage in both rivalry and cooperation. OTAs demonstrate greater efficiency in product display, sales, service assurance, and promotional cost savings, while airlines, through collaboration with OTAs, integrate the tourism supply chain to generate revenue beyond ticket sales. The founder of Tongcheng Tourism acquiring a stake in an airline exemplifies this competitive-cooperative dynamic. The partnership between HENAT and LUBAO implements a dual-hub strategy, developing logistics hubs for Europe and Asia respectively at Luxembourg Airport and Zhengzhou in China. Consequently, relationships among airlines are complex and dynamic, characterised by both fierce competition and extensive collaboration. Through cooperation, airlines can share resources, mitigate risks, and expand markets, thereby gaining competitive advantages. However, collaboration does not equate to abandoning competition; airlines must maintain a degree of competitiveness within partnerships to safeguard their market positions and profitability.

5.3.2 The Impact of Government Policies and Regulations on Civil Aviation Enterprises

Government policies and regulations exert a wide-ranging influence on civil aviation enterprises, manifesting not only through financial support and tax relief but also encompassing industry oversight, the promotion of green development, and the optimisation of the business environment. Firstly, the government provides support to civil aviation enterprises through fiscal and tax policies. Central and local governments increase financial backing to alleviate enterprises' tax burdens, including tax incentives, reductions in landing and take-off charges, parking fees, air route charges, and the purchase-sale price differential for aviation fuel. Secondly, the government regulates the operational activities of civil aviation enterprises through the formulation or amendment of relevant regulations. For instance, the revision of the 'Administrative Provisions on the Joint Restructuring and Reform of Civil Aviation Enterprises and Airports' standardises the joint restructuring and reform activities of civil aviation enterprises and airports, promoting the establishment of a fair and orderly market competition environment. Concurrently, the update to the 'Interim Measures for the Management of Integrity-Based Operations Evaluation for General Aviation Enterprises' fosters integrity-driven operations within the general aviation sector. Thirdly, the government plays a pivotal role in steering the civil aviation sector towards green development. It is committed to establishing and refining corporate accountability frameworks and supply systems. Through monitoring and administrative guidance, it encourages aviation enterprises to proactively adapt to green development requirements. Large-scale aviation enterprises are incentivised to assume leadership roles by formulating green development plans and action programmes to achieve carbon peaking and carbon neutrality.

5.4 Technological Innovation in Civil Aviation Enterprises

5.4.1 The Development and Application of Aviation Technology

Advancements in aviation technology and their application within civil aviation enterprises exhibit distinct differences and characteristics both domestically and internationally. Regarding technological adoption, both YIWUY and NAFAK actively embrace new technologies to enhance operational efficiency and safety. YIWUY expands its market share through acquisition strategies, while NAFAK elevates its market position by increasing flight hours and passenger throughput. YIWUY maintains that acquiring other airlines constitutes the most expedient route to expansion, forming one of its primary development strategies. This approach is supported by the government and the Civil Aviation Administration of China, which endorse the restructuring of loss-making carriers by financially sound airlines. Through this strategy, YIWUY achieved substantial growth in 2000. In 2023, NAFAK delivered remarkable performance, with the group completing 2,841,600 flight hours—an 82.6% year-on-year increase; achieved total transport turnover of 29.792 billion tonne-kilometres, an 81.9% year-on-year increase; transported 142.2014 million passengers, a 127.0% year-on-year increase; and handled 1.585 million tonnes of cargo and mail, a 19.5% year-on-year increase. These figures demonstrate NAFAK's substantial progress in both the scale and efficiency of its air transport operations (Fig. 5.5).

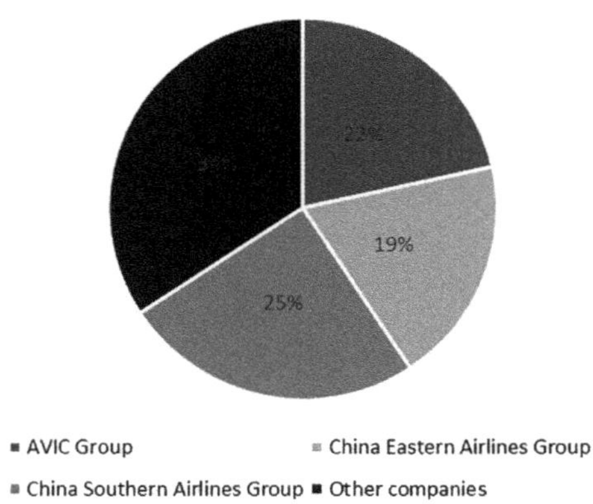

Fig. 5.5 Share of total passenger traffic turnover by airline, 2019–2023. *Data Source* 2023 Statistical bulletin on the development of the civil aviation industry

5.4.2 *The Role of Information Technology in Civil Aviation Enterprises*

Information technology assists the civil aviation industry in navigating uncertainties within complex systems through digital transformation. By leveraging big data, robust algorithms and high computing power, it enables rapid responses and optimal adjustments to competitive environments, customer demands, product lifecycles and technological shifts, thereby enhancing corporate capability value and profit margins. The implementation of smart services—such as developing intelligent airports, promoting electronic ticketing, self-service check-in and end-to-end baggage tracking—has significantly improved operational efficiency across civil aviation enterprises. Next-generation information technologies—including 5G, the Internet of Things, artificial intelligence, cloud computing, and blockchain—are reshaping global factor inputs, industrial structures, and competitive landscapes. Their application not only stimulates innovation within aviation enterprises but also propels the sector's high-quality development. By integrating cloud computing, IoT, big data, and AI, smart aviation initiatives achieve interconnected, interoperable, and shared information systems, thereby elevating overall operational efficiency.

Moreover, advancements in information technology provide civil aviation enterprises with a robust technological foundation, enabling cost savings and service quality improvements. As a leading provider of civil aviation information systems, China Air Information (CAIR) possesses a deep understanding of aviation operational logic and practical requirements, delivering tailored solutions for specific application domains. Concurrently, initiatives such as Tencent's Smart Aviation Solution leverage multiple information technologies to offer intelligent, convenient innovations in passenger services, flight optimisation, and safety management. The application of information technology within civil aviation enterprises not only enhances coordination efficiency across all stages but also enables more effective collaboration throughout operational processes. The 'Smart Aviation Joint Solution' launched by Qianfang Technology and Alibaba Cloud focuses on three core requirements: safety, punctuality, and efficiency. This initiative accelerates the digital and intelligent development of the civil aviation sector. Through continuous technological innovation and application, it has become a significant driving force for industry advancement.

5.4.3 *Future Technological Trends*

The application of future technological trends within civil aviation enterprises is progressively becoming a reality, demonstrating immense potential and prospects across multiple domains such as unmanned aerial vehicles (UAVs) and supersonic flight technology. Significant progress has already been achieved in the deployment of UAVs within civil aviation enterprises. LIKOF has secured approval from the Civil Aviation Administration of China to manage airspace, UAVs, flight plans, and operators, enabling the cluster management of multiple aircraft within the same airspace to enhance operational safety and reliability. Moreover, UAVs are increasingly deployed across diverse sectors including express logistics, agricultural and forestry pest control, geographic surveying, and emergency firefighting. By the end of 2023, China had over 1.26 million registered UAVs. This expansion of application scenarios not only boosts operational efficiency but also injects fresh momentum into the UAV industry's development. Regarding hypersonic flight technology, HUYKE stands as China's sole high-tech enterprise dedicated to the research, development, and commercial application of hypersonic flight products. Hypersonic flight technology is regarded as the new pinnacle of aerospace advancement in the twenty-first century, possessing immense commercial value. By 2040, annual sales for hypersonic travel are projected to reach US$800 billion, potentially revolutionising conventional air travel. This technological advancement could significantly reduce journey times while enhancing both comfort and cost-effectiveness in air travel.

Taking BYNSS as an example, the company's independently developed large unmanned transport aircraft has successfully completed its maiden flight, becoming I's first large unmanned transport aircraft with a payload capacity exceeding 3 tonnes. This achievement marks a major breakthrough for domestic enterprises in the large UAV sector, with numerous companies actively entering this market. The integration of UAV technology with emerging technologies such as 5G and artificial intelligence is making UAV applications increasingly diverse and intelligent. Future applications of drone technology will continue to expand across logistics, agriculture, surveying, and emergency response sectors. Supersonic flight technology also holds promise to transform the landscape of air travel within the coming decades.

5.5 Economic Analysis of Civil Aviation Enterprises

5.5.1 Revenue and Cost Structure

The revenue of civil aviation enterprises primarily derives from the following sources, as shown in Table 5.1.

Table 5.1 Primary sources of revenue for civil aviation enterprises

Income type	Specific details
Route revenue	Domestic and international flight ticket sales
Aviation revenue	Revenue from general aviation services such as helicopters and business jets
Airport service revenue	Aircraft parking fees for take-off and landing, baggage handling charges, etc.
Freight revenue	Revenue from freight transport and courier services

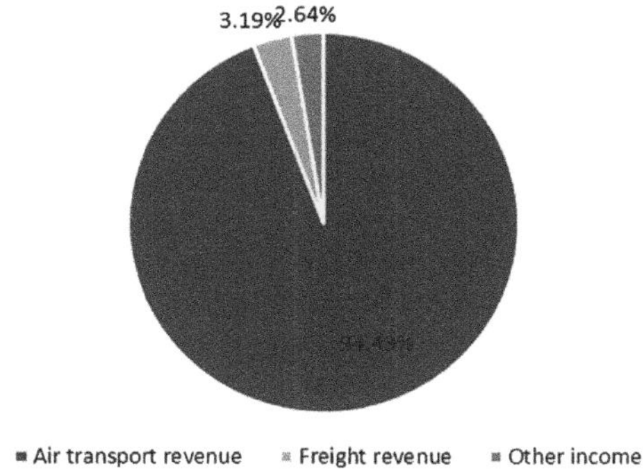

Fig. 5.6 Proportion of business revenue for China Eastern Airlines

Taking China's three major state-controlled airlines as specific examples, GUJIK's revenue structure is predominantly passenger-driven. In 2023, it achieved operating revenue of RMB 141.1 billion, representing a year-on-year increase of 166.74%. Passenger revenue reached RMB 130.517 billion, accounting for the majority of its core business income with a year-on-year growth of 240.81%. This growth was primarily driven by increased operational capacity and higher passenger load factors. However, GUJIK's revenue growth remains relatively moderate, reflecting its status as a mature enterprise. DOFAK recorded total operating revenue of RMB 113.771 billion in 2023, representing a year-on-year increase of 145.63%. Its revenue structure is similarly dominated by passenger transport, with passenger revenue reaching RMB 104.576 billion in 2023, accounting for 94.43% of aviation transport revenue. Domestic route passenger revenue constituted the bulk at RMB 83.986 billion, while cargo revenue and other business income accounted for relatively minor shares of 3.19 and 2.64% respectively. International route revenue recovered to 52% of its 2019 level in 2023, representing 17.05% of passenger revenue. In the first half of 2024, Air China's international market seat-kilometre capacity and passenger revenue both recovered to over 90% of 2019 levels, indicating a significant improvement in international route passenger revenue (Fig. 5.6).

For NAFAK, its total operating revenue for 2023 reached RMB 159.929 billion, representing an 83.7% year-on-year increase. Passenger transport revenue amounted to RMB 136.17 billion, accounting for 88.07% of its core business income, with freight revenue and other income streams also contributing to varying degrees. Furthermore, NAFAK achieved operating revenue of RMB 44.601 billion in the I quarter of 2024, representing a year-on-year increase of 30.97%, primarily driven by the continued recovery of the air passenger market.

Overall, all three major airlines recorded substantial revenue growth in 2023, with passenger revenue serving as the primary income source for each. However, freight and other operations also accounted for a proportionate share. GUJIK and DOFAK generated relatively modest freight revenue, whereas NAFAK derived a certain amount from freight alongside its passenger income (Fig. 5.7).

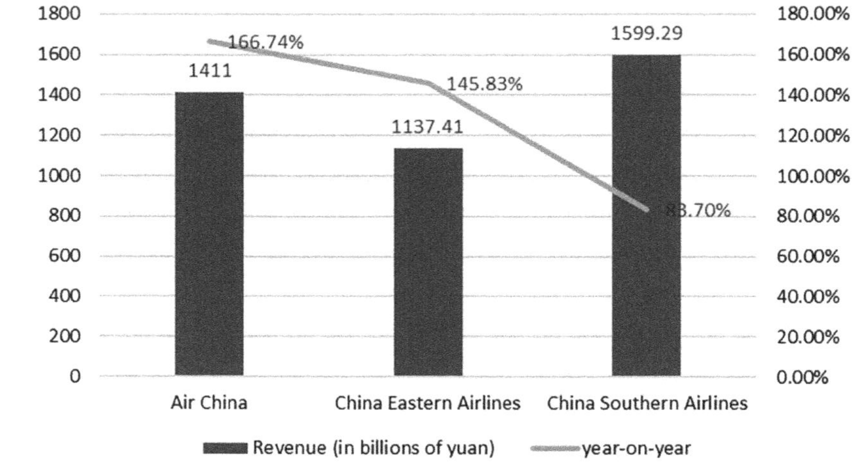

Fig. 5.7 Revenue growth of the three major state-controlled airlines in 2023

The three major airlines all face high proportions of fixed costs and fuel expenses, which are significantly affected by fluctuations in international oil prices. Additionally, aircraft depreciation, staff remuneration, landing and take-off charges, and maintenance costs constitute major components of their expenditure. Against the backdrop of intensifying competition within the aviation sector, airlines must enhance profitability through measures such as optimising debt structures and controlling operational costs (Table 5.2 and Fig. 5.8).

Table 5.2 Primary sources of costs for civil aviation enterprises

Cost category	Specific details
Aviation fuel costs	Fuel constitutes one of the largest costs for airlines
Aircraft maintenance costs	Including scheduled maintenance and unexpected repair expenses
Take-off and landing operational costs	Airport landing and take-off charges
Airport ground handling charges	Boarding bridge and baggage handling fees
Utility and power expenses	Energy costs required for airport and airline operations
Personnel costs	Wages and salaries, etc.

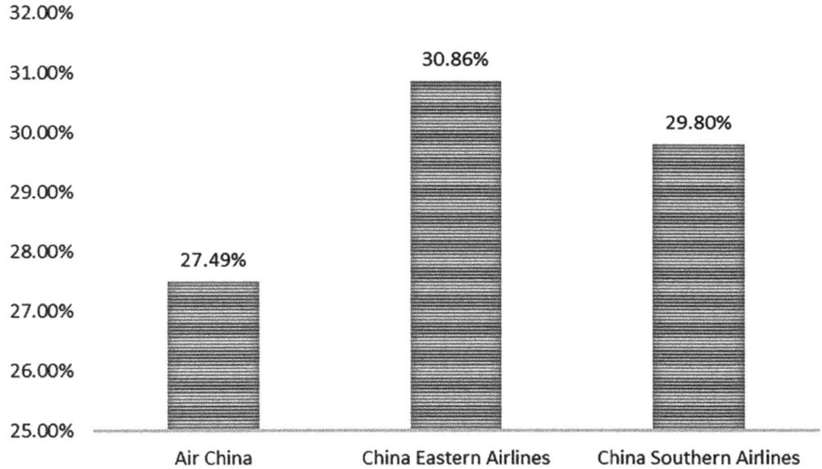

Fig. 5.8 Ratio of fuel costs to total costs for the three state-controlled airlines in 2022

5.5.2 Pricing Strategy and Revenue Management

5.5.2.1 Pricing Strategy

CHQUK has implemented a sophisticated pricing strategy, offering both affordable economy class and luxurious premium cabins while maintaining low operational costs and retail prices. This approach has secured the company a significant position within the fiercely competitive market. The airline also meticulously studies passengers' price expectations and purchasing habits to set fares accordingly. Employing flexible dynamic pricing methods, they adjust fares in response to market conditions and competitor movements, gaining deep insight into passengers' affordability and preferences to better meet consumer demands. This approach not only enhances fare elasticity but also improves seat utilisation and revenue. Concurrently, the airline monitors fluctuations in market demand, adjusting fares across different cabins accordingly. This ensures passengers willing to pay more secure seats, thereby strengthening the company's profitability.

5.5.2.2 Revenue Management

The core of revenue management lies in accurately forecasting demand and controlling inventory. Airlines employ complex statistical models and historical data to predict passenger demand for each flight, thereby formulating corresponding pricing strategies. Through overbooking and sub-class inventory restrictions, companies can achieve a 4–6% revenue uplift without increasing operational costs. A multi-tiered fare structure is also pivotal to revenue management, satisfying diverse consumer needs through differentiated fare levels while optimising seat allocation. Technological advancements, particularly digital transformation and sophisticated analytics, have profoundly reshaped airline revenue management. These tools enable deeper consumer behaviour insights and the implementation of personalised pricing strategies. Moreover, as customer expectations evolve, airlines must continually refine their revenue management approaches to enhance satisfaction and loyalty.

CHQUK leads China's low-cost carrier sector, employing flexible pricing strategies to maximise revenue. Other carriers are also exploring novel yield management approaches, employing demand forecasting, inventory control, multi-tiered fare structures, and personalised pricing to enhance seat utilisation and revenue. These measures cater to diverse consumer requirements while elevating customer satisfaction and loyalty. Differentiated pricing, flexible pricing, and demand-differentiated pricing strategies assist airlines in adapting to market dynamics and consumer behaviour. Concurrently, yield management tools facilitate optimal seat allocation and revenue maximisation.

5.6 Challenges and Opportunities for Civil Aviation Enterprises

5.6.1 Key Challenges Faced

5.6.1.1 Oil Price Fluctuations

Aviation fuel constitutes a significant portion of airline costs, typically accounting for 30–50% of total expenditure. Fluctuations in fuel prices directly impact airlines' fuel surcharge revenues and operational performance. Taking the first half of 2021 as an example, aviation fuel costs at NAFAK and GUJIK increased by 46.87 and 45.55% respectively compared to the same period the previous year. Confronted with rising fuel costs, airlines have been compelled to increase fuel surcharges to alleviate cost pressures. Furthermore, fluctuations in the Renminbi exchange rate exert significant influence on airlines, impacting not only financing costs but also potentially affecting expenditures on overseas purchases of aircraft, equipment, and fuel. For instance, the depreciation of the Renminbi in 2020 notably affected NAFAK's financial position. Substantial fluctuations in international oil prices similarly exert considerable influence on aviation fuel costs and fuel surcharge revenues, thereby further impacting airlines' operating results. By 2023, oil prices initially declined before rising again, leading to increased ex-factory prices for jet fuel and further exacerbating airlines' cost burdens.

5.6.1.2 Safety Incident

Within the aviation sector, the occurrence of air disasters and other safety incidents exerts a detrimental impact on companies' market demand and profitability. Unforeseeable natural disasters, wars, and regional conflicts may lead to flight cancellations and reduced passenger numbers, whilst simultaneously increasing expenditure on safety and insurance. To enhance civil aviation safety management and performance, China's civil aviation industry implements safety audits to bolster security. Nevertheless, safety concerns remain a principal challenge for the civil aviation sector, demanding sustained attention and continuous improvement. DOFAK has implemented measures such as adjusting fuel surcharges to counteract oil price fluctuations, yet substantial price volatility continues to exert considerable pressure on its performance. China's civil aviation enterprises confront numerous challenges in managing oil price volatility and flight safety issues. Airlines must adopt multiple strategies, including increasing fuel surcharges, enhancing safety management, and optimising route structures, to maintain stable performance and financial health.

5.6.2 Development Opportunities

As a technology-intensive sector, the aviation industry places exceptionally high demands on the research, development and application of new technologies. Technological innovation stands as one of the pivotal drivers propelling the advancement of China's civil aviation enterprises. The implementation of artificial intelligence technologies presents opportunities for cost reduction and efficiency gains, further catalysing the industry's digital transformation. Applications such as facial recognition systems, intelligent security screening technologies, and drone-based inspection and surveillance not only enhance safety and operational efficiency but also markedly elevate the passenger experience. With growing market demand, the civil aviation sector faces vast development prospects. Projections indicate that over the next two decades, China's aviation market will receive 9,250 passenger aircraft seating 50 or more, representing a market value of approximately US$1.4 trillion. The value of China's aviation services market is forecast to reach US$54.1 billion, growing at a compound annual growth rate of 5.8%. Concurrently, governmental support remains a crucial pillar underpinning civil aviation development. Through increased infrastructure investment, enhanced construction standards for airports and air transport facilities, and proactive implementation of cost-reduction measures to alleviate corporate burdens, China's general aviation equipment supply capacity and industrial innovation capabilities are projected to achieve significant advancement by 2027. Chinese civil aviation enterprises enjoy favourable development opportunities across multiple fronts, including technological innovation, growing market demand, and policy support. These opportunities will propel the high-quality development of the civil aviation sector and enable it to occupy an increasingly prominent position within the global aviation market.

5.6.3 Future Outlook

Amidst the tide of globalisation, China's civil aviation enterprises will assume an increasingly pivotal role. According to the Thirteenth Five-Year Plan for Civil Aviation in China, the term 'international' appears 67 times, underscoring the centrality of internationalisation. Chinese airlines must strengthen the development of aviation hubs, clarify their strategic positioning, cultivate an international brand image, and transform their business models to enhance core competitiveness. According to BOINN projections, China is poised to become the world's largest aviation passenger market within the next two decades, with an anticipated annual growth rate of 5.9%—surpassing the global average of 4.7%. This expansion will drive the growth of China's civil aviation fleet. By 2030, China is projected to require 1,575 new wide-body aircraft, establishing the world's largest fleet of wide-body airliners. By 2035, the development of smart civil aviation will achieve comprehensive coverage across all elements, processes, and scenarios. This will fully realise the objectives of 'one facial recognition for travel, one document for logistics, one customs clearance, one operational network, and one regulatory platform,' with the capacity to handle 30 million annual aircraft movements and an expected flight punctuality rate exceeding 85%.

Within the general aviation sector, the operational market size is projected to reach RMB 83.2 billion by 2029, with an average annual growth rate of 22%. Concurrently, the deepening implementation of reforms to streamline administration and delegate powers will drive significant development in general aviation during the 14th Five-Year Plan period. BOINN forecasts indicate China's civil aviation fleet will more than double over the next two decades, reaching 9,740 aircraft, positioning the nation to become the world's largest aviation market. This projection underscores the immense growth potential and vast scope within China's civil aviation sector. Moving forward, Chinese aviation enterprises will sustain rapid expansion through technological innovation, intelligent development, green transformation, and international route network expansion, thereby enhancing industry competitiveness and sustainable development capabilities.

China is actively expanding its international route network and attracting more foreign travellers through the implementation of a 144-h visa-free transit policy, which will positively impact the development of international aviation hubs. Technological innovation and intelligent development represent key future trends for the civil aviation sector, with innovative technologies such as artificial intelligence playing increasingly vital roles in airport management and flight operations. Furthermore, green transformation stands as one of the primary future development directions for the industry. The National Development and Reform Commission and the Civil Aviation Administration of China have launched pilot projects for Sustainable Aviation Fuel (SAF) to advance green and low-carbon development in civil aviation. Over the next five years, key operational indicators—including total transport turnover, passenger traffic, and cargo/mail volume—are projected to comprehensively surpass pre-pandemic levels. Passenger traffic is anticipated to achieve double-digit year-on-year growth, while cargo/mail volume is expected to see modest expansion. Overall, China's civil aviation sector is entering a new phase of rapid and healthy development. As market entities gradually recover and production and operational order normalises, momentum for industry growth is gathering, with supporting conditions for high-quality development continuously improving.

References

1. Shin Wan Hongyuan Research, Yan Hai, Luo Shi. Special Report on the Air Transport Industry: Global Civil Aviation Enters a New Normal, Focusing on Long-Term Industry Shifts [EB/OL]. https://new.qq.com/rain/a/20230717A011OR00, 17 July 2023.
2. Wang Di. Civil Aviation Industry Advances Towards Quality Enhancement and Efficiency Improvement Phase [EB/OL]. http://www.news.cn/fortune/20240112/29cc94c76e5a4eac8747aa1e4c4cec44/c.html, 12 January 2024.
3. Liu Jie. Multiple Indicators Hit New Highs: Civil Aviation Industry to Enter New Cycle of Sustained Rapid and Healthy Development in 2024 [EB/OL]. https://news.cctv.com/2024/02/29/ARTIa5NAF0orZV1XK7DerQ6F240229.shtml, 29 February 2024.
4. Wang Xiao. Analysis of the Current Status and Prospects of China's Civil Aviation Industry [J]. State-owned Assets Management, 2010, (I): 63–65.
5. Ma Wenhua. China to Become Aviation Powerhouse by 2035 [EB/OL]. https://www.gov.cn/zhengce/202403/content_6935404.htm?ddtab=true, 2024-03-01.
6. Xinhua News Agency. Xinhua Finance | Civil Aviation Industry Returns to Natural Growth; 2024 Airline Profits on the Horizon [EB/OL]. https://news.qq.com/rain/a/20240117A05CIM00.2024, 2024-01-17.
7. Zhu Ziyue. Century-Long Commercial Warfare in European and American Aviation: Implications for China's Participation in International Competition [EB/OL]. http://att.caacnews.com.cn/mhfzzcgjyxb/mhfzzcgjyxb5th/202111/t20211101_59843.html#,2021-11-01.
8. Cui Guoqiang, Zhang Jun. Building a Super Carrier: China's Civil Aviation Contributed Over 25% to Global Growth Last Year [EB/OL]. https://www.thepaper.cn/newsDetail_forward_2726050, 2018-12-11.
9. Li Yansheng, Guo Xiaoting. From 37th to 2nd: China Emerges as a Global Aviation Powerhouse [EB/OL]. https://www.gov.cn/gzdt/2008-12/04/content_1168516.htm, 2008-12-04.
10. Banliang Finance. Air China's 2023 Revenue Reaches 141.1 Billion Yuan, Up 166.74% Year-on-Year_Tencent News [EB/OL]. https://new.qq.com/rain/a/20240329A06LGT00, 2024-03-29.
11. Xiaolang Financial News. China Eastern Airlines reports first-half revenue of RMB 64.199 billion, up 29.89% year-on-year; net profit of -RMB 2.768 billion, up 55.70% year-on-year; sales expenses increase by 35.26% year-on-year | China Eastern Airlines_Sina Finance_Sina.com [EB/OL]. https://finance.sina.com.cn/stock/aiassist/yjbg/2024-08-31/doc-incmnhnt9855144.shtml, 31 August 2024.
12. Zhang Sunmingshuo. China Eastern Airlines reduces losses by RMB 29 billion in 2023, with international route revenue recovering to 50% | Financial Report Analysis [EB/OL]. https://finance.ifeng.com/c/8YLyhgcfw1E, 2024-03-29.
13. China Southern Airlines. China Southern Airlines Co., Ltd. 2023 Annual Report Summary [N]. 2023.
14. BT Finance. China Southern Airlines I-Q1 2024 Report: Comprehensive Financial Data Analysis [EB/OL]. https://xueqiu.com/2509537588/293493003, 2024-06-13.

Chapter 6
From Infrastructure to Innovation: The Trajectory of China's Pipeline Enterprises

Jianlin Wang and Xinmin Zhang

Abstract This paper focuses on the development of China's pipeline transportation enterprises. Starting from the development history of the pipeline transportation industry, it takes ZSHYO, ZSHHA, and ZHAYY as research objects and elaborates on their basic profiles respectively: ZSHYO has made outstanding achievements in the construction of oil and gas pipeline transportation networks; ZSHHA has unique advantages in chemical product transportation pipelines; and ZHAYY specializes in offshore oil and gas transportation. This chapter details the current development status of China's pipeline transportation enterprises, focuses on data such as pipeline mileage and transportation volume to present development achievements, and uses relevant data to analyze their market share and explore the competitive status of the three enterprises. Finally, it discusses the challenges faced by pipeline transportation enterprises, such as fluctuations in the international energy market and the breakthrough of technical bottlenecks. This paper aims to provide comprehensive and in-depth reference materials for industry professionals, facilitating further research and decision-making.

6.1 Introduction

Pipeline transportation occupies an important position in the modern transportation system. With its unique advantages, it plays a crucial role in China's energy and resource allocation as well as economic and social development. Driven by the continuous growth of China's economy and the increasing demand for energy, the pipeline transportation industry has undergone profound changes and achieved rapid development. Over the past few decades, China's pipeline transportation enterprises have grown from scratch and evolved from weak to strong, gradually building a widely covered and technologically advanced pipeline network. From the initial small-scale

J. Wang · X. Zhang (✉)
Anhui University of Finance and Economics, Bengbu City, Anhui Province, China
e-mail: 15148635083@163.com

J. Wang and M. Song (eds.), *The Green Path of Chinese Transportation Enterprises*, Resources, Climate and Sustainable Development,
https://doi.org/10.1007/978-981-95-5289-4_6

experimental pipeline construction to today's large-scale pipeline systems that crisscross the country and span thousands of kilometers, pipeline transportation enterprises have forged ahead through continuous exploration, providing solid support for national energy security and making outstanding contributions to the coordinated development of regional economies.

Currently, China is at a critical juncture of economic restructuring and energy transition, where pipeline transportation enterprises are faced with both opportunities and challenges. The urgent domestic demand for clean energy and the optimization of energy consumption structure have opened up new development directions for enterprises; however, fierce market competition, increasingly stringent environmental protection requirements, and pressure from technological innovation all require enterprises to accelerate their transformation to keep up with the new trends of the times.

This report aims to conduct an in-depth study on the development history, current status, and future trends of China's pipeline transportation enterprises. Through detailed analysis of relevant data, case studies of typical enterprises, and accurate judgment of industry development trends, it reveals the important role of pipeline transportation enterprises in the national energy strategy, analyzes the problems and difficulties encountered in their development, and puts forward forward-looking and operable development suggestions. It is expected to assist pipeline transportation enterprises in strategic decision-making and provide recommendations for promoting the healthy and sustainable development of China's pipeline transportation industry.

6.2 Overview of the Pipeline Transportation Industry

6.2.1 Development History of China's Pipeline Transportation Industry

Pipeline transportation is a special form of transportation. It uses pipelines as tools and relies on pressure differences to transport substances such as liquids, gases, or powdered solids from one place to another. Its prominent features are: large-scale continuous transportation can be realized by designing pipe diameter and pressure on demand; high continuity and stability, which can operate continuously and stably when equipment is normal; low energy consumption and cost, with small material flow resistance and no need for frequent loading, unloading and transshipment; high safety, as the closed environment reduces external impact, making it especially suitable for the transportation of dangerous substances; less land occupation, which occupies relatively less land resources; easy to realize automatic control, and real-time monitoring and remote operation management can be carried out with the help of monitoring and control systems, improving efficiency and accuracy (Fig. 6.1).

As one of the important energy transportation methods since the founding of the People's Republic of China, its development process can be divided into the

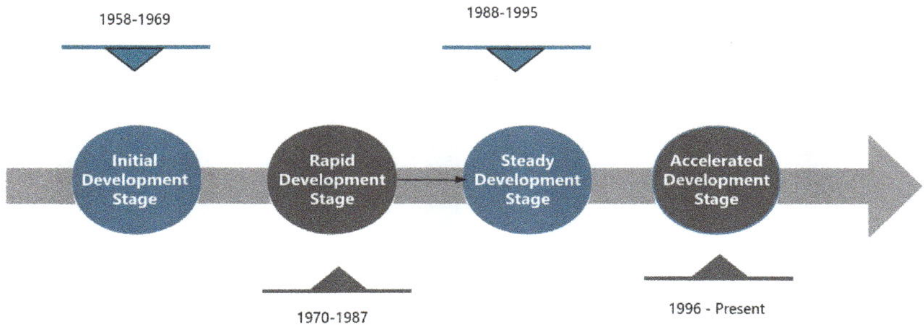

Fig. 6.1 Development history of China's pipeline transportation industry

following stages: In the initial development stage, with the increase of domestic energy demand, China began to build crude oil pipelines and natural gas pipelines. Due to the limitation of funds and technology, the scale was relatively small and the layout was relatively scattered. With the continuous increase of domestic energy demand, the government increased investment in pipeline construction, the scale of pipelines expanded rapidly, and the layout gradually improved. In the steady development stage, due to the relatively large scale and relatively perfect layout of pipelines after the previous large-scale construction, the main task of this stage is to maintain and upgrade the existing pipelines to improve the stability and safety of pipeline transportation. Finally, with the rapid development of domestic economy and the transformation of energy consumption structure, the government further increased investment in pipeline construction, the scale of pipelines continued to expand, and the layout became more perfect. At the same time, with the emergence of new technologies, the automation and intelligence level of pipeline transportation has been significantly improved. Therefore, China's pipeline transportation development has entered the accelerated development stage since 1996.

6.2.2 Market Scale and Current Situation of China's Pipeline Transportation Industry

In recent years, with the continuous national investment in infrastructure construction and the steady growth of energy consumption demand, the pipeline transportation industry has ushered in new development opportunities, and its market scale has steadily expanded. On one hand, the national strategic layout and reserve demand for energy sources such as oil and natural gas have driven the large-scale advancement of pipeline construction. On the other hand, the acceleration of urbanization and the improvement of environmental protection requirements have led to increasing demand for pipeline transportation in fields such as urban gas supply, water supply and drainage. Meanwhile, technological innovation and industrial upgrading have also injected new vitality into the pipeline transportation industry. Intelligent and

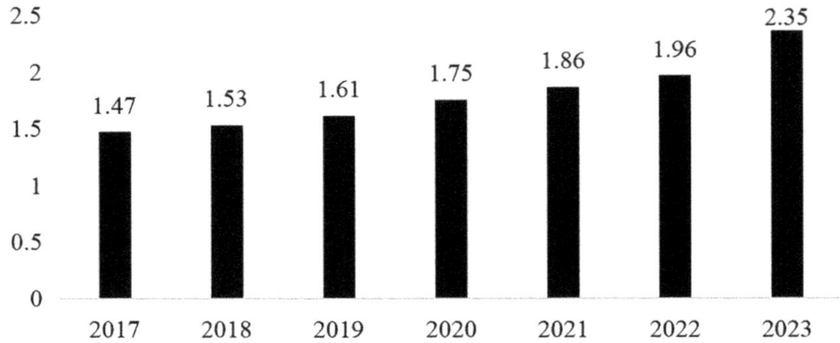

Fig. 6.2 2017–2023 China's pipeline transportation industry market scale unit: trillion yuan

networked pipeline management systems have significantly improved the transportation efficiency and safety of pipeline transportation, while reducing operating costs.

As shown in Fig. 6.2, with the strong support of the state and government and the continuous drive of market demand, the market scale of China's pipeline transportation industry increased from 1.47 trillion yuan in 2017 to 1.96 trillion yuan in 2022. Since 2023, with the recovery of the domestic economy, the market demand for natural gas, oil, water supply and drainage has further increased, driving the continuous growth of the industry's market scale. The annual market scale of the pipeline transportation industry in 2023 was approximately 2.35 trillion yuan [1].

According to data released by the National Bureau of Statistics, by the end of 2022, the total length of China's oil and gas pipelines exceeded 180,000 km. Specifically, the length of crude oil, product oil and natural gas pipelines reached 28,000, 32,000 and 120,000 km respectively. This huge pipeline network provides a solid guarantee for the stable supply of energy in China. In terms of oil transportation, pipeline transportation accounts for a relatively large proportion. For example, a large part of the crude oil produced by major oil-producing areas such as Daqing Oilfield and Shengli Oilfield is transported to major refineries through pipelines. According to relevant reports in ZSHYO Petrochemical magazine, pipeline transportation accounts for about 80% of crude oil transportation, which effectively reduces transportation costs and losses and improves transportation efficiency. In the field of natural gas, pipeline transportation plays a leading role. With the rapid growth of China's natural gas consumption demand, pipeline construction has been accelerating. Taking the famous West-to-East Gas Pipeline Project as an example, this project is a landmark achievement in the development of China's pipeline transportation. The first phase of the West-to-East Gas Pipeline Project was officially put into operation on October 1, 2004, with a total length of about 4,200 km and an annual gas transmission capacity of 80 billion cubic meters. It continuously transports the abundant natural gas resources in the west to the economically developed areas in the east. The subsequent second and third phases of the West-to-East Gas Pipeline Project further expanded the gas transmission scale and coverage, greatly increasing the proportion of natural gas in the

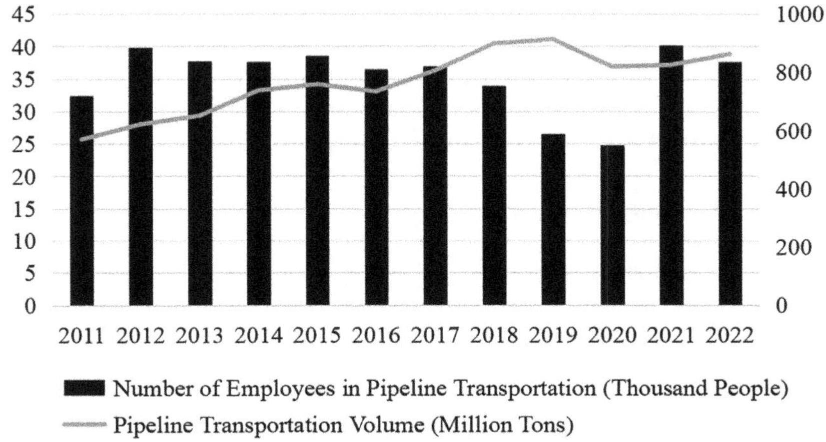

Fig. 6.3 2011–2022 number of employees and transportation volume in pipeline transportation

energy consumption structure and improving the energy supply and environmental quality in the areas along the pipeline [2] (Fig. 6.3).

According to the Global Macro and Micro Economic Database, the number of employees in China's pipeline transportation industry stood at 32,458 in 2011. It rose in 2012 and maintained a relatively stable level until 2017. In 2018, the number of employees decreased compared with previous years, standing at 33,883, and continued to decline until reaching a low of 24,930 in 2020. Relevant data indicate that the possible reasons include: the development of industrial automation and intelligentization has reduced labor demand; the improvement of production efficiency and optimization of business models have eliminated unnecessary intermediate links and labor input; industrial structure adjustment and market competition have intensified the screening of labor force, eliminating workers with unskilled technology [3]. The number of employees increased in 2021 compared with 2020, mainly due to the economic recovery after the impact of the epidemic, which led various countries to increase demand for new energy, as well as the advancement of industry investment and construction projects. In 2022, the number of employees decreased compared with 2021, possibly due to the sharp fluctuations in energy prices and adjustments in energy structure in 2022, which changed the demand for human resources. In addition, according to data from the National Bureau of Statistics, China's pipeline transportation volume has been on a continuous rise since 2011, reaching 862.604 million tons in 2022. This not only indicates the sustained growth in China's demand for energy such as oil and natural gas, but also reflects the trend of China's sustained economic growth [4].

6.3 Research on Major Pipeline Transportation Enterprises in China

6.3.1 Representative Enterprise 1

6.3.1.1 Basic Enterprise Overview

ZSHYO is a comprehensive international energy company with extensive and in-depth business sectors. In the field of oil and gas exploration and production, ZSHYO conducts oil and gas exploration and development operations in multiple large onshore oil- and gas-bearing basins in China, such as the Tarim Basin and the Ordos Basin, as well as in multiple overseas countries and regions. By 2023, ZSHYO's domestic crude oil and natural gas outputs had reached 773.7 million barrels and 4.939 trillion cubic feet, respectively. In terms of refining and chemical industry, it owns a number of large-scale integrated refining and chemical bases, equipped with advanced refining technologies and a rich range of chemical product lines. Its annual refining capacity is approximately 936 million tons per year, with a refinery operating rate of around 77%, and its chemical products cover synthetic resins, synthetic rubber, synthetic fibers and other categories. The sales business includes the sales of refined oil, natural gas and non-oil products, with a huge gas station network covering urban and rural areas across the country. In the field of pipeline transportation, ZSHYO operates the largest oil and gas pipeline network in China, including crude oil pipelines, product oil pipelines and natural gas pipelines. By the end of 2017, the total length had reached 85,600 km, with a huge annual oil and gas transportation volume, providing a solid guarantee for the stable supply of energy [5].

ZSHYO adopts an integrated operation model, realizing the coordinated development of the entire industrial chain from upstream oil and gas exploration and development, to midstream refining and chemical industry, and then to downstream sales terminals. Upstream, through advanced exploration technologies and geological research, it continuously seeks new oil and gas resources and adopts efficient extraction technologies to increase oil and gas production. Midstream, relying on advanced refining and chemical processes and equipment, it realizes the optimal utilization of resources and produces high-quality products that meet market demand. In the downstream sales stage, it meets the needs of various customers by building an extensive sales network and a complete service system. At the same time, ZSHYO attaches importance to technological innovation and digital transformation, using big data, artificial intelligence and other means to improve operational efficiency and management levels. In terms of international business, it actively carries out cooperation and investment, participates in the development and operation of international oil and gas projects, and enhances its international competitiveness.

6.3.1.2 Competitive Advantages and Development Status

ZSHYO Pipeline Company has a top domestic pipeline storage and transportation research institution, holding internationally leading core technologies such as transportation of waxy and high-viscosity crude oil and pipeline integrity management. The institution has complete simulation means, ranging from small indoor simulation devices to industrial test loops; it has also successfully developed unsteady-state simulation software for crude oil pipelines, which can accurately simulate crude oil flow and pipeline transportation characteristics under complex working conditions, reaching the international advanced level. A department of ZSHYO covers the entire industrial chain. After the pipeline construction has reached a certain scale, it can separate construction and operation. Currently, it is adjusting from line-based management to regional management, and may realize the complete separation of pipeline operation and oil and gas sales in the future.

As early as 2011, the total length of ZSHYO's oil and gas pipelines had increased from less than 30,000 km at the end of the "Tenth Five-Year Plan" to more than 50,000 km. Crude oil pipelines reached 13,000 km, accounting for approximately 70% of the national total; product oil pipelines exceeded 8,000 km, accounting for about 50% of the national total; natural gas pipelines were nearly 30,000 km, accounting for almost 90% of the national total [6]. Today, the natural gas pipeline network has connected Sichuan and other major gas regions as well as Central Asia. It is connected to more than 1,000 large users in the main consumer markets of 25 provinces and cities in China, benefiting nearly 500 million people. In addition, Yunnan-based natural gas enterprises such as ZSHYO Natural Gas Sales Yunnan Company achieved a gas supply volume exceeding 2 billion cubic meters for the first time in 2023, a year-on-year increase of 15%. In early December 2023, the Yunnan Honghe Branch Line, which took more than 2,000 days and nights to construct, was officially put into operation, bringing Honghe Prefecture into the era of pipeline natural gas, with an expected medium- and long-term consumption volume of 220 million cubic meters per year [7]. Since the commissioning of the China-Myanmar Natural Gas Pipeline ten years ago, ZSHYO has put 19 branch lines into operation, including Xiangyun and Tengchong, giving full play to the role of Myanmar gas entering Yunnan, densifying Yunnan's natural gas pipeline network and injecting strong green impetus into Yunnan's natural gas supply guarantee and stable economic growth.

6.3.1.3 Challenges and Future Outlook

According to the International Energy Agency (IEA) report, international oil prices have fluctuated more sharply in recent years. In the first half of 2023, oil prices once exceeded the $80 per barrel mark, leading to a sharp rise in ZSHYO's procurement costs. Oil price fluctuations are affected by various factors such as the global economic situation, geopolitics, and oil supply and demand relations. The uncertainty of these factors makes it difficult for ZSHYO to stably control costs and plan production and operation strategies. At the same time, with the deepening of globalization and the intensification of market competition, foreign giants have entered the oil industry with strong capital strength, advanced management concepts and technological innovations, and have expanded rapidly in the gas station market. For example, some foreign oil companies have increased their service points by about 20% in the past five years by acquiring local small gas station networks. They also use information technology, big data analysis and artificial intelligence to optimize supply chain management, improve operational efficiency and service quality, and attract consumers with environmental protection services and new energy technologies. Their business growth rate in new energy vehicle services reaches 30% annually, far exceeding the growth rate of traditional oil and gas services, bringing huge competitive pressure to ZSHYO. Moreover, domestic private oil enterprises are gradually growing, competing with ZSHYO in market share, customer resources and other aspects, making market competition more diversified and fierce. In addition, new energy has developed rapidly worldwide, and sales of new energy vehicles in China have shown explosive growth in the past few years. According to data from the China Association of Automobile Manufacturers, sales of new energy vehicles in the Chinese market increased by more than 140% year-on-year in the first quarter of 2023. This has slowed down the growth in demand for traditional petroleum energy markets. ZSHYO must accelerate its own energy structure transformation and explore new energy service models such as charging piles and hydrogen energy stations to adapt to market changes; otherwise, its traditional business may face the risk of gradual shrinkage in market share [8].

Despite the challenges, ZSHYO's future development still has the following positive trends:

First, with the development of the global economy, the demand for oil market continues to grow, especially in developing countries, where there is still huge room for growth in oil demand. ZSHYO can rely on its own strength and experience to further expand its international market share by strengthening overseas exploration and building transnational oil pipelines. For example, it can carry out energy cooperation projects in countries and regions along the "Belt and Road" to enhance its influence in the international energy market. Second, ZSHYO should comply with the trend of energy transformation and increase investment in the new energy field. For instance, it fully built 507 charging and swapping stations in 2023, plans to build more than 1,000 charging and swapping stations in 2024, and acquired Putian New Energy to restructure and establish Kunlun Wanglian Electric Energy Co., Ltd., striving to gain a foothold in the new energy market and realize the coordinated development of traditional energy and new energy businesses. Third, ZSHYO can continue to increase investment in technology research and development to improve the level of oil exploration and development technology, enhance refining efficiency and oil product quality, and reduce production costs. For example, making breakthroughs in key technical fields such as deep and ultra-deep oil and gas exploration and development, and enhancing oil and gas recovery efficiency to strengthen the core competitiveness of the enterprise. At the same time, it should actively explore the integrated development of new energy technologies, such as carrying out research and development and application of hydrogen energy technology and energy storage technology. Fourth, ZSHYO should use digital technology to improve production and operation management efficiency, optimize supply chain management, and realize intelligent production and services. By establishing a digital platform, integrating internal enterprise data resources, it can achieve precise monitoring and optimal scheduling of production, sales, logistics and other links, and improve the scientificity and timeliness of enterprise decision-making.

6.3.2 Representative Enterprise 2

6.3.2.1 Basic Overview of the Enterprise

ZSHYO Pipeline Storage and Transportation Co., Ltd. is a wholly-owned subsidiary of ZSHHA Co., Ltd., undertaking the important responsibility in ZSHHA's professional field of crude oil storage and transportation. ZSHHA is an energy and chemical enterprise with significant global influence.

In terms of business scope, ZSHHA's operations are extensive and diversified. Its core businesses include oil exploration and development, oil refining, chemical production, and the sales of oil products and chemical products. In the field of oil exploration and development, ZSHHA conducts oil and gas resource exploration and exploitation in multiple domestic and overseas regions, committed to increasing oil and gas production and ensuring national energy security. In the oil refining business, it possesses advanced oil refining technologies and large-scale refining equipment, capable of producing high-quality oil products that meet various standards. In chemical production, its product portfolio covers a wide range from basic chemical raw materials to high-end new chemical materials, such as synthetic resins, synthetic rubber, and synthetic fibers. The sales business markets oil products and chemical products through an extensive network of gas stations and sales channels to meet the needs of consumers and industrial users. Additionally, ZSHHA is actively expanding into new energy sectors, including hydrogen energy and biofuels, to adapt to the trend of energy transition.

ZSHHA adopts an integrated operation model. Upstream, it acquires oil and gas resources through technical means such as geological exploration and drilling, and conducts preliminary processing. Midstream refineries and chemical plants convert crude oil and primary products into various oil products and chemical products, while emphasizing energy conservation, emission reduction, and improvement of production efficiency. Downstream, the sales segment provides customers with high-quality products and services through multiple channels including gas stations, sales outlets, and e-commerce platforms.

Throughout the operation process, ZSHHA attaches great importance to technological innovation, continuously invests in R&D resources to enhance its technical capabilities and product quality. Meanwhile, it strengthens cooperation with domestic and foreign partners to achieve resource sharing and complementary advantages.

6.3.2.2 Competitive Advantages and Current Development Status

Firstly, ZSHHA boasts a large and well-established supply chain network covering all links such as oil exploitation, transportation, storage, processing, and sales. It has established numerous oilfields, oil depots, and gas stations nationwide, achieving vertical integration of the industrial chain. For example, as of 2023, ZSHHA has completed and put into operation 13 long-distance pipelines with a total length of 5,700 km. The complete supply chain enables ZSHHA to efficiently allocate resources, coordinate supply and demand, and gain obvious advantages in product supply and market response speed, thereby effectively reducing operating costs and enhancing market competitiveness.

Secondly, through large-scale production and resource integration, ZSHHA has effectively reduced production and operating costs. For instance, it adopts advanced production processes and equipment to improve production efficiency and product quality stability; strengthens internal management, optimizes production processes and organizational structures, and enhances employee work efficiency and teamwork capabilities. In 2023, its cash operating cost for oil and gas was 755.2 yuan/ton, a year-on-year decrease of 2.3%. Cost control and efficiency improvement have enabled ZSHHA to maintain strong competitiveness in the fierce market competition and achieve sustainable development [9].

In addition, ZSHHA has made technological breakthroughs in fields such as long-distance dense-phase pipeline transportation of carbon dioxide. On July 11, 2023, ZSHHA officially announced the commissioning of China's first high-pressure, normal-temperature dense-phase carbon dioxide transportation pipeline with a scale of one million tons and a length of 100 km. With a total length of 109 km, this pipeline can transport 1.7 million tons of carbon dioxide captured from Qilu Petrochemical's production to the underground oil reservoirs of Shengli Oilfield for enhanced oil recovery and sequestration annually, playing a demonstration and leading role in large-scale carbon dioxide pipeline transportation in China [10].

6.3.2.3 Challenges and Future Outlook

Firstly, countries around the world are attaching increasing importance to clean energy. Currently, ZSHHA's main energy sources are oil and natural gas, both of which are non-renewable. With the rapid global energy consumption rate, the risk of resource depletion is growing. Over-reliance on traditional energy sources is not conducive to the long-term development of the enterprise, so there is an urgent need to adjust the energy structure to adapt to the trend of energy development. Secondly, foreign giants have entered the domestic market with strong capital, advanced technology, and management experience, changing the supply–demand relationship in the oil market and intensifying competition. This is not only price competition but also an all-round competition in technology and services. Domestic enterprises such as ZSHHA need to reposition themselves and enhance their strength in the competition. Moreover, domestic oil refining capacity continues to increase with the construction of integrated refining and chemical projects. It is estimated that the national new oil refining capacity will be approximately 160 million tons per year from 2020 to 2025, and the total oil refining capacity will reach 1 billion tons per year by 2025. The contradiction of overcapacity in oil refining has become increasingly prominent, and some chemical products such as polystyrene, polyvinyl chloride, and polyester are also seriously overcapacity, which squeezes the profit and development space of ZSHHA's refining and chemical business. Finally, in the process of production and operation, ZSHHA may face problems such as increased difficulty in resource acquisition and rising costs. For example, the exploration and development of high-quality oil and gas resources have become more difficult, and access to other key raw materials or technologies may be subject to certain restrictions [11].

Despite the challenges, there are many positive trends in ZSHHA's future development:

First, ZSHHA is actively deploying new energy. For instance, it plans to become China's largest hydrogen energy company by 2025, increasing investment in the hydrogen energy field, including the construction of hydrogen refueling stations and the promotion of green hydrogen demonstration projects. At the same time, it is also focusing on other renewable energy fields such as solar energy and wind energy. By investing in R&D and production of new energy technologies, it can reduce reliance on traditional energy sources and lay the foundation for future energy transition. Second, by introducing advanced foreign production technologies and equipment, and promoting the application of energy-saving technologies—such as improving equipment energy efficiency and enhancing energy conversion efficiency in the production process—energy consumption can be reduced, energy utilization efficiency improved, and enterprise competitiveness strengthened. Third, as an energy-importing country, China relies on ZSHHA as one of its main energy import channels. Under the pressure of international market competition, ZSHHA can achieve mutual benefit and win–win results by cooperating with other energy companies to jointly address global energy issues. It can also expand overseas markets and carry out energy cooperation with other countries to reduce instability and risks in energy supply. Fourth, with the improvement of global environmental awareness, green development has become a mainstream trend. As an energy enterprise, ZSHHA actively promotes green development, increases R&D and application of environmental protection technologies, and improves the environmental protection level of production—such as deploying carbon capture, utilization, and storage (CCUS) technology. This helps reduce environmental impact, meets the requirements of sustainable development, and enhances corporate image and social recognition. Fifth, continuous increase in investment in key core technology research, such as breakthroughs in deep and ultra-deep shale gas exploration theories and technologies, and shale oil development technologies in faulted basins such as Jiyang and Subei, will help enhance the enterprise's core competitiveness. This enables cost reduction and efficiency improvement, as well as improved product quality and added value in fields such as oil and gas exploration and development, and refining and chemical engineering, better responding to market changes and industry competition.

6.3.3 Representative Enterprise 3

6.3.3.1 Company Overview

ZHAYY has a registered capital of 113.8 billion yuan and owns 5 holding listed companies. Its business layout spans domestic and foreign capital markets, demonstrating strong strength. Its main businesses cover oil and gas exploration and development, professional technical services, refining and sales, natural gas power generation, financial services, and other fields, while also actively expanding into new energy businesses such as offshore wind power [12].

In the first three quarters of 2023, ZHAYY stepped up its exploration and development efforts, driving steady growth in net oil and gas production and strengthening cost competitiveness. During this period, the company achieved a net production of 49.97 million barrels of oil equivalent, an 8.3% increase year-on-year, setting a new record for the same period in history and reaching the highest level. Specifically, domestic net production reached 34.55 million barrels of oil equivalent, a year-on-year increase of 6.7%, supported by increased output from oil and gas fields such as Kenli 6-1 and Lufeng 15-1; overseas net production was 15.41 million barrels of oil equivalent, a sharp year-on-year increase of 11.8%, driven by increased production from Guyana's Liza Phase II and Brazil's Buzios oilfields. In the first three quarters, the company made 8 new discoveries and successfully evaluated 21 oil and gas-bearing structures. In terms of development and production, 3 new projects including the Lufeng 12-3 oilfield development project were successfully put into operation, and other new projects are progressing steadily [13].

6.3.3.2 Competitive Advantages and Current Development Status

ZHAYY's main cost per barrel has remained at a relatively low and stable level in recent years. For instance, it dropped from $45.02 per barrel of oil equivalent in 2013 to $26.34 per barrel of oil equivalent in 2020. Although it rebounded slightly in 2022 and 2023, it generally stayed at a low level. In the first half of 2023, the main cost per barrel was $28.17, a year-on-year decrease of 7.1%; in the first three quarters of 2023, it stood at $28.37, down 6.3% year-on-year. This endows ZHAYY with stronger resilience against oil price fluctuations and ensures stable profitability (Fig. 6.4).

Secondly, ZHAYY holds substantial offshore proven and unproven oil reserves domestically. The three trillion-cubic-meter gas zones currently under exploration and development are expected to have a total geological gas reserve of approximately 6.5 trillion cubic meters. Based on a 40% recovery rate, the recoverable reserve is about 2.6 trillion cubic meters, equivalent to 1.58 billion barrels of oil equivalent, with net equity recoverable reserves reaching around 1.49 billion barrels, representing a massive scale. Meanwhile, the company is actively expanding overseas markets and acquiring overseas oil and gas resources through mergers and acquisitions, joint

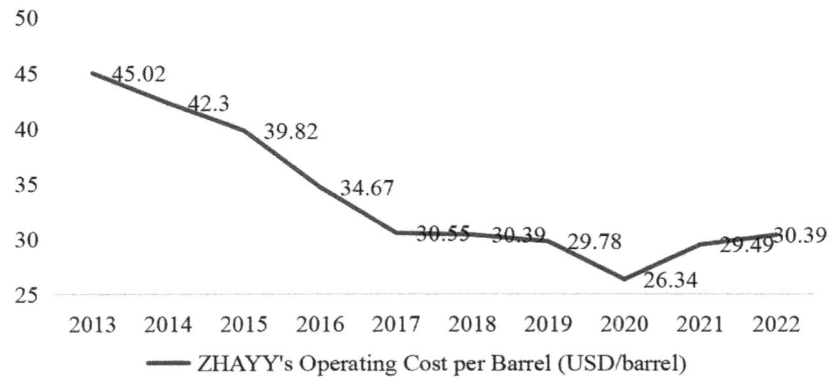

Fig. 6.4 ZHAYY's operating cost per barrel

ventures, and other means. For example, the Guyana project has delivered extremely high profit per barrel, and increased production from projects such as Guyana's Liza Phase II and Brazil's Buzios Oilfield has brought substantial returns to the company.

In addition, ZHAYY continues to invest in technological R&D and has achieved major breakthroughs in technologies such as deepwater drilling platforms, enabling efficient operations in harsh marine environments. As the world's first ultra-large deepwater operation platform with remote control production capability, "Deep Sea No. 1" has effectively contributed to the increase in oil and gas reserves. At the same time, the company is continuously advancing the construction of offshore unmanned platforms and intelligent oilfields, steadily improving production efficiency and management standards.

In the first quarter of 2024, ZHAYY's total net production rose to 180.1 million barrels of oil equivalent, a year-on-year increase of 9.9%. Among this, domestic net production reached 123.2 million barrels of oil equivalent, up 6.9% year-on-year, driven by the commissioning of new projects; overseas net production increased by 16.9% to 56.9 million barrels of oil equivalent. In the first three quarters of 2023, the net production hit a record high of 499.7 million barrels of oil equivalent for the same period in history. Domestically, net production reached 345.5 million barrels of oil equivalent, a 6.7% year-on-year increase, thanks to higher output from oil and gas fields; overseas, net production rose by 11.8% year-on-year to 154.1 million barrels of oil equivalent, supported by increased production from specific oilfields.

In Q1 2024, the company's net profit attributable to shareholders of the parent company was 39.72 billion yuan, a year-on-year increase of 23.7%; operating revenue reached 111.4 billion yuan, up 14.1% year-on-year; and oil and gas sales revenue stood at approximately 89.98 billion yuan, a year-on-year growth of 21.3%. In 2023, oil and gas sales revenue was 151.7 billion yuan in the first half, and 238.4 billion yuan in the first three quarters, with net profit attributable to the parent company amounting to 97.6 billion yuan. ZHAYY has stepped up efforts in natural gas exploration and development, enhanced unconventional gas development capabilities, and strengthened the construction of liquefied natural gas (LNG) supply security capacities, continuously improving the supply level of clean energy [14].

China has achieved remarkable results in the LNG transportation sector. The first vessel "Lü Neng Ying" of the country's largest LNG carrier construction project has been successfully delivered and put into operation. Since 2000, ZHAYY has focused on the construction of large domestic LNG carriers, and 10 vessels have been completed and delivered so far. According to plans, these carriers can transport up to 16 million tons of LNG to China annually, greatly facilitating the replenishment of domestic natural gas resources. Meanwhile, China's first deep-sea floating wind power platform "Hai You Guan Lan Hao" has successfully connected to the grid for power generation; the country's first offshore carbon dioxide storage demonstration project has been officially put into operation; and the first onshore centralized photovoltaic power generation project has also completed grid connection.

6.3.3.3 Challenges and Future Outlook

In 2023, the average international crude oil price fell significantly, leading to a decline in ZHAYY's performance. Its operating income decreased by 1.3% year-on-year, and the net profit attributable to the parent company dropped by 12.6% year-on-year. Moreover, with the improvement of environmental awareness and advances in science and technology, renewable energy sources such as solar and wind energy have developed rapidly, posing a threat to the market position of the traditional oil industry. ZHAYY needs to find its correct positioning amid changes and cope with the competitive pressure from the new energy industry. Additionally, oil resources are limited, and as extraction volume increases, the issue of resource depletion has gradually emerged. While ensuring current extraction volume, ZHAYY must actively seek new exploration areas and improve exploration success rates to ensure the company's sustainable development in the future. Although ZHAYY has overseas business layouts, it may also face political, economic, legal and other uncertainties in different countries and regions, as well as geopolitical risks [15]. For example, ZHAYY once received a notice from a customer in the Middle East to suspend operations of four drilling platforms, which affected its business expectations in the Middle East region.

In the future, ZHAYY will continue to focus on "increasing reserves and boosting production" as its development priority. In 2024, the company has set a net production target of 700–720 million barrels of oil equivalent, representing an increase compared with 2023. ZHAYY has also set a net production target of 780–800 million barrels of oil equivalent for 2025 and 810–830 million barrels of oil equivalent for 2026, with production expected to maintain a growth trend in the next few years.

In 2024, the total capital expenditure budget is between RMB 125 billion and RMB 135 billion, roughly the same as in 2023. Among this total budget, exploration, development, and production capitalization are expected to account for approximately 16,

63, and 19% respectively [16]. The company will rely on effective project connection and high-intensity capital investment to support the growth of production and business development.

ZHAYY adheres to the strategy of "stabilizing oil output and increasing gas supply", enhances natural gas supply capacity, strengthens the layout of natural gas exploration, and steadily advances the construction of the South China Sea Trillion-Cubic-Meter Gas Zone Exploration Project, the Bohai Sea Trillion-Cubic-Meter Gas Zone Exploration Project, and the Onshore Trillion-Cubic-Meter Gas Zone Reserve Enhancement Project.

Meanwhile, making full use of its advantages in marine resources, ZHAYY focuses on strategic emerging industries such as deep-sea floating wind power and related technological R&D, and actively builds a multi-energy complementary integrated energy system. For example, the successful grid connection of China's first deep-sea floating wind power platform "Haiyou Guanlan Hao" symbolizes a key step in offshore oil and gas development towards the "green electricity era".

In addition, ZHAYY conducts a potential assessment of CCS/CCUS (Carbon Capture, Utilization, and Storage) across China's coastal waters, builds the North China Offshore CCUS Demonstration Center and the South China Offshore CCUS Demonstration Center, and continuously follows up on the CCUS cluster project in the Daya Bay area.

IV. Competitive Status of Pipeline Transportation Enterprises

In terms of total oil and gas production, ZSHYO reached 1.56 billion barrels, ZHAYY stood at 680 million barrels, and ZSHHA was relatively lower. Leveraging its extensive resource layout and strong production capacity, ZSHYO maintained a distinct advantage in total oil and gas production, which can be attributed to the stable output of its numerous large-scale oil and gas fields as well as its active exploration and development activities both domestically and internationally. As an enterprise specializing in offshore oil and gas development, ZHAYY also achieved a considerable total production volume, demonstrating strong capabilities in the offshore oil and gas sector. Although ZSHHA's total oil and gas production was lower than that of ZSHYO and ZHAYY, it still played an important role in China's oil and gas supply. Its strengths in refining, chemical engineering, and other fields enabled it to conduct in-depth processing and efficient utilization of oil and gas resources. In terms of crude oil production, ZSHYO hit 770 million barrels, ZHAYY reached 530 million barrels, and ZSHHA was 250 million barrels. ZSHYO took the lead in crude oil production, which was closely linked to its abundant onshore oilfield resources and advanced extraction technologies. ZHAYY also achieved remarkable crude oil production; its continuous new breakthroughs in offshore oilfield development have made significant contributions to the growth of China's crude oil production. ZSHHA's relatively lower crude oil production was related to the characteristics of its business layout—it focused more on the development of refining and chemical businesses, and met its production needs by importing crude oil and purchasing it from other enterprises (Fig. 6.5).

Overall, ZSHYO holds significant advantages in both total oil and gas production and crude oil production, making it an important oil and gas producer in China;

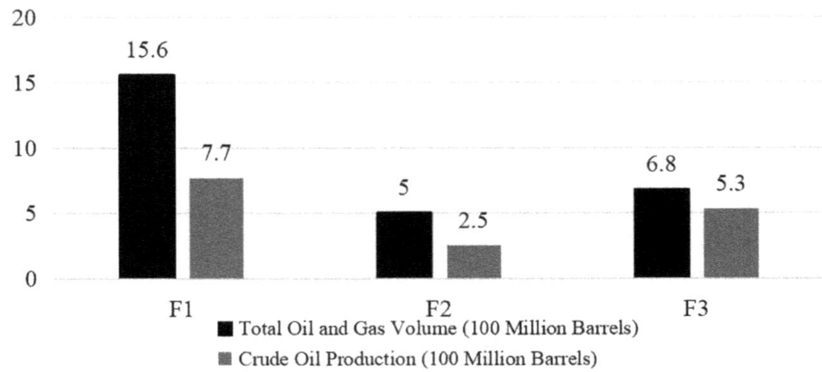

Fig. 6.5 Oil and gas production of ZSHYO, ZSHHA and ZHAYY in 2023

ZHAYY has unique strengths in the offshore oil and gas sector and also achieves relatively high crude oil production; ZSHHA, on the other hand, boasts strong capabilities in the refining and chemical industries, and enhances value through the deep processing of oil and gas resources. All three enterprises occupy important positions in China's oil and gas industry, each with distinct characteristics and advantages, and collectively contribute to China's energy supply and economic development. Their development in different fields also reflects the pattern of diversification and coordinated development of China's oil and gas industry (Fig. 6.6).

In China's domestic oil and gas market in 2023, ZSHYO demonstrated a strong dominant position. Its oil and gas production accounted for 53.85% of the country's total domestic oil and gas output, and this high proportion fully reflects ZSHYO's leading role in the domestic oil and gas production sector. Relying on its abundant resource reserves, extensive business layout, and advanced technological strength, ZSHYO has stood out among numerous domestic oil and gas production enterprises. Over the years, ZSHYO has continuously increased its investment in exploration and development, and has sustained efforts in multiple important domestic oil and

Fig. 6.6 Market share

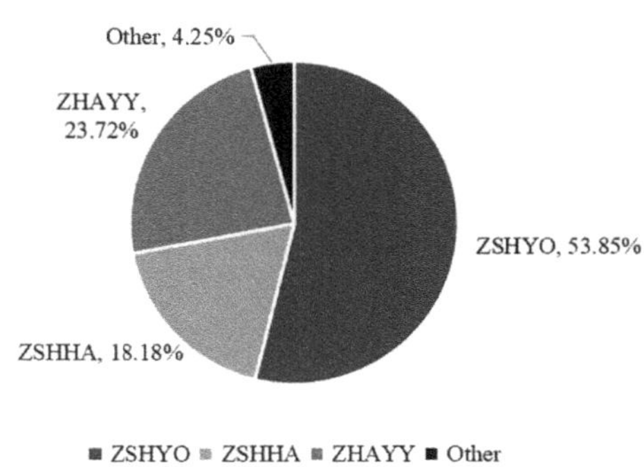

gas production areas, ensuring stable oil and gas supply and making significant contributions to national energy security.

ZSHHA and ZHAYY also hold important positions in the domestic oil and gas market. ZSHHA's oil and gas production accounts for 18.18%. As a key domestic energy and chemical enterprise, ZSHHA has strong capabilities in refining, chemical engineering and other fields, and also plays an important role in oil and gas production. ZHAYY's oil and gas production accounts for 23.72%. As an enterprise specializing in offshore oil and gas development, ZHAYY has fully leveraged its professional advantages in offshore resource exploration and development, continuously expanded its offshore oil and gas business, and contributed significantly to the growth of domestic oil and gas production. Other pipeline transportation enterprises collectively account for 4.25% of the share. Although relatively small, they play an indispensable role in the oil and gas transportation link, ensuring the efficient transportation and allocation of oil and gas resources.

Overall, ZSHYO, ZSHHA, ZHAYY, and other pipeline transportation enterprises together form an important structure of China's domestic oil and gas industry. Each enterprise plays a unique role in different fields, jointly promoting the sustainable development of the domestic oil and gas industry.

6.4 Development Suggestions and Countermeasures

For ZSHYO, it should continue to leverage its advantages in resource reserves and market share, increase investment in technological innovation, further improve the efficiency of oil and gas exploration and development, and reduce production costs. At the same time, it should strengthen cooperation with international energy enterprises, expand overseas markets, and enhance global competitiveness. In terms of pipeline transportation, it should continuously optimize the pipeline network layout and improve transportation safety and reliability. For ZSHHA, while maintaining its advantages in the chemical industry, it can increase investment in upstream oil and gas businesses to boost oil and gas production. It should focus on technological innovation to improve refining efficiency and product quality, and strengthen cooperation with other enterprises to achieve resource sharing and complementary advantages, thereby jointly promoting the development of the pipeline transportation industry. For ZHAYY, it should continue to deepen its presence in the offshore oil and gas sector and step up efforts in deep-sea oil and gas exploration and development. It needs to upgrade its technical capabilities to address the challenges posed by the marine environment, and strengthen cooperation with domestic and foreign marine engineering enterprises to jointly develop offshore oil and gas resources and improve pipeline transportation facilities.

Second, for all pipeline transportation enterprises, they should increase investment in technological research and development, improve the intelligence level of pipeline transportation, and realize real-time monitoring and precise scheduling. They should strengthen safety management, establish a sound safety risk prevention and control system, and ensure the safe and stable operation of pipeline transportation. Enterprises should actively expand their business scope and carry out diversified operations, such as venturing into new energy transportation and other fields. They should also strengthen cooperation to jointly construct and maintain pipeline infrastructure, thereby reducing operating costs. Additionally, enterprises need to enhance communication and coordination with government departments to secure policy support and promote the sustainable development of the industry. Meanwhile, emphasis should be placed on talent cultivation and introduction to improve the core competitiveness of enterprises.

References

1. China Research Industry Institute. Prospects Forecast and In-depth Research Consultation Report on China's Pipeline Transportation Industry from 2024 to 2029[R]. https://m.chinairn.com/scfx/20240926/154351324.shtml.
2. West-East Gas Pipeline: Delivered to Shanghai and Lighting up Hong Kong. (2012-10-23)[2025-03-06]. http://www.nea.gov.cn/2012-10/23/c_131923381.htm.
3. CEIC Database.https://www.ceicdata.com.
4. National Bureau of Statistics.https://www.stats.gov.cn/.
5. China National Petroleum Corporation.https://www.cnpc.com.cn/.
6. China Petroleum News. China Petroleum Circulation Industry Development Blue-book (2023–2024). https://trxf.cnpc.com.cn/content/h/content_35996.shtml.2024.4.18.
7. National Energy Administration. China Natural Gas Development Report (2024). https://www.nea.gov.cn/2024-07/23/c_1310782456.htm.2024.7.23.
8. NetEase. Foreign Giants Enter the Gas Station Market! PetroChina and Sinopec Face Challenges, and the Transformation Path is Full of Variables. November 6, 2023. https://www.163.com/dy/article/IIRPUC4U05562ZU8.html.
9. Zuo Qianming, Hu Xiaoyi. Sinopec: The Core Business is Operating Well, and the Long-term Value of the Petrochemical Leader is Highlighted. March 25, 2024. https://stock.finance.sina.com.cn/stock/go.php/vReport_Show/kind/lastest/rptid/764685921773/index.phtml.
10. Sinopec Group. Under Fierce Competition, the Integrated Advantages are Highlighted. August 25,2019. http://www.chinacpc.com.cn/info/2019-08-25/news_3333.html.
11. NetEase. What Crises is Sinopec Facing? July 24,2023. https://www.163.com/dy/article/IADBALF405529Z50.html.
12. China National Offshore Oil Corporation.https://www.cnooc.com.cn/.
13. Xueqiu. The Business Model and Competitive Advantages of CNOOC. June 7, 2024. https://xueqiu.com/6186763452/292684054.
14. Hou Ruining. CNOOC: The Natural Gas Supply and Demand are Balanced, and the Possibility of Reducing Industrial Demand is Small. November 23, 2018. https://www.jiemian.com/article/2647944.html.
15. Lin Zi. Analysis of CNOOC's Annual Report: Production Increased but Income Did not Increase Last Year. Can Diversified Development Break the Deadlock? March 22, 2024. https://news.qq.com/rain/a/20240322A071SX00.
16. Jiemian News. Oil and Gas Production Maintains Strong Growth, CNOOC Announces 2024 Business Outlook. January 26, 2024. https://www.163.com/dy/article/IPB7VFSC053469RG.html.

Chapter 7
Development and Future Trends of China's Highway Enterprises

Jianlin Wang and Xueqin Wen

Abstract Supported by national policies, China's expressway enterprises have continuously developed. Through multi-stage construction and optimization, they have gradually formed a vast and technologically advanced expressway network. With the promotion of intelligent and informatized technologies, smart transportation applications are increasingly prevalent in the expressway sector. This has not only significantly enhanced traffic management efficiency but also accelerated the digital evolution of related industries. Concurrently, the regional layout of the expressway industry is closely tied to local economic development. The eastern coastal regions feature denser road networks, while western regions are progressively strengthening infrastructure construction. Despite notable achievements, the industry still faces challenges such as financing pressures, imperfect management systems, and lagging regulatory frameworks. To achieve high-quality development, it is recommended that expressway enterprises broaden financing channels, optimize management models, strengthen legal and regulatory support, and integrate green environmental concepts with intelligent technologies. Through the Belt and Road Initiative, Chinese expressway enterprises actively participate in international infrastructure development, exporting technology and management expertise, thereby demonstrating China's significant role in global infrastructure construction.

J. Wang · X. Wen (✉)
Anhui University of Finance and Economics, Bengbu City, Anhui Province, China
e-mail: wxq18305589742@163.com

J. Wang and M. Song (eds.), *The Green Path of Chinese Transportation Enterprises*, Resources, Climate and Sustainable Development,
https://doi.org/10.1007/978-981-95-5289-4_7

7.1 Current Development Status of China's Expressway Enterprises

7.1.1 Policy Framework

In terms of national policy support for the expressway industry, the focus has gradually shifted from early trunk road construction to the establishment of a national highway system, ultimately entering a phase of transportation network refinement and enhancement. During the 8th (1991–1995) to 9th (1996–2000) Five-Year Plan periods, the central government established a development strategy led by local governments with central government support, prioritizing the construction of national highway trunk lines and the development of expressways in coastal transportation hubs. During the 10th (2001–2005) to 12th (2011–2015) Five-Year Plan periods, the focus shifted to refining the national expressway network. During the 13th (2016–2020) to 14th (2021–2025) Five-Year Plan periods, efforts focused on optimizing and upgrading the existing national expressway network through expansion and enhancement [1]. However, influenced by China's diverse geographical features and regional economic development disparities, the industry layout of expressways is closely tied to local economic progress, exhibiting distinct county-level characteristics. From a local development strategy perspective, different regions have distinct policy orientations for expressway construction: Central and Eastern China focus on refining the expressway network and improving transportation efficiency in congested areas, striving for comprehensive expressway coverage. In contrast, Western and Northern China, while pursuing increases in total expressway mileage, place greater emphasis on establishing transportation links with Central and Eastern China, particularly connecting to major urban agglomerations like the Beijing-Tianjin-Hebei region.

7.1.2 Market Scale

China embarked on its expressway construction journey in the 1980s, since when the expressway network has undergone rapid expansion with continuous mileage growth [2]. Particularly in recent years, China's expressway development has witnessed significant growth and expansion. By the end of 2023, the national expressway mileage had reached 184,000 km, an increase of 6,700 km from the previous year [3]. Compared to 2018, when the national expressway network stood at 142,600 km, this represents a substantial increase of 29.3%. The continuous expansion and optimization of China's expressway network has been accompanied by heightened national emphasis on upgrading road technical standards, reflecting strategic planning for a higher-quality, higher-standard expressway network. By the end of 2023, the total length of highways meeting Grade IV or higher standards reached 5.2701 million kilometers, an increase of 107,600 km from the previous year. This represents 96.9%

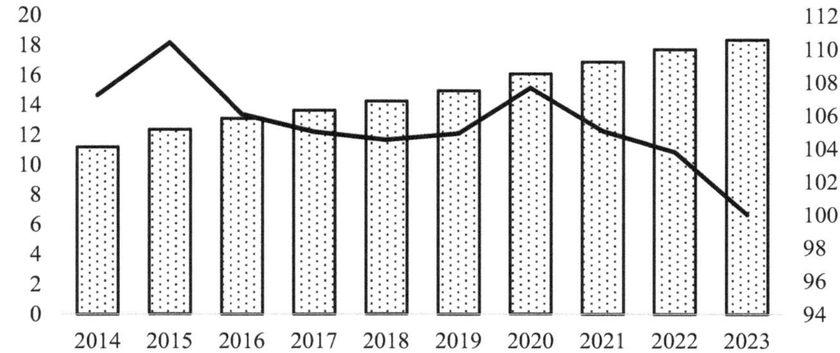

Fig. 7.1 2014–2023 expressway route mileage and mileage growth index. *Source* China Economic Data (CEIData)

of the total highway mileage, marking a 0.5% increase in proportion [4]. In the 2023 statistics, the mileage of highways meeting Grade II or higher standards totaled 762,200 km, an increase of 18,600 km compared to the previous year. This accounted for 14.0% of the total highway mileage, with the proportion rising by 0.1 percentage points [5] (Fig. 7.1).

7.1.3 Corporate Structure

7.1.3.1 Competitive Tiering

China's expressway enterprises coexist across varying scales and influence, forming distinct competitive tiers. Provincial-level conglomerates—including SADOG, JZHH, and NHAGA—typically receive provincial government support and control key expressway corridor resources within their jurisdictions. These entities generally operate extensive expressway networks and possess substantial asset portfolios. These enterprises exhibit diversified business scopes, with actual control often held by provincial State-owned Assets Supervision and Administration Commissions (SASAC) or transportation departments. Due to their scale and governmental backing, they dominate the industry. Municipal transportation construction and operation entities form the middle tier of the competitive tier, including SHGAS, NAYAJ, GAZHG, and others. Compared to provincial groups, municipal transportation construction and operation entities are smaller in scale, primarily responsible for constructing and operating regional expressways. Their service scope is more localized, yet they wield significant influence within their service areas. Finally, there are enterprises with limited highway assets, predominantly private companies or smaller state-owned units. These typically control fewer expressway resources and have relatively limited influence, though they may demonstrate flexibility and specialization in specific regions or niche markets.

7.1.3.2 Regional Development Imbalances

In terms of regional development, China's expressway construction is closely tied to regional economic growth, resulting in an uneven distribution that aligns with local prosperity levels. Coastal eastern regions, benefiting from early economic advantages, feature relatively dense expressway networks. These networks are often based on vertical highway layouts with broad industrial chains, forming well-developed transportation systems. This is particularly evident in Guangdong, Zhejiang, and Jiangsu provinces, where expressway enterprises are highly concentrated. Guangdong leads in the number of listed expressway companies, hosting multiple publicly traded entities including YGASA, YGASB, DOWAK, and SGSSS, making it the province with the highest concentration of listed expressway firms in the regional sector. Central China's expressway enterprises exhibit certain disparities in capital strength and market competitiveness compared to their eastern counterparts, yet their overall development remains relatively balanced. Companies like SAXIL and ZHYAG play crucial roles in the regional transportation networks of central China. Due to geographical constraints and economic development levels, western enterprises such as Sichuan Chengyu in Sichuan Province lag behind in scale, technology, and capital compared to their eastern and central counterparts. However, recent years have seen significant progress in western highway construction driven by the sustained implementation of the Western Development Strategy. Taking the Jixig Expressway as an example, its full opening has drastically reduced travel time between Beijing and Xinjiang, providing a powerful impetus for economic growth and regional connectivity in Northwest China [6].

7.1.4 Technological Advancements

7.1.4.1 Intelligent Development

China's intelligent highway development primarily manifests in the advancement of the intelligent transportation industry. This sector encompasses numerous subfields, including urban multimodal transportation hubs, parking facilities (covering both indoor parking lots and on-street parking spaces), and intelligent management systems for urban road networks. Furthermore, the industry extends to broader application scenarios such as autonomous driving technology and smart urban logistics. The growth of the intelligent transportation sector is driven by national policy support, technological progress and application, as well as increasing market demand. From a policy perspective, the national government has introduced a series of measures to support and drive the digital transformation of intelligent transportation and highways. Technologically, China's Electronic Toll Collection (ETC) system has achieved widespread adoption, intelligent infrastructure construction has steadily strengthened, and the development of smart highway systems has matured, demonstrating significant achievements in intelligent applications within the expressway sector. Market demand analysis indicates frequent financing activities within the intelligent transportation industry. In the first half of 2023, the sector witnessed 20 financing cases totaling RMB 3.631 billion. These Figures demonstrate the industry's capital attractiveness and the capital market's confidence in intelligent transportation, further confirming the expanding market demand within this sector. Several enterprises have already achieved notable milestones, such as ZXKEZ's C-V2X vehicle-road-cloud collaboration technology, HUWEA's cloud-network integrated tolling solution, smart highway edge integrated machine solution, F5G all-optical communication network, and cloud-network-endpoint integrated security technology.

7.1.4.2 Technology Integration and Innovation

China's expressway sector has successfully enhanced traffic network efficiency and safety by integrating advanced technologies including cloud computing, big data, IoT, mobile internet, blockchain, and artificial intelligence. Cloud computing provides data storage and computational capabilities for digital devices; big data tools extract critical insights from vast traffic datasets to support decision-making in traffic management and deliver real-time information, forming the foundational information architecture of the digital economy [7]. IoT technology connects physical objects to the internet via sensing devices for information exchange; mobile internet enables anytime, anywhere access to traffic information, enhancing service convenience [8]; blockchain technology ensures data security and transparency through distributed networks and asymmetric encryption algorithms; Artificial intelligence technology plays a pivotal role in intelligent transportation systems, processing data through algorithms to provide decision support for applications such as traffic flow forecasting and intelligent signal control. Notable enterprises in this field include: JZHH, which integrates leading-edge technology and management practices to innovatively develop the Collaborative Intelligent Transportation Management System (C-ITS) using big data, IoT, traffic simulation, and digital twin technologies; BESFJ collaborated with NIHUG to establish Beijing's first smart expressway; ZHSHG's intelligent traffic cones enhance safety levels and efficiently guide drivers and passengers through smart road condition information.

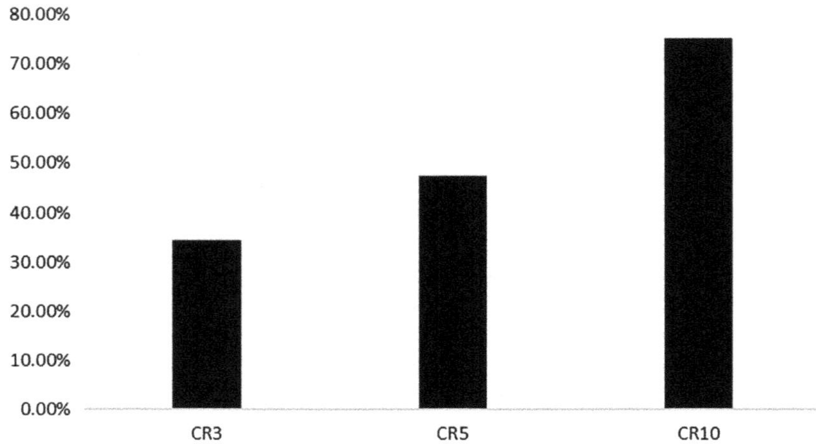

Fig. 7.2 2023 SGJ China expressway industry market concentration (by expressway business revenue). *Data Source* Qianzhan Industry Research Institute

7.1.5 Industry Competitive Environment

7.1.5.1 Market Competitive Landscape

The market competition landscape facing China's expressway enterprises is relatively complex. The expressway industry is a typical capital-intensive sector. Although entry barriers are not high, the industry exhibits a highly oligopolistic market structure due to factors such as lengthy construction cycles, slow capital recovery, stringent national technical standards, and substantial capital requirements [9]. A handful of enterprises dominate the market with strong control, limiting opportunities for new entrants. Data from the first half of 2023 shows the top three companies held a combined market share of 34.22%, with CR5 at 47.37% and CR10 at 74.94% [10]. Among them, SADOG and JZHH held the largest shares at 14.08 and 12.27% respectively. Other major players including ZHSHG, SCHCY, GAYUG, NIHUG, FUJIG, and SHGAS demonstrated strong competitiveness and influence within their respective regions. Major enterprises hold substantial market shares, exerting significant influence on industry development and market share allocation. Furthermore, the performance of expressway enterprises is inversely correlated with market concentration and internationalization levels, indicating issues of insufficient competition and inefficiency in the market (Fig. 7.2).

7.1.5.2 Core Competitive Advantages

Key directions for enhancing the core competitiveness of China's expressway enterprises include leveraging geographical advantages, improving emergency response mechanisms, implementing service value chain management, and cultivating innovation capabilities. Regional advantages have consistently played a role in the development of China's expressway enterprises, and current operators continue to rely on these advantages to develop core competitiveness in different regions. For instance, Shandong Expressway, operating in Shandong Province—a key transportation hub connecting North China, East China, and the Central Plains—leverages its coastal location along China's eastern seaboard to provide logistics and passenger transportation services within the region. Regarding emergency response mechanisms, expressway enterprises are investing more resources in enhancing security technology and emergency handling capabilities to ensure stable transportation system operations. For instance, FUGAO established a "5 + N" collaborative mechanism, bringing together the Fujian Provincial Department of Transportation, Provincial Public Security Bureau Traffic Police Corps, Provincial Emergency Management Department, Provincial Fire and Rescue Corps, Provincial Expressway Group, and relevant departments like health and meteorology. This enables centralized command, coordination, and cooperation, enhancing expressway safety operations and rapid emergency response capabilities while strengthening overall contingency support. In service value chain management, highway enterprises continuously elevate service standards and operational efficiency. The 2023 National Highway Traffic Civil-Military Joint Emergency Drill, successfully held in Loudi, Hunan Province, demonstrated FUGAO's cutting-edge concepts and emergency rescue technologies in highway traffic emergency management, showcasing its highly coordinated response capabilities within the service value chain. To cultivate innovation capabilities, highway enterprises have increased investments in technologies integrating cloud computing, big data, the Internet of Things, mobile internet, blockchain, and artificial intelligence to achieve intelligent traffic management and service quality enhancement [11].

7.1.6 International Competitive Environment

Under initiatives such as the Belt and Road Initiative (BRI), Chinese highway enterprises have actively engaged in international infrastructure development, enhancing their global competitiveness and achieving significant results. For instance, a subsidiary of SADOG participated in constructing Bangladesh's Dhaka Airport Elevated Expressway. This project represents not only Bangladesh's first elevated expressway but also its largest public–private partnership (PPP) initiative, demonstrating Chinese enterprises' competitive edge across the entire infrastructure industry chain. Similarly, ZHXIJ constructed Algeria's East–West Highway, spanning 17 northern provinces over 1,216 km. Upon full completion, this highway will greatly facilitate local transportation and boost domestic economic interaction [12]. These cases demonstrate that China's expressway enterprises have achieved notable successes in international cooperation, creating a win–win situation both domestically and globally. While showcasing China's formidable professional capabilities and collaborative strength, Chinese expressway enterprises have also exported substantial technical expertise and management experience to foreign counterparts. For instance, in the Barilla Oil Field project in Sudan, SADOG not only transferred construction technologies but also provided expertise in investment financing, financial management, and project administration, thereby shaping the "Spirit of Sudan." Additionally, New Era Company's cotton cultivation initiative in Sudan has spurred local economic growth, created employment opportunities for residents, and elevated the quality of life for Sudanese citizens. Chinese enterprises have also constructed the ring expressway around Serbia's capital. Spanning approximately 20.34 km and divided into three sections, this highway will reduce travel times, lower traffic accident rates, and mitigate harmful gas emissions and noise pollution within the city center upon completion. DILIJ participated in the construction of the S14 ring highway in Lodz, Poland, making Lodz the first city in Poland with a complete ring highway network, thereby improving the logistics distribution efficiency of the China–Europe freight train. Meanwhile, TIJIG undertook the renovation and expansion of the Vanuco-Vayanga Highway in Peru. This 236.63-km route traverses the Andes Mountains, and the completion of the Wactawaru Tunnel laid a solid foundation for the smooth progress of subsequent highway projects.

7.2 Development Trends of China's Expressway Enterprises

7.2.1 Government Policies

7.2.1.1 Financing Structure Optimization and Support

Highway enterprises predominantly adopt the "borrow to build, toll to repay" model. As the revenue-expenditure gap widens, corporate debt burdens intensify. On one hand, revenue margins are compressed: rising land prices and escalating costs for materials and labor continuously increase highway project construction expenses. Simultaneously, the rapid development of high-speed rail and civil aviation has diverted passenger traffic from highways, while policies like "shifting freight from road to rail" have impacted expressway freight revenue. On the other hand, financing difficulties have intensified. Strict national controls on local government debt and slowing growth in land transfer revenues have constrained local governments' financing capacity, leading to higher financing costs for expressway enterprises [13]. A typical example is China Merchants Highway (001965.SZ). According to its 2023 annual performance report, despite achieving significant growth in operating revenue and net profit, the rise in financing costs still exerted pressure on the company's profitability. To address these financing challenges, the government has implemented a series of policies, including optimizing financing structures, deepening investment and financing reforms, supporting private enterprise participation, promoting expansion and renovation projects, and strengthening coordinated funding guarantees. Regarding financing difficulties, enterprises are encouraged to adopt diversified financing methods, including bond issuance, asset securitization, and real estate investment trusts (REITs) in the infrastructure sector, to reduce financing costs [14]. Concurrently, the Implementation Opinions on Supporting Private Enterprises in Participating in the Development of Transportation Infrastructure Construction (a policy document issued by the Ministry of Transport and the National Development and Reform Commission) was released. This document aims to eliminate market access barriers, ensure private enterprises can enter the market on an equal footing, incentivize private investment, and broaden financing channels. To increase revenue, existing highways undergoing expansion or renovation may have their operating periods reassessed, extending toll collection durations to enhance investment returns. Regarding funding safeguards, the Ministry of Transport proposes improving the government-led public finance support system for transportation infrastructure, strengthening fiscal funding guarantees. Additionally, transportation authorities at all levels must accelerate the establishment of linkage mechanisms between highway development planning and funding guarantees, while reinforcing risk prevention and control measures for highway construction project financing.

7.2.1.2 Toll Policy Adjustments

China's expressway operators are progressively establishing a more flexible, efficient, and user-friendly toll adjustment mechanism. Toll rates are periodically revised based on factors such as traffic volume, operational costs, and debt repayment to adapt to evolving transportation demands and economic conditions. While long-term toll policies have consistently faced public scrutiny, expressway companies universally grapple with financial pressures. NIHUG's overall toll road revenue in 2023 remained largely unchanged from the previous year, yet operating costs increased by 10.9%, revealing significant toll collection pressures. To address toll revenue challenges, the government has implemented several adjustment measures. These include differentiated toll strategies based on road sections, vehicle types, time periods, entry/exit points, travel directions, and payment methods. Vehicle classification standards for toll road fees have also been revised to align with the new charging model. At the user level, the government actively promotes preferential policies for ETC electronic payments to enhance public acceptance of the toll system. Additionally, auxiliary tools like map navigation have improved route recommendation accuracy and toll cost estimation capabilities, thereby enhancing toll transparency and providing the public with better travel services. Concurrently, the government has strengthened publicity and toll transparency, intensifying efforts to explain and interpret differentiated toll policies for expressways. Concurrently, the government will make efforts to ensure the sustainable development of highway enterprises. Toll adjustments for these enterprises will also consider debt repayment and maintenance costs, ensuring the sustainable operation of toll roads and maintaining good road conditions.

7.2.1.3 Industry Regulation and Management

In recent years, highway enterprises have faced challenges such as inadequate service area facilities, subpar service quality, and public scrutiny over toll standards. To address these issues, the Chinese government has implemented multiple measures, including strengthening market access rules, price regulation, service quality management, and optimizing the reform management framework. Strengthening market access regulations involves vetting companies entering the expressway market, intensifying oversight of the market environment for expressway enterprises, standardizing market order, and ensuring highway construction quality. Enhanced price regulation adjusts toll rates and implements differentiated toll policies to reduce the public's economic burden while ensuring the sustainable development of expressway enterprises. Designated service quality regulations incentivize enterprises to improve service standards by evaluating service area quality levels. Optimize management structures: The government will drive digital transformation in the highway sector, accelerate smart highway construction, and enhance digitalization across the entire lifecycle of highway design, construction, maintenance, and operation. Improve laws and regulations: By revising regulations such as the "Regulations on the Management of Toll Highways," clarify toll collection periods and rates while strengthening oversight of toll revenue usage and management.

Based on the Chinese government's policies in recent years, it is evident that in the future, China's expressway sector will continue to strengthen market access regulations, pricing regulations, and service quality regulations while improving the reform management system. Furthermore, the government will provide legal support for standardized expressway management from a legal perspective. Regarding the debt challenges faced by expressways, the government may create corresponding financial opportunities, such as loan refinancing, to smooth maturity structures and alleviate debt burdens. Concurrently, it will promote the investment value of smart expressways and encourage private investment. China's government is expected to achieve more standardized and efficient management in the industrial regulation and oversight of expressway enterprises.

7.2.2 Corporate Ecosystem

7.2.2.1 Environmental Ecology and Green Development

The release of the "14th Five-Year" Development Plan for Green Transportation (2021–2025) underscores the importance of ecological conservation and environmental friendliness, reflecting the commitment to advancing green development in highway construction. On one hand, the concept of ecological highways emphasizes the seamless integration of engineering structures with the ecological environment throughout all stages—planning, design, construction, and operational management. This signifies that future highway construction will increasingly prioritize the use of green building materials, noise and pollution reduction, and biodiversity conservation. Additionally, future highway development must explore further protective measures to achieve sustainable development goals. This includes strengthening research on slope ecological protection and emphasizing the study and practice of road ecology. The Shandong Jiwei Expressway Project, as China's first "zero-carbon" highway, effectively controlled carbon emissions by adopting green construction and low-carbon environmental principles. The project utilizes photovoltaic power generation systems to save substantial annual carbon dioxide emissions. Simultaneously, slope greening through plant carbon sequestration injects green energy into "zero-carbon" construction. This initiative demonstrates the deep integration of green, low-carbon practices with modern intelligent technology applications. The Guangxi Leye-Baise Expressway incorporates multiple green construction technologies in its design, including photovoltaic power generation, green plant glare-reduction greening for bridges, shade canopies, and fog zone guidance systems. XYSJ's "Innovative Low-Carbon Emission Reduction Project for MTC Paper Invoice Digitization Based on Multi-Dimensional Data Fusion Technology" was recognized as a "2023 Exemplary Case of Green and Low-Carbon Expressway Development." Through digital transformation, this initiative has eliminated vast quantities of unnecessary paper invoices across the industry, significantly enhanced vehicle throughput efficiency on manual lanes, and reduced carbon emissions.

7.2.2.2 Economic Ecology

The key to the sustainable development of the expressway industry lies in the efficient implementation of ecological construction. From the perspective of expressway enterprises, future development will increasingly focus on reducing economic losses while enhancing environmental and social benefits [15]. Enterprises must adopt scientific economic-ecological development strategies to establish an eco-friendly, sustainable expressway enterprise model. From the perspective of smart highway development trends, smart highways serve as a critical platform for the deep integration of new technologies and the transportation sector. They bear the mission of enhancing highway transportation efficiency, incubating emerging industries, and driving new digital economic growth. Future highway development will increasingly rely on cutting-edge technologies such as big data, cloud computing, and artificial intelligence to achieve functions like holistic perception, proactive management, and companion services, thereby enhancing road safety and operational efficiency. Analyzing the dual-circulation development pattern, highways serve as the primary conduit linking domestic and international circulation. Their industrial development is crucial for ensuring smooth economic flow and industrial connectivity. Future highway development will prioritize effective integration with domestic and international markets while amplifying its role in advancing the dual-circulation framework. TIJIG's application of digital innovation achievements for expressways based on a big data middle platform has enabled the aggregation and scientific management of expressway data resources. Through this data middle platform, the group achieves comprehensive control over expressways, promptly identifying and addressing various anomalies while providing scientific basis for resolving challenges encountered during project construction. This not only reduces economic losses but also demonstrates increased social benefits. Additionally, the group has developed a smart expressway operation management system and digital twin applications. These enable realistic simulations of expressway infrastructure, surrounding environments, and vehicles, thereby enhancing operational efficiency and service levels while improving road safety and operational effectiveness.

7.2.3 *Technical Capabilities*

7.2.3.1 Digitalization, Informatization, and Intelligence

Thanks to the widespread application of technologies such as big data and cloud computing, coupled with the public's demand for faster travel speeds and improved vehicle throughput efficiency, highways will evolve toward greater networking and informatization. According to the Ministry of Transport's "Guidelines on Promoting Digital Transformation of Highways and Accelerating the Development of Smart Highways," significant achievements in highway digital transformation are projected by 2027. This will establish a unified digital model and unified data covering the entire lifecycle—from design and construction to maintenance and operation—comprehensively enhancing highway operational efficiency, service quality, and traffic flow capacity. Furthermore, as a key direction for future expressway development, intelligence is being realized through the introduction of advanced management technologies and equipment. During its digital transformation, SADOG actively adopted BIM + GIS technology to achieve integrated data management and optimize survey and mapping processes. JZHH is constructing a high-precision digital foundation to integrate and govern comprehensive data resources for expressways. This includes data collection using technologies such as drone LiDAR surveying, oblique photography, high-resolution remote sensing, and BeiDou positioning, alongside BIM + GIS technology for integrated data management. Qianxun Positioning Network Co., Ltd. has launched the world's largest-scale BeiDou high-precision positioning road test, which will conduct algorithm verification on all national expressways and major urban expressways. China's expressway enterprises are continuously enhancing service levels and operational efficiency through digital, informational, and intelligent technological innovations to meet future society's high-standard demands for the expressway network.

7.2.3.2 Electrified Expressway System

Under the national goals of "carbon peaking and carbon neutrality," the proposal and implementation of electrified highway systems have become a necessary strategic direction for China's highway enterprises. Such systems not only drive energy structure reform but also enable efficient utilization of electricity while reducing emissions of polluting gases, holding significant importance for building an environmentally friendly society. By utilizing electric power to drive vehicles, the electrified highway system reduces reliance on fossil fuels and lowers emissions of pollutants such as carbon monoxide and carbon dioxide. Furthermore, employing electric power for vehicle propulsion enhances the effective utilization rate of electricity. Currently, China has made considerable progress in electrified highway technology, accelerating the development of such systems. ZGZHO launched China's first demonstration project for electrified highways and mining transportation systems in Datong, Shanxi Province. By constructing electrified highways, developing heavy-duty electrified transport equipment, and establishing an integrated operations and dispatch platform, this project has formed a comprehensive electrified highway transportation system, achieving a green transformation of traditional heavy-duty freight transportation. SAYIJ participated in constructing China's first electrified highway test line in Zhuzhou, jointly developed by SAYIJ, ZHOZH, Tsinghua University, and other institutions. This two-lane, bidirectional test line features a 7-m-wide roadway with 1.8 km of overhead contact lines and box-type substations. It provides diverse simulated scenarios to support vehicle performance and system reliability testing.

7.2.3.3 Communication Technology

The application of communication technology in expressways represents another key direction for future development. Specifically, highway communication systems deploy fiber-optic digital transmission systems and programmable digital switching systems to build fully digital integrated communication networks. These networks provide essential voice services and data/image transmission channels for operational management, monitoring, and toll collection systems. With the advancement of technologies like 5G, highway communication capabilities will significantly improve. The application of these technologies will enhance data transmission and processing power, supporting more efficient traffic management and control.

7.2.4 International Cooperation

7.2.4.1 Smart Expressways

The development of smart highways represents a crucial reference point for China's highway enterprises in international cooperation. As a prominent symbol of deep integration between new technologies and the transportation sector, smart highways play a pivotal role in enhancing transportation efficiency, fostering emerging industries, and driving digital economic growth. Developed nations such as the United Kingdom and the United States have achieved significant accomplishments in smart transportation research and construction, accumulating extensive experience particularly in variable speed limit signs, automated traffic signal control, traffic flow sensing, and video surveillance systems. China's expressway enterprises are just beginning their journey in smart highway construction. By learning from and adapting successful international practices to their own developmental realities and needs, they can forge a uniquely Chinese path for smart highway advancement. This process of introducing and assimilating international best practices will help China avoid detours in modernizing its smart transportation systems and accelerate alignment with global leading standards.

7.2.4.2 The New Dual-Circulation Development Pattern

Under the new dual-circulation development paradigm, the international cooperation trends of China's expressway enterprises are also evolving and deepening. Entering this new development stage, the Central Committee has proposed a new development pattern centered on the domestic cycle while promoting mutual reinforcement between the domestic and international cycles. This presents greater opportunities for China's expressway enterprises in technological innovation and industrial integration. The "Long-term Development Plan Outline for Scientific and Technological Innovation in the Transportation Sector (2021–2035)" explicitly states the need to accelerate the deep integration of cutting-edge technologies such as artificial intelligence, new materials, and new energy with transportation, break through key core technologies, and enhance transportation science and technology innovation capabilities. To achieve this goal, China's expressway enterprises must strengthen cooperation with international counterparts. Through technology introduction and joint R&D, they should continuously achieve breakthroughs in technological innovation and management system innovation, thereby providing robust support for the new dual-circulation development paradigm. International collaboration not only facilitates the introduction of advanced technologies but also enhances the capacity for integrated industrial development through joint construction, ultimately strengthening the overall competitiveness of China's expressway industry.

7.2.4.3 Collaborative Development of an Environmentally Friendly Society

Facing global challenges such as climate change, population growth, and land resource scarcity, China's expressway enterprises should actively collaborate with the international community to jointly promote the construction of green expressway systems. By drawing on international best practices in environmental protection and green infrastructure development, these enterprises can introduce innovations in technology, design, and materials. This will enable the achievement of more environmentally friendly and low-carbon infrastructure construction goals, thereby making positive contributions to global environmental protection.

7.2.4.4 International Cooperation Under the Belt and Road Initiative

The Belt and Road Initiative provides a vital international cooperation platform for China's expressway enterprises. Within this framework, these enterprises actively participate in infrastructure development—particularly highway projects—along the routes, leveraging their comprehensive strengths in design, construction, and management. Such collaboration not only advances regional economic integration in Asia but also provides crucial support for the socioeconomic development of participating nations. Moving forward, China's expressway enterprises will continue advancing infrastructure cooperation while further strengthening policy coordination and strategic alignment. They will actively engage in green development partnerships and explore new areas of collaboration. For instance, in fields such as new energy infrastructure, smart transportation, and digital management, these enterprises can broaden the scope and depth of international cooperation, laying a solid foundation for building sustainable and mutually beneficial global partnerships.

7.3 Challenges Facing China's Expressway Enterprises

7.3.1 Financing Structure

As a capital-intensive industry, expressway projects inherently involve large capital requirements, extended operational lifespans, and lengthy payback periods. Currently, however, China's expressway financing primarily relies on indirect loans such as bank loans and government investments. This approach not only lacks diversified financing channels but also suffers from underdeveloped indirect financing markets. The unique nature of expressways and the limited financing channels have led to a high dependence on indirect financing methods like bank loans for expressway projects. This reliance has resulted in financing issues for China's expressway enterprises, including high debt levels, high financing costs, and low investment efficiency

[16]. Moreover, existing investment and financing mechanisms, along with management systems, exhibit certain distortions—such as toll road policies violating the principle of earmarked funds for highway construction. Additionally, deficiencies in fiscal systems, horizontal and vertical institutional frameworks, and policy regulations contribute to the challenges in highway investment and financing. JIAJK serves as a typical case study. JIAJK sought to broaden its financing channels by issuing REITs (Real Estate Investment Trusts), but it faces refinancing challenges. Publicly offered REITs require funds to distribute dividends at least once annually, with a payout ratio of no less than 90%, thereby constraining its refinancing capacity [17]. Additionally, JIAJK expanded its scale through acquiring external road assets to address issues such as the high cost of new road construction and the expiration of toll collection periods for existing assets.

7.3.2 Operational Management

Highway operations management suffers from insufficient specific regulations and unclear responsibilities within the administrative system, primarily manifesting in four aspects. First, lagging regulations and inadequate policies. The existing Regulations on the Management of Toll Highways no longer effectively support the sustainable development of toll highways. Unreasonable toll rates and concession periods make it difficult to achieve investment returns. Furthermore, issues such as an incomplete responsibility system in administrative law enforcement, conflicts of authority and responsibility, and excessive discretionary power reflect deficiencies in the legislative framework. Second, unreasonable functional allocation. The highway management system suffers from blurred boundaries between government and business, monopolistic operations, and inefficiency, resulting in unclear demarcation between government and market roles and inefficient resource allocation and utilization. Administrative enforcement entities for highways vary across provinces, with enforcement models ranging from unified to decentralized to hybrid. This diversity may instead lead to management chaos and inefficiency [18]. Third, the oversight system is inadequate. Weak supervision of administrative law enforcement in expressway management has rendered the existing oversight framework ineffective. Fourth, institutional arrangements are irrational. Duplication of expressway management bodies is widespread. Many provinces and municipalities have established separate expressway administration bureaus or expressway project offices alongside existing highway administration bureaus. Some have even created agencies directly authorized and managed by provincial governments. This irrational institutional setup impedes the smooth implementation of transportation policies.

7.3.3 Regulatory Development

Currently, there are several shortcomings in the regulatory framework governing China's expressway management model. First, due to the vast geographical expanse of China, provinces and municipalities employ differing approaches to expressway project construction and investment financing models. This has resulted in a lack of unified policy guidance and a consistent management structure within China's expressway management system. Such regional variations constrain the development of expressway enterprises. Second, while China's expressway network has expanded rapidly in recent years, legislative development has lagged behind emerging issues. Moreover, the complex problems faced by expressway enterprises have inevitably exposed shortcomings in existing legislation, including legal conflicts and suboptimal legislative quality. The existing Regulations on the Management of Toll Highways no longer adequately support toll road operations or sustainable development. Unreasonable toll rates and durations make it difficult to achieve investment returns. Hubei Province's 2023 toll road revenue-expenditure balance resulted in a deficit of −1.044 billion yuan, representing an overall loss of 1.044 billion yuan. Furthermore, the emergence of new government-enterprise cooperation models, such as the highway PPP model, has created ambiguity in the government's dual role as both partner and regulator. The absence of explicit legal provisions has blurred these responsibilities, dampening private capital participation and hindering project health. These new cooperation models and emerging challenges have inevitably created regulatory gaps in the government's oversight framework and legal regulations for highway enterprises.

7.3.4 Cost Management

Issues in cost management among China's expressway enterprises primarily manifest in the following areas: budgets lacking strategic guidance, cost control lacking incentive mechanisms, absence of systematic cost management systems, opaque or untimely cost information, and insufficient ongoing evaluation and feedback on cost control effectiveness. During the budget management phase, while many enterprises have attempted to establish modern budget management systems and achieved some initial results in implementing comprehensive budget management, numerous operational issues persist. Budget management lacks sufficient strategic guidance [19]. During the cost control phase, while highway projects have clear cost control targets, the absence of effective incentive mechanisms makes it difficult for project teams and individuals to gain sufficient motivation to achieve these goals. Consequently, even when cost control measures are established, they often fail to be effectively implemented. In the cost management system phase, the lack of a unified and systematic cost management framework hinders the integration of cost control measures with the enterprise's overall strategy. In the cost information phase, some expressway enterprises suffer from insufficient cost transparency or untimely updates to cost data, directly undermining the accuracy and effectiveness of cost control. In the evaluation and feedback phase, continuous assessment and feedback mechanisms remain inadequate. Even when cost control issues are identified, they may remain unresolved due to the absence of timely adjustments and improvements.

7.4 Recommendations for the Development of China's Expressway Enterprises

7.4.1 Diversify Financing Channels

Given the financing challenges faced by China's expressway enterprises—including high debt levels, elevated financing costs, and low investment efficiency—it is essential to expand and innovate financing channels. This can be achieved through methods such as Public–Private Partnerships (PPP), bond issuance, and attracting private investors to diversify funding sources. The PPP model can attract broader investor participation, helping projects secure greater social capital investment. This effectively addresses funding gaps in traditional financing approaches and enhances projects' fiscal sustainability. Corporate bonds, as a form of direct financing, can provide essential capital support, particularly when traditional bank loans are constrained. Although the volume and total scale of bonds issued by China's listed highway companies remain extremely low at present, corporate debt financing holds long-term potential to optimize financing structures and reduce costs. Introducing private investors can diversify existing financing methods, lower the proportion of loans, and alleviate debt pressure.

7.4.2 Reforming Operational Management Models

Addressing deficiencies in specific regulations and unclear responsibilities within the management system, China's highway enterprises are optimizing through top-level design reforms of operational management models. This includes streamlining organizational structures and functional allocations, strengthening oversight systems, and reforming management frameworks. First, organizational structures and functional allocations should be optimized by clearly delineating the respective responsibilities of government and market entities, ensuring proper functional distribution, and establishing a scientifically viable management model where both parties fulfill their roles effectively in the development of highway enterprises. Furthermore, an effective oversight system must be established to ensure rigorous supervision of administrative law enforcement in highway operations. Leveraging current digital, information, and intelligent technologies, while introducing third-party oversight mechanisms, will help develop a scientific and effective supervision system that enhances efficiency and transparency. Finally, to address issues of government-enterprise entanglement, monopolistic operations, and inefficiency, restructuring is needed in legal framework development, management function configuration, and operational model selection. This includes implementing enterprise concessions and achieving economies of scale in expressway operations to improve resource allocation and utilization efficiency.

7.4.3 Strengthening Regulatory Development

In accordance with the directives of the Fourth Plenary Session of the 18th CPC Central Committee, highway management must achieve decision-making, supervision, and administration in accordance with the law. All decisions and actions in highway planning, construction, and operation must strictly adhere to national laws, regulations, and rules. Strengthening regulatory frameworks for the expressway sector not only enhances safety and reliability, safeguarding public life and property, but also builds societal trust and satisfaction with expressway services. This is also a crucial means to prevent and reduce illegal activities while ensuring the normal operation and healthy development of expressway enterprises. To this end, improving road administration management is paramount, requiring the lawful fulfillment of duties to protect road assets and enforce road rights, while reasonably regulating the behavior of road users. Furthermore, establishing a comprehensive standardized management system for expressway construction projects is vital. This system should incorporate clear operational mechanisms, execution controls, and feedback adjustments to ensure smooth project implementation and high-quality completion. Concurrently, updating management philosophies and standardizing operational practices through scientific, standardized management methods and robust enforcement mechanisms will drive the modernization and efficiency of expressway management.

7.4.4 Improving the Financial Management System

To enhance the budget and cost management capabilities of expressway operators, the financial management system must be refined across multiple dimensions. First, strengthen the strategic guidance of budget management by rationally scheduling budget preparation timelines and establishing a timely reporting system to ensure accuracy and timeliness. Simultaneously, develop a comprehensive budget management mechanism tailored to industry characteristics based on the enterprise's actual conditions. Second, design reasonable incentive mechanisms, such as performance evaluation and reward systems, to effectively boost employee motivation and accountability, thereby aiding in achieving cost control objectives. Building upon this foundation, a cost management system covering the entire project lifecycle must be established. This system should encompass all stages—from bidding and construction management to project completion—ensuring cost control aligns with the company's overall strategic objectives. To further enhance management efficiency, developing an effective information system is crucial. This system will guarantee timely updates and transparent sharing of cost information. Third, an efficient evaluation and feedback mechanism should be implemented. Regular assessments of cost control effectiveness, coupled with timely adjustments and optimizations based on feedback, will ensure continuous improvement.

7.5 Conclusion

Driven by sustained national policies, China's expressway enterprises have evolved from the infrastructure construction phase to a new stage of technological innovation and intelligent transformation. This evolution has solidified their competitive edge through expanded market scale and enhanced technological capabilities. Concurrently, these enterprises play a pivotal role in building industrial ecosystems and advancing environmentally sustainable societies. However, confronting financing constraints, operational management challenges, regulatory gaps, and cost management issues, companies must adopt diversified financing channels, reform operational models, strengthen regulatory frameworks, and establish robust financial management systems to overcome obstacles and achieve sustainable development. With the deepening implementation of the Belt and Road Initiative, China's expressway enterprises have demonstrated remarkable vitality and potential in international cooperation, contributing Chinese wisdom and strength to global transportation infrastructure development. This not only reflects their international influence but also showcases their capacity to drive infrastructure construction and development worldwide.

References

1. Jiang Yanfei. Research on the Operation Management System and Toll Policies of Expressways [D]. Doctoral Dissertation, Chang'an University.
2. 2023 Statistical Bulletin on the Development of the Transportation Industry.
3. Liang Shihao. Research on Strategies to Improve Rural Road Spatial Planning in Xi'an under Ecological Orientation [D]. Master's Thesis, Chang'an University, 2019.
4. Wang Chunyao. Analysis of Operating Performance and Influencing Factors of Listed Expressway Companies [D]. Master's Thesis, Dalian Maritime University, 2021.
5. Li Fuyong. System Evaluation Research on Expressway Traffic Safety Facilities [D]. Master's Thesis, Chang'an University, 2006.
6. Summary and Interpretation of 2021 Expressway Industry Policies in China and 31 Provinces – OFweek Smart Manufacturing.
7. Chen Yazhu. Application of Data Warehouse Technology in China's Housing Management [D]. Master's Thesis, University of International Business and Economics, 2006.
8. Zhang Yan. Challenges and Countermeasures of Information Security in the FinTech Era [J]. China Financial Computer, 2017.
9. https://finance.sina.cn/2024-02-07/detail-inaheuik8258972.d.html.

10. Yan Kai. Construction of Financial Information Systems in Expressway Enterprises [J]. Public Investment Guide, 2019.
11. Wang Tianyu, Shen Xueqing, Fan Dalu. 2017 China Commercial Logistics Bank Research and Industry Development Report [M]. Beijing: China Economic Publishing House, 2017.
12. Li Hongmei. International Economic Cooperation [M]. Beijing: China Social Sciences Press, 2013.
13. Wang Ruchao. Research on the Impact Path and Issues of Media Attention on Corporate Performance [J]. Modern Auditing and Accounting, 2014.
14. Jiang Xuqin. Research on the Market-Oriented Transformation of Local Government Financing Platforms in Jiangsu Province — A Case Study of A Urban Investment Company [D]. Master's Thesis, Anhui University of Finance and Economics, 2023.
15. Wang Yong. Exploration on the Economic Ecological Development Strategies of Expressway Enterprises [J]. 2013.
16. Final Report (Abridged) of the Global Environment Facility Project: "Reducing Carbon Emissions by Mitigating Congestion in Large Cities". Beijing Jiaotong University et al., April 30, 2016.
17. Long Zhigang Column. This Article Clearly Explains the Current Status and Issues of Expressway Public REITs | Tencent News.
18. Li Jing. Research on Administrative Law Enforcement Issues of Expressways [D]. Master's Thesis, Hebei Normal University, 2016.
19. Ministry of Transport — Admits that the Current Toll Road Policy is Still Incomplete (http://news.xinxunwang.com/china/sdbd/8960.html).

Chapter 8
Smart and Green Ports: Strategic Pathways for Sustainable Development of Port Enterprises

Jianlin Wang and Xue Xing

Abstract As a critical nexus in international trade and logistics systems, the strategic position of ports is becoming increasingly significant. This study focuses on exploring the current development status, challenges faced, and future development trends of port enterprises in China. It first delves into the latest achievements of port enterprises in areas such as infrastructure upgrading and the construction of smart ports. Furthermore, this research emphasizes how port enterprises adapt flexibly and strengthen risk management mechanisms to ensure long-term sustainable development. The final conclusion points out that domestic port enterprises need to continuously strengthen international cooperation to promote the optimization and upgrading of the industrial structure. Simultaneously, these enterprises should be driven by innovation to guide themselves towards a stage of high-quality development, thereby contributing positively to the prosperity of global trade.

8.1 Introduction

With the deepening development of global economic integration, the strategic position of ports has become increasingly prominent, serving as vital transportation hubs connecting sea and land and radiating inland. As the world's second-largest economy and a key participant in global trade, the development of port enterprises in China is not only crucial for the steady growth of the national economy under the "Belt and Road" initiative but also constitutes an essential link in the construction of international logistics corridors. In recent years, given the complexities of the international trade environment, Chinese port enterprises have actively sought transformation and upgrading, relying on technological innovation, improved management efficiency, and the implementation of green and low-carbon development strategies to continuously enhance their service quality and international competitiveness. The

J. Wang · X. Xing (✉)
Anhui University of Finance and Economics, Bengbu City, Anhui Province, China
e-mail: 19956325055@163.com

J. Wang and M. Song (eds.), *The Green Path of Chinese Transportation Enterprises*, Resources, Climate and Sustainable Development,
https://doi.org/10.1007/978-981-95-5289-4_8

core objective of this study is to comprehensively analyze the current development status, challenges, potential opportunities, and future trends of Chinese port enterprises, thereby providing theoretical support and practical strategies to promote their sustainable development.

8.2 Overall Development Situation of Chinese Port Enterprises

8.2.1 Introduction to the Development of Chinese Port Enterprises

8.2.1.1 Industry Overview

A port refers to an important transportation hub located along the coasts of oceans, rivers, lakes, and reservoirs, providing facilities and conditions for ships to enter, leave, and berth. The port industry, as a vital pillar of the national economy, plays an indispensable foundational role in promoting economic development. This industry encompasses not only the construction and maintenance of berths, wharves, warehouses, and yards but also includes a series of services such as cargo transshipment, loading and unloading, storage, and distribution. Ports are core intersections for global cargo transportation and trade, as well as crucial links connecting inland areas and the oceans, bridging domestic and international markets. In recent years, the rapid development of international trade and the recovery of the global economy have significantly elevated the status of Chinese ports within the global port network.

8.2.1.2 Development History

Since the reform and opening-up, China's port industry has achieved remarkable accomplishments through relentless efforts, making significant contributions to the nation and society. With the gradual maturation of global trade and logistics systems and the sustained strong growth of the Chinese economy, China's port industry is expected to maintain a robust development trajectory. The development history can be detailed into the following stages.

After the founding of the People's Republic of China, due to the impacts of historical events such as the War of Resistance against Japan and the Liberation War, most ports in China suffered severe damage. To restore the national economy and promote foreign trade, the state began to promote the recovery and development of the port industry. By the early 1970s, China's port throughput achieved a significant leap from an initial 10 million tons to a historic breakthrough of 100 million tons, marking the preliminary revival of China's port industry.

After the reform and opening-up, China's economy entered a phase of rapid growth, leading to a substantial increase in foreign trade and shipping demand. Simultaneously, the introduction of the "Port Law" and the decentralization of national port management authority brought new development opportunities for China's port industry. Port construction gained rapid momentum, infrastructure was quickly improved, and handling capacity and scale expanded rapidly. By that time, the number of productive berths in major coastal ports had increased to 311, and cargo throughput grew to 200 million tons.

Since entering the 21st century, China's shipping demand has maintained a continuous growth trend. The number of productive berths in coastal ports increased to 5,461, and cargo throughput climbed to a new height of 9.48 billion tons. Based on this, China's port industry has established a leading position in the global port network, becoming one of the largest and busiest port network systems, holding a significant position in global port throughput and container throughput rankings. Ports have also made significant progress in intelligent and green transformation, continuously enhancing their service quality and international competitiveness. In the future, Chinese port enterprises will further promote the industry's transition towards high-quality development, deepening the construction of intelligent and green ports, aiming to comprehensively enhance the comprehensive service efficiency of ports.

8.2.2 Current Development Status of China's Port Industry

8.2.2.1 Continuous Expansion of Port Throughput

Currently, the market size of China's port industry continues to expand. According to the latest data, by the end of 2023, the national port cargo throughput reached 16.973 billion tons, an increase of 8.2% compared to the previous period. Among this, the cargo throughput of inland river ports was 6.139 billion tons, and that of coastal ports was 10.835 billion tons, representing year-on-year growth of 10.5% and 6.9%, respectively (Fig. 8.1).

Relying on the Yangtze River Delta, situated at the confluence of the golden waterway, and adjacent to the main international shipping routes, Shanghai Port is China's largest coastal hub port. Latest data shows that Shanghai Port's container throughput reached 49.158 million TEUs in 2023, ranking first in the world for the 14th consecutive year. In July 2024, Shanghai Port's container throughput is expected to exceed 4.8 million TEUs, setting a new monthly record. In the future, Shanghai Port will further expand its capacity to help open up new space for high-quality development (Fig. 8.2).

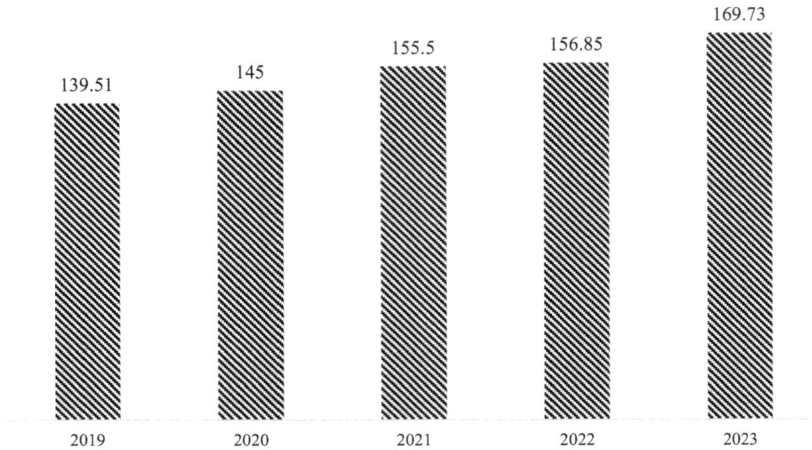

Fig. 8.1 Annual port cargo throughput completed from 2019 to 2023. *Data source* Ministry of Transport

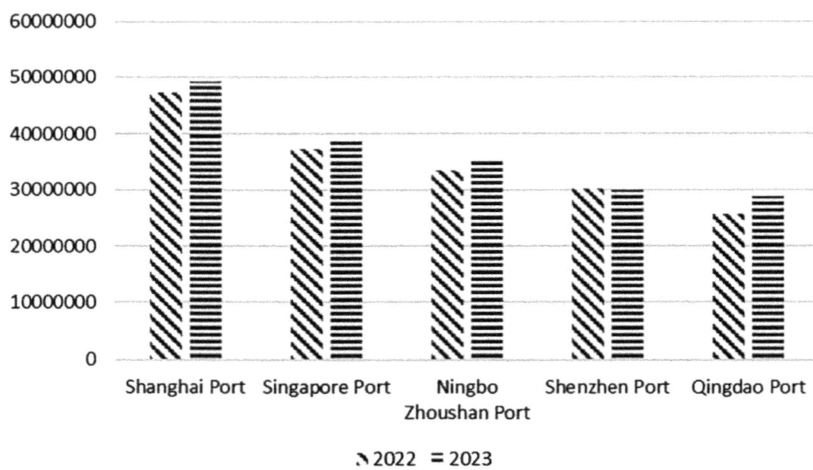

Fig. 8.2 Global top five container throughput in 2022 and 2023. *Data source* Logclub

8.2.2.2 Increasingly Improved Port Infrastructure

Currently, China's port infrastructure is undergoing a crucial phase of comprehensive upgrading and improvement. This significant progress is mainly reflected in the critical transformation and optimization of port facilities, channel conditions, and ship support facilities. By the end of 2023, the total number of operational quay berths in national ports had climbed to 22,023, achieving steady growth compared to the previous year, with the specific addition of 200 new berths. This fully demonstrates

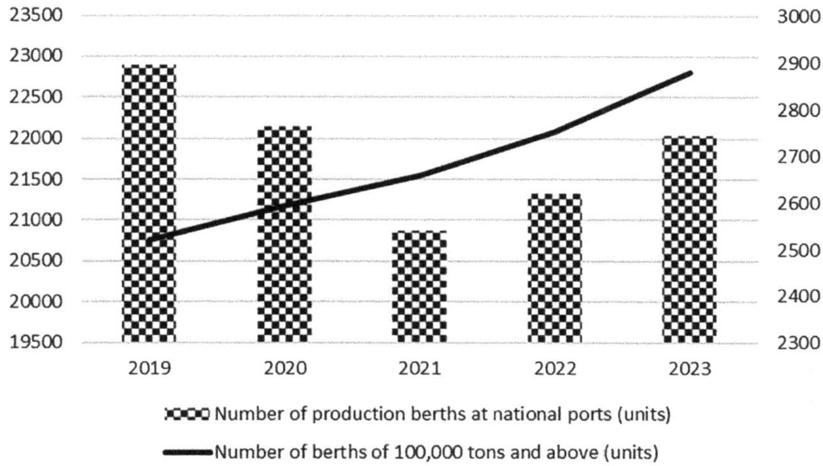

Fig. 8.3 Number of berths in national ports from 2019 to 2023. *Data source* Ministry of Transport

the continuous increase in investment and the efficient progress in port infrastructure construction in China. These newly added and optimized berths not only greatly enhance the cargo handling capacity and operational efficiency of ports but also build a more solid security line for the smooth circulation of international trade. Meanwhile, the number of large berths of 10,000 tons and above also showed a significant growth trend, steadily increasing from 2,520 in 2019 to 2,878 in 2023, a net increase of 358 over four years. This increase directly reflects the continuous expansion of China's foreign trade scale, especially the rapid growth in demand for the transportation of bulk commodities such as iron ore, crude oil, and containers, profoundly reflecting China's strategic deployment in response to changes in the global trade landscape. The increase in large berths allows ports to accommodate more and larger vessels, effectively improving logistics efficiency and providing strong support for Chinese port enterprises in the global market competition (Fig. 8.3).

8.2.2.3 Accelerated Construction of Smart Ports

In recent years, the deep application of cloud computing and artificial intelligence technologies is driving the port sector through a profound transformation towards automation and intelligence. Taking Shenzhen's Mawan Port as an example, it has taken the lead domestically in completing the transition from a traditional terminal to a 5G low-carbon smart terminal. Port equipment, through the 5G RTG (Rubber-Tyred Gantry crane) remote control system, achieves remote operation, significantly improving Mawan Port's operational efficiency while effectively reducing labor costs. On the other hand, Mawan Port has successfully activated the unmanned driving function of intelligent tractors, which can autonomously drive to the yard according to the system-planned route before the automated rail cranes complete container yard operations. Mawan Port uses AI technology to conduct real-time safety monitoring and early warnings for tractor drivers' behavior, further enhancing port operational safety. The achievements in building Mawan Smart Port not only enhance the port's operational efficiency but also set an example for the sustainable development of the port industry, accelerating the process of intelligent transformation of ports in Shenzhen and nationwide. Since Mawan Smart Port began operations, the number of calling routes has continuously increased, and throughput has grown steadily. On the first anniversary of the official operation of Berths 3 and 4 at Mawan Smart Port, the cumulative handling throughput exceeded 1.53 million TEUs, demonstrating strong growth momentum [1].

8.2.3 Analysis of the Current Status of Chinese Port Enterprises

8.2.3.1 Current Development Status of Chinese Port Enterprises

In China's port industry, leading enterprises in terms of competitiveness include SGJ, NBOG, and XMG, primarily engaged in port container and bulk cargo handling, storage, and ship towing services. In the fiscal year 2023, SGJ ranked first in operating revenue, reaching 37.552 billion RMB, a year-on-year increase of 0.7%, followed by NBOG and XMG. Leveraging superior geographical locations, comprehensive collection and distribution networks, and excellent service quality, NBOG and XMG have achieved significant accomplishments in container and bulk cargo transportation, with operating revenues both exceeding 20 billion RMB, showing strong growth momentum and becoming key drivers of the vigorous development of China's port economy. Additionally, the operating revenues of LGGF and TJG also exceeded 10 billion RMB, making important contributions to the prosperity of China's port industry. Looking ahead, with the continuous deepening of global trade, Chinese port enterprises will continue to leverage their advantages, strengthen international cooperation, and promote port logistics towards higher levels and quality, aiming to make more prominent contributions to the prosperity and development of the global economy (Fig. 8.4).

8.2.3.2 Introduction to Key Enterprises

(1) SGJ was established in 2006, and its cargo throughput and container handling capacity rank among the top globally. SGJ mainly comprises two major ports: Shanghai Port and Yangshan Port. The group's core business focuses on four areas: port handling operations, port logistics services, port support services, and port commercial activities. It also emphasizes technological innovation and intelligent

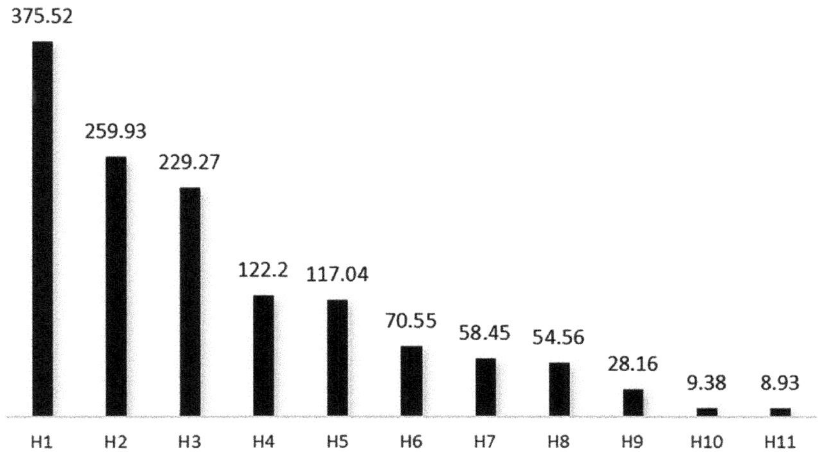

Fig. 8.4 Operating revenue of Chinese port enterprises in 2023. *Data source* Annual reports of respective port enterprises

development, effectively enhancing port operational efficiency through the introduction of advanced intelligent equipment. Compared to other enterprises, SGJ has several major advantages: First, a superior geographical location. As one of the largest comprehensive ports in the Far East, Shanghai Port is located at the confluence of the Yangtze River estuary and China's south coastline of the North Sea, possessing a unique geographical advantage. This location not only makes Shanghai Port a bridge connecting the inland and the sea but also greatly promotes the prosperity and development of domestic and foreign trade. Second, well-developed infrastructure. SGJ possesses advanced port facilities and equipment, including large container terminals, equipped with automated yards capable of handling large-scale, high-density container operations, intelligent dispatching systems, and efficient quay cranes and yard cranes, greatly improving loading and unloading efficiency and operational accuracy to meet the handling needs of different types of cargo. SGJ has also established a comprehensive logistics network including various transportation modes such as highways, railways, and waterways, achieving cargo collection, distribution, and rapid transshipment.

As of now, SGJ owns a total of 121 productive berths, with a throughput capacity of 139.81 million tons. Furthermore, the group possesses high-grade dedicated container terminals that maintain a water depth of 15.5 meters year-round, holding an important position in the global port industry. According to the latest data, the group's operating revenue has been continuously increasing in recent years. In 2023, the operating revenue reached 37.552 billion RMB, net profit attributable to the parent company reached 13.1 billion RMB, container throughput completed 49.158 million TEUs, an increase of 3.9% compared to the previous period, ranking first globally for 12 consecutive years. Port cargo throughput reached 564 million tons, a year-on-year

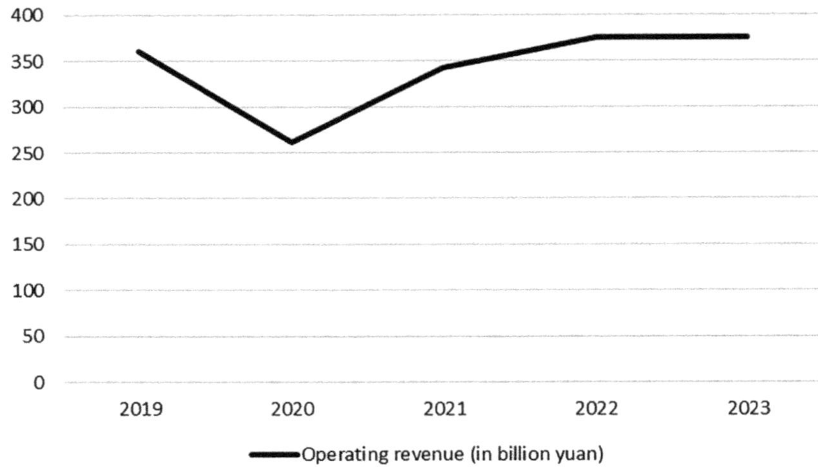

Fig. 8.5 SGJ operating revenue from 2019 to 2023. *Data source* Company annual report

increase of 9.7%. These achievements fully demonstrate the group's strong capabilities and exceptional competitiveness in the global logistics and transportation field, also laying a solid foundation for its future development (Fig. 8.5).

(2) NBOG adopts a comprehensive, efficient, and innovative operational model. By integrating various port resources, it provides all-round services including storage, transportation, distribution, and customs declaration to meet diverse customer needs. The company is also equipped with modern port facilities, including deep-water berths, and advanced equipment, such as efficient handling systems and storage facilities, all ensuring operational efficiency and high-quality customer service, positively contributing to enhancing the overall operational benefits of the port. Additionally, the company is committed to introducing advanced technology and management concepts, continuously optimizing operational processes, and placing high importance on technological innovation and management model innovation, enabling it to maintain an industry-leading position in the fierce market competition. NBOG's operating revenue has maintained a stable growth trend in recent years. In 2023, the company's total revenue reached 25.993 billion RMB, a year-on-year increase of 0.12%, demonstrating strong profitability. Meanwhile, net profit attributable to the parent company also showed a steady upward trend, increasing by 10.45% year-on-year to 4.668 billion RMB. As of the first half of 2024, the company continued its strong development momentum. Its operating revenue rapidly climbed to 14.442 billion RMB, with a year-on-year growth rate as high as 18.11%, far exceeding market expectations. This reflects both the strong recovery of port business and the high efficiency of the company's logistics supply chain operations. Net profit attributable to shareholders of the listed company achieved a year-on-year increase of 1.87%, reaching 2.233 billion RMB, fully demonstrating the group's excellent operational and profitability capabilities.

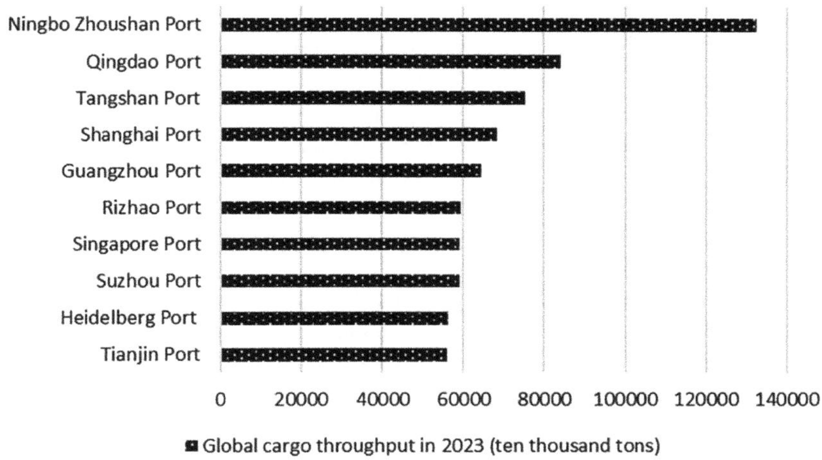

Fig. 8.6 Global top 10 cargo throughput in 2023. *Data source* Logclub

The company primarily covers multiple ports including Ningbo-Zhoushan Port, Wenzhou Port, Jiaxing Port, Taizhou Port, and Yiwu Inland Port. Among them, Ningbo-Zhoushan Port's cargo throughput ranks first in the world. It consists of 19 port areas including Zhenhai, Beilun, and Daxie, currently possessing over 620 productive berths. According to the latest data, by the end of 2023, Ningbo-Zhoushan Port's cargo throughput reached 1.324 billion tons, a year-on-year increase of 4.94%; container throughput reached a new high of 35.301 million TEUs, a year-on-year increase of 5.85%. Consequently, the port has maintained its leading position as the world's number one in cargo throughput for 15 consecutive years and ranked third globally in container throughput for 6 consecutive years (Fig. 8.6).

(3) Since its establishment in 1997, XMG has transformed from an enterprise focused on traditional port handling and logistics support services into a comprehensively developed group, becoming a core driving force for the construction of the Xiamen International Shipping Center and the city's industrial economic platform. The group's business areas are extensive, such as: port operations (covering port freight and passenger services, shipping logistics services, and port-related support), engineering construction sector (involving the development, construction, and professional consulting of port engineering), environmental protection business (handling marine oil pollution, water resource recycling, and environmental monitoring), information technology services (R&D and sales of information products, information consulting, technical services, and development and implementation of information engineering projects), and other diversified businesses (real estate, hotel operations, property management, tourism, and commodity trade). As of now, the group owns 53 productive berths, capable of accommodating 200,000-ton container ships and bulk cargo ships of the same tonnage. In recent years, the group has actively promoted the "Xiamen Port Smart Port Construction Project," relying on

cutting-edge technologies such as cloud computing, the Internet of Things, artificial intelligence, and mobile internet to extend information applications from the port core to the entire logistics service chain, promoting multi-party collaborative operations and significantly enhancing the ability to integrate and optimize the allocation of logistics resources. In 2023, the group's cargo, container, and bulk cargo throughput increased to 158 million tons, 10.523 million TEUs, and 34.6148 million tons, respectively. Financially, according to the group's latest financial report, XMG's financial situation shows a fluctuating trend. In the first quarter of 2024, the company achieved revenue of 14.010 billion RMB, a year-on-year increase of 7.70%; net profit decreased by 3.73% to 155 million RMB. This change reveals that while maintaining revenue growth momentum, the company is facing pressure and challenges in profitability. In the future, the group needs to optimize its business composition and increase the proportion of high-value-added businesses in the overall operations to enhance corporate profitability.

8.3 Development Environment for Chinese Port Enterprises

8.3.1 Internal and External Environment for the Development of Chinese Port Enterprises

8.3.1.1 Policy Environment

In recent years, China has successively issued guiding documents regarding port development. On the one hand, the government has released a series of policy documents to specifically guide the development of port enterprises. The "Outline of the National Comprehensive Three-dimensional Transportation Network Plan" proposes efforts to build a multimodal transport network integrating air, water, ground, and underground transportation; the "Action Plan for New Infrastructure Construction in the Transportation Sector (2021–2025)" also emphasizes strengthening the construction of smart waterways to enhance waterway operation support capacity and comprehensive service capability; the "Opinions on Strengthening the Planning and Construction of Coastal and Inland Waterway Ports and Channels in the New Era" further explicitly proposes strengthening the infrastructure construction of ports and channels, optimizing channel layout, and further enhancing the throughput capacity and service level of ports to meet the growing demand for cargo transportation.

On the other hand, to accelerate the market-oriented reform of the port industry, the government, while granting port operating enterprises more autonomy, has gradually implemented market-adjusted prices for fees including port facility security fees and storage fees, to reduce the logistics costs of port enterprises and stimulate market vitality. Additionally, to alleviate the economic pressure on import-export and enterprises, the government has adopted a series of measures such as simplifying fee items, reducing government-priced projects, and lowering relevant fee standards, promoting further improvement in port enterprise competition and service quality. Through the implementation of the above reform measures, the government creates a more favorable competitive environment for port enterprises, helping them possess stronger market competitiveness and sustainable development capabilities [2].

8.3.1.2 Economic Environment

In recent years, the overall economic situation in China has shown a trend of stable recovery and continuous improvement, injecting strong driving force into the development of the port industry. By the end of 2023, the total cargo throughput of national ports had climbed to 17.0 billion tons, an increase of 8.2% compared to the previous year. This significant growth not only marks the position and influence of Chinese port enterprises in the international maritime landscape but also vigorously promotes the prosperity of domestic commercial activities. Under the backdrop of globalization, the future direction of the global economy remains shrouded in numerous uncertainties. The intensification of international trade protectionism and the escalation of geopolitical conflicts mean that the global port logistics industry is facing unprecedented severe tests. In this specific context, the development path of Chinese port enterprises is fraught with diverse risks and challenges. Therefore, Chinese port enterprises need to maintain high vigilance, closely monitor fluctuations in the macroeconomic situation, while strengthening internal operational management, improving service quality, and optimizing resource allocation, thereby further comprehensively enhancing the enterprises' overall competitiveness and ability to withstand external shocks.

8.3.1.3 Technological Environment

In recent years, technological progress has driven port enterprises towards intelligence and automation. Faced with continuously rising trade demands, numerous port enterprises widely integrate cutting-edge technologies such as the Internet of Things (IoT) and big data to enhance operational efficiency and safety. Firstly, by extensively applying IoT technology, enterprises closely connect facilities, goods, and human resources within the port, constructing a real-time, dynamic information exchange network that ensures accurate data collection and high-speed transmission, enabling seamless integration of various port operation stages. Furthermore, the application of big data technology allows for in-depth analysis of massive data, revealing valuable insights and operational patterns hidden within the data. By analyzing historical data to forecast future cargo flow, facility demand, and potential risk points, it provides forward-looking strategic guidance for port management. Additionally, the integration of artificial intelligence (AI) technology further enhances the intelligence and automation level of ports. Port enterprises widely use advanced facilities such as

Automated Guided Vehicles (AGVs), intelligent robots, and even unmanned port areas, achieving automated operations in container handling and horizontal transportation, significantly improving operational efficiency, reducing human errors, and ensuring the safety and stability of port operations. Taking TJG as an example—as a leading comprehensive major port in the western Bohai Sea region of Northern China and a key window for national foreign trade, TJG actively adopts advanced technologies such as IoT and big data to optimize port operational efficiency and service quality, driving the port's transformation towards intelligence and automation. In recent years, TJG has achieved significant results in smart port construction, with both cargo and container handling volumes showing continuous growth. According to statistics, by the end of 2023, TJG's total cargo handling volume reached 559 million tons, while container handling volume exceeded 22.19 million TEUs. In the future, TJG will continue to strengthen its integrated development strategy with the inland hinterland, further expanding the port scale to enhance the breadth and depth of port services.

8.3.2 Challenges Facing the Development of Chinese Port Enterprises

8.3.2.1 Infrastructure Remains Weak in Some Ports

On the one hand, some port infrastructure built initially did not fully anticipate the development needs of modern container logistics during the planning and construction period, resulting in limited capacity, low operational efficiency, and difficulty in accommodating the requirements of large vessel berthing. Additionally, some port enterprises lack the application of contemporary information technologies such as automated handling systems and intelligent dispatching technology support, further inhibiting the potential for improving port operational efficiency. Meanwhile, the inadequacy of supporting service facilities in ports, such as insufficient storage capacity, aging transportation vehicles, and low efficiency of handling machinery, constitutes an important factor restricting the improvement of port logistics efficiency. These issues affect the smoothness of port logistics operations, thereby increasing the logistics and time costs for port enterprises.

On the other hand, some port enterprises face challenges such as insufficient deep-water channels and large-scale comprehensive port facilities, making it difficult to cope with the continuously growing logistics demand. Although some ports possess deep-water conditions, they still have multiple deficiencies in functional layout, operational procedures, and service support, failing to fully leverage the potential advantages of deep-water ports. TJG lags behind in logistics infrastructure construction, with a shortage of specialized berths and unreasonable functional area planning, leading to congestion in some areas and idle resources in others. This imbalance in resource allocation not only significantly reduces the overall efficiency of port resources but also directly increases the complexity and cost pressure of port operations. Therefore, TJG can enhance the port's operational efficiency and market competitiveness by optimizing the port's logistics infrastructure, increasing the supply of specialized berths, and scientifically and reasonably adjusting the functional area layout [3].

8.3.2.2 Overcapacity and Structural Imbalances Are Becoming Increasingly Prominent

In recent years, with the growth of international trade demand and the promotion of regional economic development, China's port enterprises have accelerated the construction of port infrastructure. However, this rapid expansion process has not kept pace with the actual steps of market demand, leading to the aggravation of overcapacity and structural imbalance problems. Specifically, some ports face situations such as rising berth vacancy rates, declining yard utilization rates, and severe equipment idleness, prompting a significant reduction in the effectiveness of resource use. According to statistics, the average utilization rate of major coastal port areas in China is as low as 60.3%, with this situation being more prominent especially in the Northeast region. Overcapacity not only leads to a massive waste of resources but also significantly increases enterprises' operational costs, including maintenance fees, labor costs, and financial expenditures, posing a major threat to the long-term sustainable development of ports [4]. Simultaneously, the irrationality of the port structure has become another issue that cannot be ignored. Some ports, while pursuing comprehensiveness, neglect clear professional positioning, resulting in weak competitiveness in specific cargo handling and business areas, making it difficult to build unique competitive advantages. This widely distributed and professionally insufficient layout pattern limits the port's ability to deepen services and improve efficiency in specific fields, indirectly affecting the port's service quality and market competitiveness [5].

8.3.2.3 Intensifying Competition Among Port Enterprises

Under the macro trends of a recovering global trade environment and structural adjustments, the competitive landscape among ports is becoming increasingly fierce. On the one hand, competition within regions has intensified sharply, and the rivalry among domestic port terminal enterprises has reached a white-hot level. Port enterprises in close geographical proximity with overlapping service areas are facing direct and severe competitive challenges. These enterprises need to compete for resources, customers, and market share within a limited market space. On the other hand, the continuous expansion of global trade volume is promoting competition among international ports to become increasingly intense. This is not only a contest of throughput but also a comprehensive test of operational efficiency, intelligence level, service quality, and other aspects, setting higher standards for port throughput and operational efficiency. As an indispensable part of global trade, Chinese port enterprises urgently need to strengthen their own capabilities to meet challenges in order to secure a stable position in the world port arena.

8.3.2.4 Environmental Pollution Issues Are Becoming Increasingly Severe, Presenting Greater Challenges to Sustainable Development

With the rise in global environmental awareness and the increased emphasis on sustainable development, governments worldwide are continuously strengthening environmental regulations and standards, imposing stricter requirements on port enterprises regarding emission control and waste management, among other aspects. As a responsible major country, China is also continuously strengthening environmental legal construction and law enforcement effectiveness to ensure that port enterprises comply with environmental standards and sustainable development principles. However, as areas of intensive economic activity, ports inevitably generate pollutants such as noise and wastewater during their operations, adversely affecting the surrounding environment. Noise not only disturbs the daily lives of nearby residents but may also harm their hearing health and psychological state with long-term exposure; exhaust gases contain harmful components such as sulfur dioxide and nitrogen oxides, seriously threatening air quality; the discharge of untreated wastewater pollutes the ocean and disrupts the balance of aquatic ecosystems. Furthermore, resource waste in port operations cannot be ignored, such as inefficient use due to improper shoreline resource planning, and idle facilities and equipment caused by technological iteration or management neglect, all of which increase enterprise operational costs and violate the basic principles of the circular economy and sustainable development. To address these challenges, port enterprises need to take active action, strengthen the construction of environmental protection facilities, promote clean energy, and implement energy-saving and emission-reduction strategies; simultaneously, the government should increase environmental supervision, promote technological innovation and industrial upgrading, and jointly support the sustainable development of the port industry.

8.4 Future Development Trends and Strategic Recommendations

8.4.1 Accelerate Green Port Construction to Promote Sustainable Port Development

Green port construction is a key path to achieving the sustainable development of port enterprises and an inevitable choice in response to global environmental calls. Against the backdrop of continuously rising global environmental awareness, green port construction has leapt from a marginal issue to a core component of port development strategies leading port enterprises towards a more environmentally friendly and sustainable future.

8.4.1.1 Actively Promote the Wide Application of Clean Energy

To significantly reduce reliance on fossil fuels and cut greenhouse gas and air pollutant emissions, port enterprises must actively promote the widespread application of renewable energy sources such as solar and wind power. Taking the Luojing Container Port Area of the Shanghai International Shipping Center as an example, this port area actively introduces cutting-edge equipment such as quay cranes and rail cranes powered purely by electricity, not only significantly reducing noise levels during operation but also greatly reducing the exhaust pollution previously emitted by fuel-powered equipment. Simultaneously, the port area has also implemented a supporting photovoltaic power generation system that effectively converts solar energy into electricity, further promoting the optimization and upgrading of the energy structure and providing reliable, environmentally friendly green energy support for various facilities in the port area.

8.4.1.2 Increase Efforts in Energy Saving and Emission Reduction

To enhance port area efficiency and reduce energy consumption, port enterprises should commit to continuously improving operational processes and equipment layout, proactively adopting various advanced energy-saving technologies and devices such as frequency conversion adjustment technology, as well as potential energy feedback systems. Simultaneously, promote the use of port transportation vehicles powered by clean energy, such as electric forklifts and electric tractors, to further reduce carbon emissions within the port. Implementing these measures can effectively lower the port's operating expenses and has a very positive effect on shaping the port's green image and enhancing its market competitive advantage. Therefore, these strategies are particularly necessary for current efforts towards efficient energy use and environmental protection.

8.4.1.3 Promote the Establishment of Intelligent Pollution Monitoring Systems

Adopting intelligent sensing technology and big data analysis tools enables port enterprises to achieve real-time monitoring, data collection, and in-depth analysis of the surrounding environment. By deploying sensing facilities such as air quality monitoring stations and water quality monitoring devices, enterprises can capture environmental data in real time. Once potential pollution sources or environmental abnormalities are detected, the system quickly triggers warning signals and automatically activates corresponding emergency response procedures. Simultaneously, enterprises can use technologies like big data and cloud computing to comprehensively mine and analyze this real-time data, aiming to plan more scientific and reasonable environmental protection measures for the port [6].

8.4.2 Improve Port Infrastructure, Optimize Port Layout and Collaboration

Improving port infrastructure and optimizing port layout and collaboration are key measures for building a modern and efficiently operating port system. They are of positive significance for enhancing the comprehensive service capability of ports and promoting the long-term development of port enterprises.

8.4.2.1 Increase Yard Storage Capacity

To meet the growing demand for container handling, port enterprises can adopt a series of measures to enhance their storage capacity. Specifically, enterprises can consider expanding yard areas outside the port, optimizing yard structure layout, and upgrading yard facilities. Geographically, enterprises can choose suitable locations around the port to add external yards, thereby expanding the yard area, reducing the waiting time for internal container trucks, and thus enhancing the overall operational efficiency of the port. Additionally, reasonably planning the yard layout and upgrading yard facilities can also help enterprises enhance yard storage capacity. For example, by implementing efficient zone division and container management strategies, introducing advanced handling equipment and automation technology, and constructing modern storage facilities and cargo stacking areas, the utilization efficiency of yard space can be significantly improved, container turnover speed accelerated, and storage potential enhanced.

8.4.2.2 Optimize Port Layout and Functional Positioning

The key to addressing the shortage of deep-water ports and large-scale port facilities lies in implementing scientific planning and rational allocation of port resources to adapt to long-term regional economic development needs. Port enterprises need to conduct in-depth analysis of the regional economic composition, industrial distribution characteristics, current logistics demand, and future development trends, and clearly define the role and position of each port within the regional logistics system. Based on this, implement detailed planning of port shoreline resources to ensure they fully meet modern logistics needs and reserve sufficient development space to cope with future challenges. Furthermore, enterprises should establish a port system with complementary functions, clear responsibilities, and collaborative operation to avoid the drawbacks of duplicate construction, positioning deviations, and resource wastage. Through this optimization strategy, the port's radiating force and driving force on the regional economy can be enhanced, thereby substantively improving the port's operational efficiency and service quality.

8.4.2.3 Strengthen Division of Labor and Collaboration Among Ports

Implementing division of labor and collaboration among ports requires consideration from multiple dimensions. Primarily, it is necessary to clearly define the functional role and development path of each port based on factors such as geographical location, resource conditions, and transportation networks. Large comprehensive ports can focus on high-end service areas such as container transportation and ocean shipping; while small and medium-sized ports can focus on characteristic businesses such as bulk cargo handling and inland waterway shipping. Through scientific division of labor layout, homogenized competition among ports can be effectively avoided to achieve complementary advantages and synergistic progress among ports. Secondly, strengthening information sharing and optimized resource allocation among ports is crucial. By building port cooperation alliances, signing cooperation agreements, and other means, promote information flow and resource integration among ports aimed at improving the utilization efficiency of port resources. Simultaneously, enhance the comprehensive competitiveness of ports by promoting business linkage among ports, such as jointly developing shipping routes and co-building logistics infrastructure.

8.4.3 Extend the Industrial Chain to Achieve Diversified Development

Against the current backdrop of continuously advancing globalization, Chinese port enterprises are in a critical stage of transition from traditional service models to modernized, diversified business models. This transformation is crucial for enhancing the comprehensive competitive strength of port enterprises and promoting their sustainable development.

8.4.3.1 Promote the Extension of Logistics Business

Port enterprises must not only consolidate their foundation in traditional handling and storage services but also actively expand deeply into the upstream and downstream of the logistics industry chain, creating a comprehensive logistics service system. This system not only encompasses the flexible combination of sea, land, and multimodal transport, ensuring seamless connection of goods from origin to destination, providing efficient and safe transportation solutions, but also focuses on optimizing warehouse management processes, utilizing cutting-edge warehousing technology and advanced information systems to achieve precise inventory control, rapid response, and intelligent management. Port enterprises can also launch customized distribution services based on specific customer needs, ensuring the safe and timely arrival of goods, designing personalized transportation solutions including special areas like cold chain transport, and adding value-added services such as packaging, sorting, and labeling to enhance the market competitiveness of port enterprises.

8.4.3.2 Achieve Diversified Service Expansion

To enhance corporate competitiveness, port enterprises need to go beyond the single scope of logistics services and actively expand into high-end service areas covering shipping finance, insurance, etc., achieving service diversification and integration, thereby bringing more significant convenience and security to port enterprises in terms of capital operation and risk management. At the level of shipping finance, port enterprises should cooperate with financial institutions such as banks and insurance companies to provide solutions for financing, credit guarantees, insurance services, etc., for enterprises, Accelerating the flow of funds in enterprises aims to effectively manage risks. In the field of information services, create a cutting-edge information technology platform integrating functions such as port operations, logistics tracking, and market trend analysis, providing customers with instant data resources to assist them in making wise decisions. Port enterprises can also venture into the operational management of transportation vehicles, covering leasing services, maintenance, scheduling, etc., for ships, containers, and logistics vehicles, thereby establishing a complete network of logistics equipment to further optimize the overall operational efficiency of the enterprise.

8.4.4 Promote Industrial Integration and Innovative Development

Currently, port enterprises are actively embracing the trend of industrial integration and innovative development, driving ports and surrounding areas towards high-quality development with a more open and forward-looking vision. Through industrial integration and innovation strategies, Chinese port enterprises strengthen their core competitive advantages. While providing strong momentum for the transformation, upgrading, and high-quality development of the regional economy, they also contribute valuable wisdom and strength to the global port industry.

8.4.4.1 Actively Promote Linkage with Port-Proximate Industries

Port enterprises are well aware that a single industry can no longer adapt to the needs of the diversified modern economic system. Chinese port enterprises urgently need to strengthen synergistic development with related industries such as port-proximate industries and commerce, attract port-proximate industries like high-end manufacturing and modern services, and rationally layout port facilities. Simultaneously, efforts should be made to implement targeted development strategies for port-proximate industries, aiming to create industrial clusters with complete chains and tight upstream-downstream connections, thereby promoting harmonious coexistence of multiple industries. Port enterprises also need to deepen integration with urban development, promote seamless connection of port economic zones to achieve efficient resource sharing and complementary advantages. In terms of trade, port enterprises, leveraging their logistics hub status, can build a connecting platform between international trade and the domestic market, thereby accelerating commodity circulation and trade activities; in terms of tourism, they can rely on the port's unique natural scenery and cultural heritage to develop tourism projects with port characteristics, thereby enhancing the city's brand image and popularity [7].

8.4.4.2 Promote Technological Innovation and Application

Technology is the primary productive force. While introducing top-tier technology from domestic and international sources, Chinese port enterprises must focus on their own innovation and absorption, increase investment in technological innovation, and thereby promote the continuous upgrading and optimization of port equipment and technology. From the perspective of port equipment, to improve the level of automated operations and work efficiency, enterprises should adopt the most advanced automated terminal systems, unmanned container trucks, intelligent cranes, and other cutting-edge equipment; In terms of technology application, it is necessary to fully utilize information technologies such as big data and the Internet of Things to achieve precise scheduling, smart supervision, and remote control. Additionally, port enterprises must translate the software achievements of technological innovation into practical applications, by establishing smart port and green port demonstration projects, leading the entire industry towards a higher level of development [8].

8.5 Conclusion

Chinese port enterprises play a core role in the international port sector. As important bridges connecting domestic and international markets, they exert a significant influence on national economic growth. As key facilities for cargo handling and container transshipment, port enterprises are drivers promoting regional economic integration and active participants on the global supply chain competition stage. Facing a complex and ever-changing domestic and international environment, Chinese port enterprises urgently need to demonstrate forward-looking strategic vision and agile adaptation strategies. The primary task is to actively overcome internal challenges such as intensified competition, overcapacity, and structural imbalances, promote industrial innovation through technological reform, advance the automation and intelligence of port operations, enhance resource utilization efficiency, and reduce operating costs.Furthermore, Chinese port enterprises should precisely seize external opportunities such as global economic recovery, trade growth, and the rise of emerging markets, strengthen synergy with global ports, shipping companies, and related upper and lower logistics chains, and work together to establish an efficient, safe, and environmentally friendly global logistics system.

References

1. Li Ding. Development Outlook of China's Port Industry [J]. Port Technology, 2017, (01): 1–3.
2. Li Lan, Baoxin Qin, Enchang Deng, et al. Practice and Exploration of Comprehensive Planning in the Management of Port Enterprises [J]. Energy Technology, 2024, 22(01): 18–22.
3. Yihong Chen. Problems and countermeasures in the management of fixed assets of port enterprises [J]. Caixun, 2024, (12): 114–116.
4. Ni Zhang. Current Status and Transformation Path of Financial Digitalization in Port Enterprises [J]. Taxation, 2024, 18(18): 61–63.
5. Xuanrui Yu. Financial Management Problems and Countermeasures in the "Belt and Road" Cooperation of Port Enterprises [J]. Land Bridge Vision, 2024, (01): 99–101.
6. Feiyun Wu. Analysis of production and intelligent safety management strategies of port enterprises [J]. Water Safety, 2024, (12): 4–6.
7. Jingjing Gu. Suzhou innovates international logistics supply chain model [J]. China Ports, 2023, (05): 52–55.
8. Minghui Qu. Exploration on the construction of core competitiveness of port enterprises based on the value chain [J]. Shandong State-owned Assets, 2024, (Z1): 130–131.

Chapter 9
Competitiveness and Challenges of Chinese Airport Enterprise in the Global Market

Jianlin Wang and Mengjie Zhu

Abstract Airport enterprises serve as pivotal transport hubs connecting the global economy and fostering local development, occupying a strategically significant position. This paper delves into the current state of China's airport enterprises, covering their historical evolution, global and domestic development trajectories, and contributions to the global economy. It further examines management and operational approaches within airport enterprises, addressing organisational structures, safety oversight, flight operations, customer service management, and financial controls. Through case studies of SHOGU and SPUGJ, the profitability mechanisms of airport enterprises are examined, followed by a discussion on future development trends and policy recommendations. The paper proposes that intelligent advancement, personalised services, and cross-industry collaboration and integration represent primary future development trends for airport enterprises. Policy recommendations to foster their growth include refining regulatory frameworks, enhancing policy support, optimising airport planning and layout, and boosting the competitiveness of airport enterprises.

9.1 Analysis of the Current Situation of Airport Enterprises in China

9.1.1 The Historical Transformation of Airport Enterprises

The development history of airport enterprises can be traced back to the early 20th century, primarily divided into the following stages (Fig. 9.1).

J. Wang · M. Zhu (✉)
Anhui University of Finance and Economics, Bengbu City, Anhui Province, China
e-mail: zikw2859091185@163.com

J. Wang and M. Song (eds.), *The Green Path of Chinese Transportation Enterprises*,
Resources, Climate and Sustainable Development,
https://doi.org/10.1007/978-981-95-5289-4_9

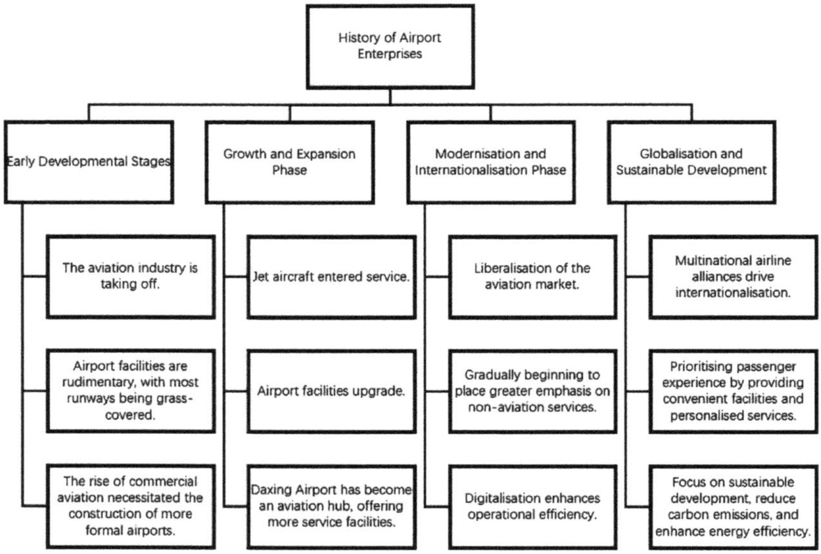

Fig. 9.1 Historical transformation of airport enterprises

9.1.2 Current Situation of Airport Enterprises

9.1.2.1 Current State of Global Airport Enterprise Development

Airport enterprises refer to companies that manage and operate airports, with their operations encompassing aviation support services and a broad range of commercial activities. The global development landscape of airport enterprises exhibits trends towards diversification and rapid expansion. As aviation serves as a vital bridge connecting the world, its significance continues to grow. With the accelerated advancement of global economic integration, the aviation industry maintains steady growth momentum. International routes are continually expanding, domestic route networks are becoming increasingly comprehensive, and the coverage and service quality of air transport are steadily improving.

The development characteristics of global airport enterprises are primarily concentrated in Europe, Asia-Pacific, and North America. Notably, regions like Europe and America, which implemented open skies and aviation liberalisation policies earlier, have fostered a number of internationally competitive airport groups. FLKJC stands as one of the 'global-level players' in the aerotropolis industry; its Frankfurt airport handles over 140,000 passengers and more than 80 aircraft movements daily [1]. Over 50 airport groups now manage five or more airports globally, collectively accounting for 51% of worldwide passenger throughput. Many airports have further catalysed regional economic growth by developing airport economic zones. HSHJI and HAREC, for instance, have attracted numerous enterprises and logistics centres through such zones, significantly boosting local economies.

Certain airports have also established themselves as major global air cargo hubs through the construction of freight terminals. Memphis International Airport, for instance, has become the world's largest cargo airport thanks to its FedEx Superhub. Its airport economic zone has generated 166,000 jobs and $20.7 billion in direct economic output for Memphis, cementing its status as a pivotal global aviation logistics centre. The success of MEFEG lies in its concentric development model. It has evolved beyond a mere logistics hub into a more comprehensive airport city, spearheading the formation of an airport metropolitan area.

9.1.2.2 Current Development of Airport Enterprises in China

Since the founding of the People's Republic of China, the nation's civil aviation sector has achieved remarkable progress, particularly during the past three decades of reform and opening-up, which saw both civil aviation and airport construction enter a new phase of rapid development. Firstly, China's airport industry has maintained steady growth in recent years, with particularly outstanding performance in passenger and cargo throughput alongside operating revenue. According to 2023 data, Chinese airports generated operating revenue of RMB 101.98 billion, representing a 71.1% increase over the previous year. Airport profitability continues to improve, with losses narrowing despite persistent deficits. Currently, the density of China's airports is gradually increasing, with service capacity steadily improving and modernisation levels continuously advancing. A fundamental framework has been progressively established, centred on major hub airports such as Beijing, Shanghai, and Guangzhou, supported by provincial capital or key city airports including Chengdu, Kunming, Chongqing, Xi'an, Ürümqi, Shenzhen, Hangzhou, Wuhan, Shenyang, and Dalian, and complemented by feeder airports in other cities. This signifies the preliminary establishment of China's civil transport airport system.

By 2022, mainland China operated 166 civil aviation airports (excluding Hong Kong, Macao, and Taiwan), comprising 165 airports with scheduled flights and 45 military-civilian dual-use airports. Of these, 165 airports maintained scheduled flight operations. The average airport density nationwide stands at 1.53 airports per 100,000 square kilometres. This comprises 25 Category 4E airports, 35 Category 4D airports, 58 Category 4C airports, and 29 Category 3C airports. By the end of 2023, mainland China (excluding Hong Kong, Macao and Taiwan) operated 259 transport airports, handling 16.8331 million tonnes of cargo and mail and recording 11.7082 million aircraft movements—representing year-on-year increases of 142.2%, 15.8% and 63.7% respectively. There were 38 transport airports handling over 10 million passengers annually and 63 transport airports handling over 10,000 tonnes of cargo and mail annually [2] (Table 9.1).

Civil aviation transport airports nationwide handled 1.26 billion passenger movements, representing a year-on-year increase of 142.2% (Fig. 9.2).

Table 9.1 Number of licensed transport airports by region in 2023

Region	Number of certified transport airports	Percentage of nationwide total (%)
Nationwide	259	100.0
Eastern region	56	21.6
Central region	44	17.0
Western region	132	51.0
Northeastern region	27	10.4

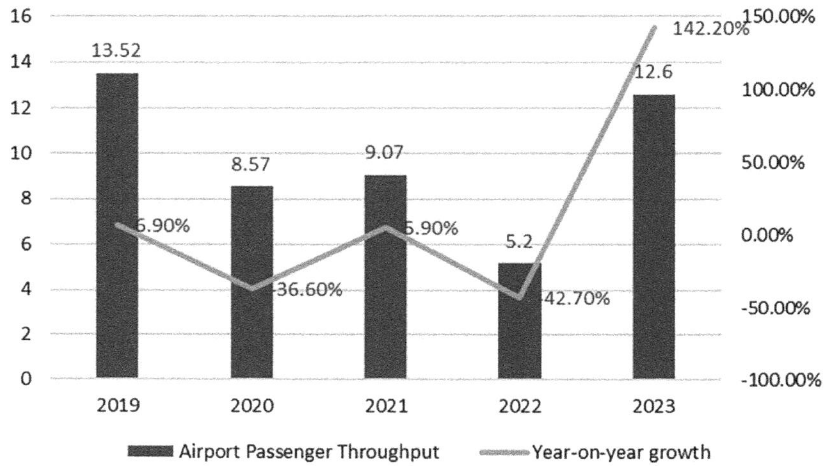

Fig. 9.2 Passenger throughput at civil aviation airports, 2019–2023. Data Source: 2023 statistical bulletin on the development of the civil aviation industry

In 2023, passenger throughput in China's eastern regions achieved a year-on-year increase of 156.8%, reaching 638 million passenger journeys; central regions recorded a year-on-year growth of 128.8% to 143 million passenger journeys; western regions saw a year-on-year rise of 129.6% to 399 million passenger journeys; while northeastern regions registered a year-on-year increase of 125.6% to 81 million passenger journeys (Fig. 9.3).

By 2023, China's civil aviation airports handled 16.8331 million tonnes of cargo and mail throughput, representing a year-on-year increase of 15.8%. Within this total, eastern regions accounted for 12.0679 million tonnes, up 12.8% year-on-year; central regions handled 1.5154 million tonnes, rising 20.3% year-on-year; the western region handled 2.6675 million tonnes, a 24.0% increase; and the northeastern region managed 582,200 tonnes, a 37.8% rise.

GBAYU, established in 2000 and listed on the Shanghai Stock Exchange in 2003, stands as one of China's three major gateway composite hub airports, holding significant strategic importance. Its operating entity has witnessed two decades of growth

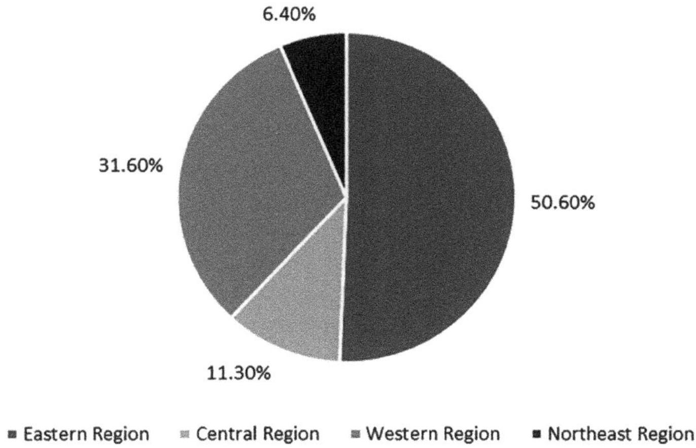

Fig. 9.3 Passenger throughput at civil aviation airports by region in 2023

and leads the industry in smart airport development; SHGJJ Co., Ltd. has similarly excelled, ranking first in passenger and cargo throughput during the first half of 2024; XJCCC Group, as the nation's second-largest inter-provincial airport operator, oversees the construction and management of 23 member airports across Northwest China, including the trunk airport XXXXJ. The group employs big data analytics to optimise operational efficiency, exemplifying smart transformation. This underscores the vital role Chinese airport enterprises play in regional economic development and intelligent transformation.

Beyond operational growth, Chinese airport enterprises prioritise non-aviation business development, such as commercial retail, advertising, and VIP services, which constitute vital revenue streams. SHOGU enhanced advertising quality and returns by partnering with internationally renowned advertising operators. SHOGU Co., Ltd. showcased exemplary cases of 'one-stop' services and aviation express products during the group's 2023 annual case review. Meanwhile, SJJJJ, a major aviation hub in East China with two principal airports (SPUGJ and HQGJJ), derives approximately 26% of its revenue from aviation-related sources. In 2023, the company's aviation business revenue rebounded rapidly, surpassing 2019 levels. The advancement of China's airport enterprises extends beyond enhanced revenue and profitability to encompass contributions to smart airport development and regional economic growth. Looking ahead, with the nation's emphasis on building a transportation powerhouse and advancing smart transportation, China's airport sector is poised to sustain its robust development momentum.

9.1.3 The Contribution of Airport Enterprises to the Global Economy

Airports, serving as rapid transit hubs for high-value logistics and passenger flows worldwide, have significantly propelled the development of international trade. As enterprises increasingly globalise, the demand for long-distance, high-frequency, high-value, and efficient logistics continues to grow, with airports playing an indispensable role in this process. Firstly, as the fundamental infrastructure of air transport, airports act as vital links connecting regions across the globe, exerting a profound influence on the global economy and cultural exchange. Against the backdrop of economic globalisation and integration, airports not only facilitate the rapid movement of high-value goods and passengers but also provide essential services and support for airlines, travellers, and cargo. Furthermore, airport development has spurred the emergence of aerotropolis zones, becoming a vital engine for regional economic growth.

Taking Chinese airports as an example, Guangzhou Baiyun International Airport (GBAYU) has achieved remarkable progress in transport operations and hub development, handling 26.11 million passengers in 2022—ranking first nationally [3]; Passenger throughput reached 63.17 million in 2023, representing a year-on-year increase of 142%; Passenger throughput for the first half of 2024 has already reached 36.6528 million, representing a year-on-year increase of 26.67%. This not only demonstrates GBAYU's significant progress in transport operations and hub development, but also reflects the pivotal role of Chinese airports within the global aviation industry. GBAYU's success can be attributed to its efficient operations and management, alongside its contribution to global aviation logistics and passenger flows (Fig. 9.4).

Moreover, as the world's fourth and Asia's first cargo-focused new airport, EZHHH is poised to become a new engine for national and regional development. The construction and operation of EZHHH will positively impact the growth of the airport economic zone, driving rapid regional economic expansion. The airport will further stimulate economic growth by promoting tourism and cargo transportation. Through its expansion project, ZGXXX is projected to contribute HK$88.9 billion annually to Hong Kong's economy (representing 5.5% of GDP), with direct contributions accounting for 57% and indirect contributions for 26%. Furthermore, DDIII contributes as much as 28% to local GDP, comprising direct contributions of US$11.7 billion and indirect contributions of US$10.4 billion. The success stories of Chinese airport enterprises such as GBAYU, EZHHH, and ZGXXX further demonstrate the vital role airports play in driving both global and regional economic development.

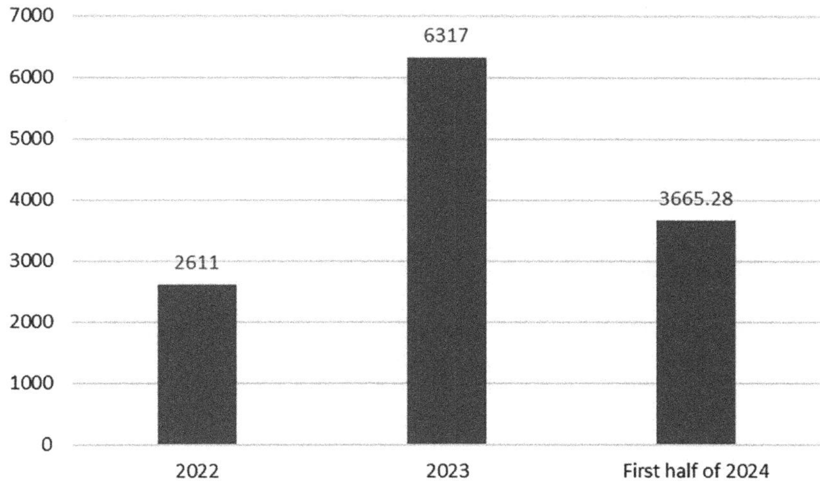

Fig. 9.4 Passenger throughput at GBAYU

9.2 Operational Management Models for Airport Enterprises

9.2.1 Organisational Structure of Airport Enterprises

9.2.1.1 Shareholding Structure

The equity structure of airport enterprises typically comprises multiple shareholders, which may include government agencies, state-owned enterprises, private enterprises, or foreign-funded enterprises. While the equity structure and actual controllers may vary across different airport enterprises, there is usually a primary controlling shareholder or actual controller. The controlling shareholder of SHOGU Co., Ltd. is the Civil Aviation Administration of China, holding 100% of the shares. Its principal secondary subsidiaries include SHOGU Co., Ltd. (56.61% stake), JIAJC Group Corporation, BBIIN, and others. The controlling shareholder of SJJJJ Co., Ltd. is SJJJJ (Group) Co., Ltd., with the Shanghai State-owned Assets Supervision and Administration Commission as the actual controller holding a 58.38% stake [4]. This control is exercised through the holding of 100% equity in SJJJJ Group. GBAYU's controlling shareholder is Guangdong Airport Management Group, holding 61.62% of shares, with the Guangdong Provincial People's Government as the ultimate controlling shareholder. XIAAO Co., Ltd.'s controlling shareholder is Zhejiang Airport Group Co., Ltd., holding 44.526% of shares [5]. JIGJJ Co., Ltd. has SHANN Group Co., Ltd. as its controlling shareholder with a 38.95% stake. SHANN Group's shareholder roster includes Shandong Provincial Financial Investment Group Co.,

Ltd. and Shandong LUXIN Investment Holding Group Co., Ltd., while its controlling shareholder is the Shenzhen Municipal State-owned Assets Supervision and Administration Commission holding 56.97% of shares. These equity structures reflect the multi-tiered holding system established within China's airport sector under the leadership of local governments and state-owned enterprises, facilitating the concentration of resources and management.

9.2.1.2 Management Levels

The management hierarchy of Chinese airport enterprises typically features a relatively standardised organisational structure (Table 9.2).

Table 9.2 Hierarchy of management levels in Chinese airport enterprises

Hierarchy of management levels in Chinese airport enterprises	Board of Directors	Chairman	Preside over the board of directors and oversee the company's strategic planning and major decision-making
		Vice Chairman	Assist the Chairman in his duties and act as his proxy in his absence
		Director	Participate in board decision-making, which may include independent directors, executive directors, and non-executive directors
	Senior management	Chief Executive Officer (CEO)	Responsible for the company's daily operations, implementing board decisions, and reporting to the board of directors
		Deputy General Manager	Assist the General Manager in their duties and oversee different business areas
		Chief Financial Officer (CFO)	Responsible for the company's financial management and capital operations
	Middle management	Department Manager	Responsible for the management of their respective departments
		Branch Manager	Where an airport enterprise has no branch offices, the branch manager shall be responsible for the day-to-day operations and management of the branch
	Front-line management	Section Chief/ Manager	Responsible for the day-to-day management and guidance of specific departments or teams
		Team Leader	On the I line, responsible for the allocation and execution of tasks within the team

The management hierarchy of China's airport enterprises has undergone multiple reforms, progressively streamlining from a complex multi-tiered structure to a more streamlined organisation. Particularly following accession to the WTO, the Civil Aviation Administration of China actively promoted sectoral institutional reforms, reducing the airport management system from three tiers to two: the Civil Aviation Administration and regional bureaus. Concurrently, the Civil Aviation Safety Supervision and Management Office was established to oversee safety supervision and market regulation. Local governments also formed management companies, introducing corporate management and operational models.

At the airport enterprise level, the SHOGU Group oversees the management and operation of 13 airports, including MEIII, FENGG, and YIICA. Concurrently, SHOGU Group assumes a trustee management role, operating airports under the provincial airport groups of Inner Mongolia and Heilongjiang. Xiamen Airport International Aviation Port Co., Ltd. has demonstrated a clear hierarchical structure from top management to operational staff. SSZZZ's recent development is equally noteworthy, having elevated its commercial operations through enhanced business planning and the introduction of direct-operated brands, earning recognition as a national 'Model Case of Integrity-Driven Business Development'. GBAYU, meanwhile, has improved service quality through its Spring Breeze service brand initiative, being selected as a national model case for state-owned enterprise brand development. Building upon streamlined management hierarchies and diversified operational models, China's airport enterprises have achieved high-quality development through corporate management practices, service brand optimisation, and enhanced commercial operations.

9.2.2 *Operational Management of Airport Enterprises*

9.2.2.1 Safety Management

The safety management of China's airport enterprises is primarily achieved through the establishment and implementation of Safety Management Systems (SMS), aimed at enhancing airport safety management standards and ensuring the security of aviation operations. In accordance with Civil Aviation Administration requirements, all transport airport (group) companies must standardise and guide the development of transport airport Safety Management Systems (SMS), driving continuous improvements in operational safety. SHOGU Corporation has undertaken a comprehensive optimisation and upgrade of its '1-2-1' intelligent safety system, introducing 23 new functionalities including emergency training, safety conduct protocols, and stakeholder safety management. This includes refining the safety accountability framework, such as organising anti-fraud initiatives at DDXXX to heighten safety awareness among all personnel. Since 2005, the XIIIJ Group has implemented a risk-based airport safety management system. By collating International Civil Aviation Organisation materials on SMS development and organising research into its implementation, they have progressively transitioned from semi-empirical, semi-systematic management towards a fully structured approach. The successes of these airport groups exemplify the efforts and achievements of China's airport enterprises in safety management. Through establishing and implementing Safety Management Systems (SMS), integrating optimised smart safety systems, and conducting specialised safety prevention initiatives, they continuously elevate safety management standards to ensure aviation operational safety.

9.2.2.2 Flight Management

Flight management at China's airport enterprises encompasses multiple facets, including flight scheduling, flight information management, flight monitoring, and flight optimisation. Firstly, the flight scheduling system constitutes a vital component of airport operational systems, responsible for arranging flight arrival and departure times alongside flight routes. Captain Sun Qixin independently developed a flight scheduling optimisation platform based on multi-objective optimisation. This system aims to enhance the refined management of flight route optimisation projects while responding to the requirements of smart civil aviation development. The flight information management system assists airlines or airport authorities in performing operations such as inputting, modifying, querying, and deleting flight data. Regarding flight operations, airlines utilise flight operations management systems to track departure/arrival times, flight paths, crew, and passenger information, aiming to enhance on-time performance. Concurrently, the smart airport solution integrates emerging technologies such as AI, video cloud, and big data to achieve automated data collection at flight support nodes and intelligent scheduling, thereby boosting operational efficiency. SPUGJ leveraged digital twin technology and low-code development to establish China's largest video analytics project for aircraft turnaround support nodes. Covering 19 support nodes across 290 aircraft parking bays, this initiative enhances ground safety through video analysis, enabling efficient integrated operations within the smart airport ecosystem. Furthermore, to address unforeseen circumstances, the concept of emergency dispatch incorporates an emergency scheduling model to safeguard civil aviation operations. This model comprehensively evaluates capacity, timeliness, and decision-making quality. The system encompasses multiple stages, including incident monitoring and intelligent dispatch, enabling rapid responses to emergencies and minimising losses. Through diverse systems and methodologies, China's airport enterprises ensure efficient and secure flight operations while continuously leveraging new technologies to enhance management efficiency and passenger experience.

9.2.2.3 Customer Service Management

In the realm of customer service management, numerous airports across the nation are continuously enhancing passenger experiences through digitalisation. GGDDD Group has achieved secure 'airport-in-the-cloud' operations by migrating vast volumes of business to the cloud, thereby elevating customer service standards. By establishing a financial digital intelligence platform, SSJJC supports smart airport development while operating a customer service centre offering round-the-clock support to handle feedback and complaints, alongside implementing corresponding management systems and operational workflows. GBAYU's 'Spring Breeze Service' brand initiative has garnered multiple accolades, reflecting its excellence in customer service delivery. Airport enterprises also employ CRM systems for refined customer relationship management, elevating service quality. Numerous operators have established customer care lounges and implemented contactless service measures. WoFeng Technology's intelligent large-scale call centre solution for HENNN has optimised the smart airport service ecosystem. Consequently, China's airport enterprises have made significant strides in customer service management. Through digital transformation, customer service centre development, innovative service models, customer relationship management, and customer experience optimisation, they continuously enhance service quality to meet customer demands.

9.2.2.4 Financial Management

Airport enterprises enhance the achievement of internal management objectives through meticulous financial management. SJJJJ improves financial analysis efficiency and strengthens operational risk assessment and daily financial management capabilities via its financial intelligence cockpit, early warning indicator analysis, and monthly intelligent reporting. NEMIN Group has elevated the modern management standards and technological innovation capacity of airport enterprises by establishing a financial data middle platform system, providing more reliable and precise data support for corporate decision-making. To counter financial risks such as market volatility, exchange rate fluctuations, and liquidity pressures, certain airport enterprises have effectively mitigated and controlled such risks through strengthened internal control mechanisms and refined risk management frameworks.

Despite these financial challenges, China's airport enterprises continue their upward trajectory. In the first half of 2024, SJJJJ achieved operating revenue of RMB 6.064 billion, representing a year-on-year increase of 24.55%. This growth primarily stemmed from increased aviation business volumes at SPUGJ and HQGJJ [6]. By the end of Q1 2024, SJJJJ's total assets stood at RMB 70.611 billion with a debt ratio of 39.83%, indicating the company maintains a relatively stable financial structure while expanding operations [7]. In 2023, SJJJJ achieved total operating revenue of RMB 11.047 billion, a year-on-year increase of 101.57%; net profit attributable to shareholders reached RMB 934 million, rising 131.19% year-on-year [8]. During the same period, SHOGU's core business revenue reached RMB 4.559 billion, growing approximately 104.3% year-on-year. Aviation-related revenue amounted to RMB 2.097 billion, representing a year-on-year increase of approximately 195.0%. Although the net loss after tax stood at RMB 1.697 billion, this represented a reduction in losses of approximately 51.9% compared to 2022, indicating that the core business continues to grow rapidly [9].

9.2.3 *Airport Enterprises' Profit Models*

9.2.3.1 Aviation revenue

The aviation-related revenue of China's airport enterprises primarily comprises fees directly associated with flight operations, including aircraft landing and take-off charges, parking fees, passenger service charges, and security inspection fees, alongside revenue from passenger and cargo services. Airport enterprises generate revenue by providing services such as aircraft take-off and landing, parking, and passenger arrival and departure. SJJJ's aviation-related income encompasses various flight-related fee revenues and passenger and cargo service revenues, accounting for approximately 26% of its total aviation business income. This segment experienced a rapid rebound in 2023, surpassing 2019 levels. SSZZZ's principal business revenue is entirely derived from aviation and related services. In the first half of 2024, it achieved operating revenue of RMB 2.251 billion, representing a year-on-year increase of 15.60%, with net profit attributable to shareholders of RMB 174 million [10]; GBAYU drove business growth through airport expansion and duty-free tenant recruitment. Its aviation service revenue dominated at RMB 4.80 billion, accounting for 77.8% of total revenue and contributing RMB 1.26 billion in net profit, representing 90.2% of total net profit [11].

9.2.3.2 Non-aviation revenue

Within the profit models of China's airport enterprises, non-aviation revenue constitutes a significant component, primarily encompassing commercial retail, advertising, catering, VIP services, leasing, and parking operations [12]. These businesses leverage airport passenger traffic and resources, generating stable income streams through market-oriented operations. Although the proportion of non-aviation revenue varies across different airports, it generally accounts for a substantial share of total income. SJJJ's non-aviation revenue constitutes nearly 60% of its total income, surpassing aviation revenue. In 2018, its non-aviation revenue reached RMB 5.34 billion, representing a 23.2% year-on-year increase. Duty-free operations contributed the majority of profits, accounting for 65% of non-aviation revenue. SJJJ further boosted duty-free sales by re-tendering contracts and increasing commission rates; Xiamen Airport's non-aviation revenue share rose to 39.3% [13]; for SHOGU, advertising, retail and catering operations accounted for 76.31% and 52.97% of that year's

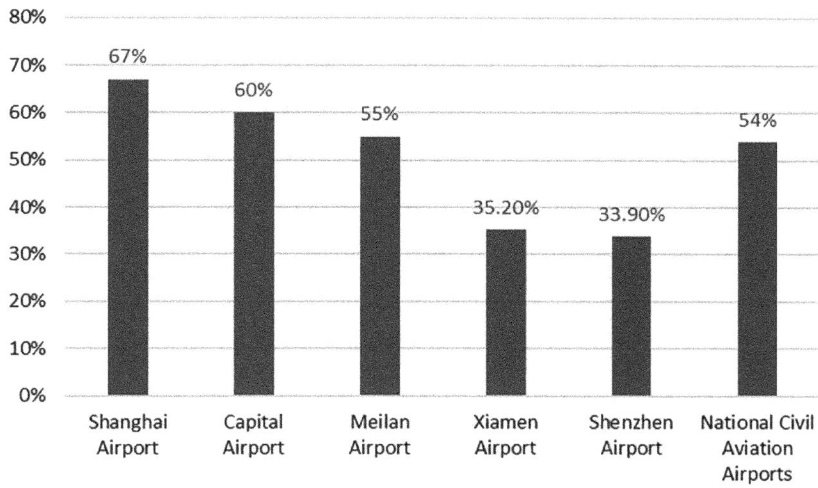

Fig. 9.5 Non-aviation revenue share

non-aviation income respectively. These operations not only provided airports with stable revenue streams but also enhanced operational efficiency through market-oriented management. Consequently, non-aviation businesses have become a vital profit source for China's airport enterprises, offering robust support for sustainable development through diversified operations and market-driven approaches (Fig. 9.5).

9.2.3.3 Government subsidies

Airports, as vital infrastructure, often require substantial capital investment for their construction and operation. Government subsidies provide essential financial backing, enabling airports to sustain operations and stimulate regional economic development. Specifically, as public facilities, airports' construction and operation not only impact civil aviation efficiency but also significantly drive local economic growth. Consequently, governments typically support airports through various means, including fiscal subsidies, route subsidies, and construction grants. These subsidies cover not only daily operational costs but also expenses related to route development and infrastructure construction. The domestic passenger flight subsidy policy jointly issued by the Civil Aviation Administration and the Ministry of Finance provides subsidies of up to ¥24,000 per hour. Additionally, local governments offer route subsidies tailored to specific circumstances to expand route networks and stimulate aviation market growth. SJJJ has leveraged government subsidy schemes, particularly those for international routes, to attract substantial international flight operations and passenger traffic. These subsidy policies have not only enhanced SJJJ's international competitiveness but also incentivised base carriers to establish intercontinental routes. This has secured growth in SJJJJ's international passenger volumes and consistently delivered high-quality, highly profitable passenger traffic.

Through securing regulatory backing, QIJII achieved a substantial reduction in subsidies for newly launched routes. Leveraging these subsidies, the airline proactively developed tourist routes, attracting more potential visitors and contributing to the growth of the local tourism industry while generating increased fiscal revenue. Concurrently, the government actively supported QIJII's development. Despite the significant reduction in subsidies for new routes, the airline increased its revenue and actively expanded its route network. XINMM's rapid expansion similarly relies on governmental backing and prioritisation. Municipal leadership annually allocates ¥120 million in route subsidies while actively securing premium connecting routes. The government further invested over ¥20 million to supplement specialised vehicle equipment, coordinate city bus costs, and advance supporting infrastructure alongside Phase II expansion works, thereby providing robust operational guarantees for the airport. To foster route development, MMMYYY has adopted flexible partnership models and tailored subsidies to market conditions. Annual subsidies have risen from RMB 9 million in 2007 to RMB 40 million presently. Specific measures include guaranteeing minimum funding for route development (minimum guarantee), providing fixed subsidies for qualifying routes (fixed subsidy), and implementing reductions or exemptions for non-compliant or underperforming routes (reductions).

9.3 Development of Typical Airport Enterprises in China

9.3.1 Typical Airport Enterprises in China 1

9.3.1.1 Development History

See Fig. 9.6.

9.3.1.2 Operational Status

As China's foremost gateway, Shanghai Pudong International Airport (SHOGU) presents its operational status through multiple dimensions. Firstly, in terms of facilities and services, SHOGU ranks among the world's most advanced airports. It currently operates three terminals connected by complimentary shuttle buses for passenger convenience, with a designed annual passenger throughput capacity of 92.36 million and an annual cargo throughput capacity of 2.15 million tonnes. Secondly, in terms of transport scale and growth, SHOGU's annual passenger throughput has surpassed 70 million, with projections indicating it may exceed 74 million this year, solidifying its position as the world's second-busiest airport. This figure not only demonstrates the airport's high capacity but also reflects its pivotal

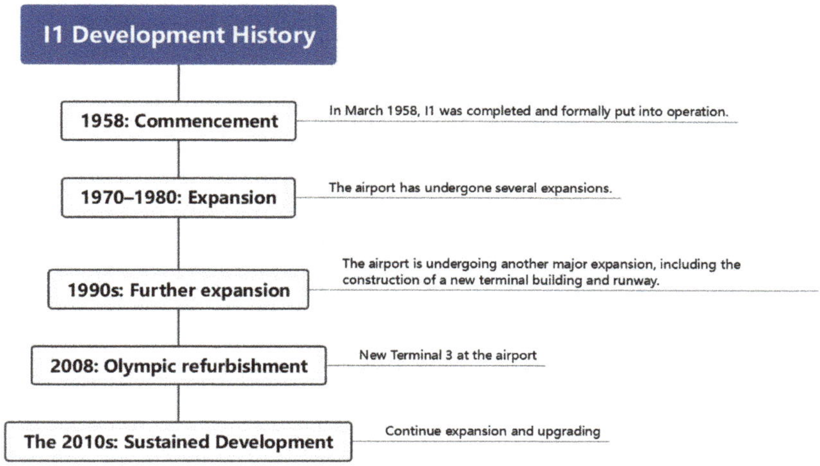

Fig. 9.6 The evolution of SHOGU

role in global air transport. In 2021, passenger throughput reached 140 million, representing a 9.7% year-on-year increase; cargo and mail throughput amounted to 2.265 million tonnes, up 12.7% year-on-year; aircraft movements totalled 1.215 million, rising 8.9% year-on-year; operating revenue reached 19.7 billion yuan, growing 5% year-on-year. In 2022, SHOGU's passenger throughput reached 12.7033 million, a year-on-year decline of 61.1%; cargo and mail throughput stood at 9.8867 million tonnes; aircraft movements totalled 157,630 [14]; In 2023, passenger throughput was 52.8792 million; cargo and mail throughput reached 1.1159 million tonnes; aircraft movements totalled 399,700 [15]. SHOGU experienced significant fluctuations between 2021 and 2023, with both passenger and cargo throughput declining notably under the pandemic's impact, though operations rebounded markedly in 2023.

From a route network perspective, SHOGU bears the vital mission of establishing an international aviation hub. Detailed analysis and optimisation recommendations concerning the airport's route network operations indicate ongoing efforts to enhance international route coverage and service quality. The airport demonstrates excellence in facilities and services, transport scale and growth, and route network development, continuously elevating its functionality and influence as an international aviation hub.

9.3.1.3 Development Strategy and Challenges

In recent years, SHOGU has consolidated its position as a global aviation hub through strategic development. By implementing three major strategies—'New Airport', 'New Hub' and 'Airport Cluster'—the airport has prioritised enhancing service quality and operational efficiency to address increasingly fierce market competition.

In its collaborative development with SSZZZ, SHOGU will concentrate on restoring and expanding international routes, particularly those to Europe, America, and countries along the Belt and Road Initiative, aiming to achieve higher international passenger throughput. However, future development faces a series of challenges. The persistent impact of the COVID-19 pandemic has severely undermined the airport's profitability, resulting in consecutive losses this year. This makes resource allocation and management efficiency pressing issues requiring urgent resolution. Furthermore, constraints on air rights and slot allocations limit international route expansion, requiring SHOGU to strike a balance between service enhancement and market development. Confronted by SSZZZ's competition, it must continually elevate service standards and passenger experience to maintain its leadership in the international route market. Nevertheless, with the gradual recovery of the international market anticipated in 2024, SHOGU projects significant growth in passenger volumes, particularly for international flights, striving to restore pre-pandemic levels. By improving flight punctuality and optimising transfer procedures, the airport aims to reclaim its competitive edge in the international aviation market over the coming years and continue serving as a vital aviation gateway for China.

9.3.2 *Typical Airport Enterprises in China*

9.3.2.1 Development History

See Fig. 9.7.

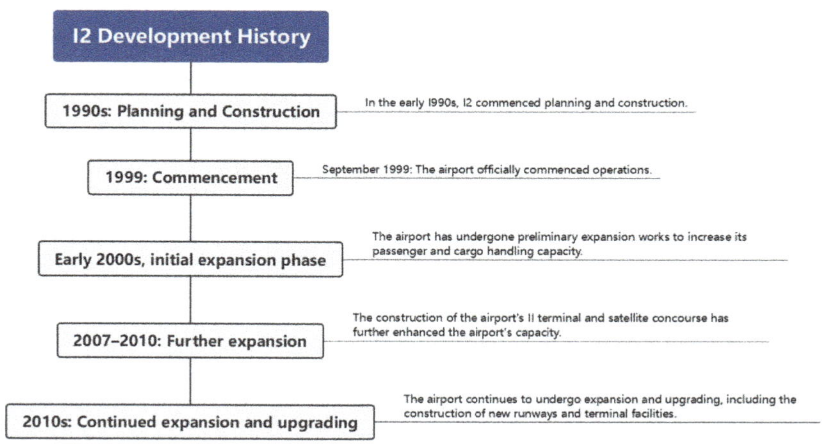

Fig. 9.7 The development history of SPUGJ

9.3.2.2 Operational Status

SPUGJ, as a vital international aviation hub in China, has achieved significant operational progress in recent years, with both flight movements and passenger throughput experiencing sustained growth over the past few years. For the full year of 2023, passenger throughput is projected to reach 54.25 million, representing a recovery to 71.24% of the 2019 level. Cargo and mail throughput is anticipated to reach 3.37 million tonnes, with aircraft movements totalling 350,000. In 2022, annual passenger throughput stood at 14.1784 million, representing a year-on-year decrease of 56.0%. Annual cargo throughput reached 3.1172 million tonnes, down 21.7% year-on-year, while total flight movements amounted to 204,400, a 41.5% year-on-year decline. In 2019, SPUGJ recorded 267,000 domestic flight movements and 240,800 international and regional flight movements. However, due to the pandemic, these figures declined to 213,400 and 252,300 movements respectively in 2020 and 2021. with international and regional flight movements falling to 107,000 and 93,000 respectively (Fig. 9.8).

The infrastructure development of Shanghai Pudong International Airport (SPUGJ) has also progressed steadily. By 2015, the airport's fourth runway became operational, significantly enhancing operational efficiency. The commissioning of this fourth runway established a four-runway configuration at SPUGJ, enabling an efficient 'dual take-off and dual landing' model. This theoretically increased peak-hour capacity to approximately 80–82 aircraft movements, with daily flight capacity surpassing 1,400 movements [16]. Positioned as a major composite international hub, the airport serves as mainland China's largest air gateway, handling the highest volume of international flights within the mainland. Although Shanghai's international passenger transport development base remains relatively underdeveloped compared to Beijing and Hong Kong, its international air passenger and cargo traffic has grown rapidly, establishing an initial aviation network. Its near-term planning targets are to accommodate 60 million passenger movements, 4.75 million tonnes of cargo and mail, and 555,000 aircraft movements. Long-term planning targets are set

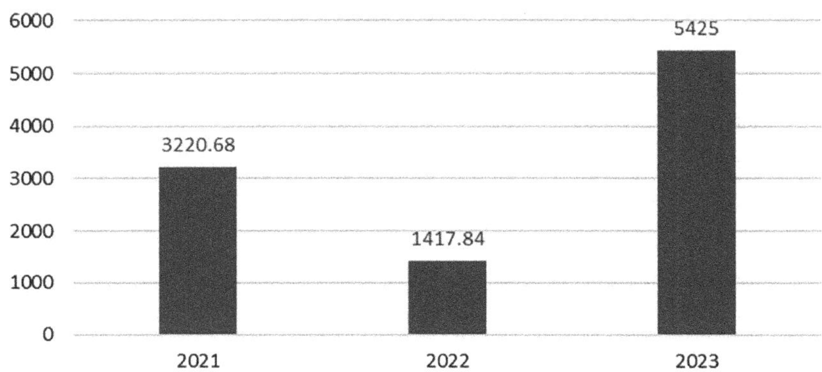

Fig. 9.8 SPUGJ passenger throughput

at 80 million annual passenger movements, 5.7 million tonnes of annual cargo and mail throughput, and 653,000 annual aircraft movements [17].

9.3.2.3 Development Strategy and Challenges

SPUGJ's development strategy centres on enhancing its status as a world-class aviation hub. Firstly, SPUGJ is committed to establishing a globally-reaching, regionally-serving international aviation hub. It will continuously strengthen facility provision capabilities, prioritising the advancement of the fourth phase expansion project. This includes constructing the new Terminal 3, cargo terminal, and integrated supporting works such as apron taxiways within the airfield area, thereby enhancing the airport's passenger and cargo handling capacity. Furthermore, the airport actively responds to the national Yangtze River Delta integration strategy by accelerating regional connectivity to serve the public's demand for high-quality travel. In commercial development, the airport has shifted from self-operation to a tenant-based model through competitive tendering, thereby increasing rental income per square metre. However, these elevated rents have impacted the introduction of premium brands and the enhancement of commercial service standards.

Although the airport's revenue in the first half of 2024 exceeded that of the same period in 2019, its profitability has yet to fully recover. Notably, international flights and passenger throughput have not yet achieved a complete structural recovery, which may impact the growth rate of the company's aviation revenue. Furthermore, the impact of high rents on commercial service standards and the demand for future passenger growth rates represent significant challenges for SJJJ. Currently, SPUGJ's Terminal 1 and Terminal 2 are operating near capacity limits. They will absorb new passenger flows for the entire Shanghai region, with passenger growth projected to maintain a 5-10% annual rate over the next decade. Concurrently, the airport must address pressures from HQGJJ's near-saturation and fulfil long-term planning objectives through its Phase III expansion project.

9.4 Development Trends and Policy Recommendations for Airport Enterprises

9.4.1 Development Trends

9.4.1.1 Intelligent Development

The intelligent development of China's airport enterprises is currently undergoing a phase of rapid advancement, primarily driven by robust national policy support and technological progress. As a vital component of Digital China's construction, the core of smart airports lies in leveraging next-generation information technologies—such as the Internet of Things, big data, and mobile internet—to enhance operational efficiency, passenger service standards, and security capabilities. The Civil Aviation Administration of China's (CAAC) Smart Civil Aviation Development Roadmap clearly outlines three phased objectives: the large-scale application of intelligent equipment, significant improvements in passenger experience, substantial increases in operational efficiency, and more effective governance capabilities. Furthermore, the CAAC emphasises the importance of enhancing key digital capabilities for smart civil aviation development, including strengthening enterprises' role as innovation leaders and focusing on digital technology innovation within civil aviation operations and service management. Currently, China's smart airport market is experiencing rapid growth, reaching approximately RMB 6.276 billion by 2022. However, the competitive landscape remains fragmented, with most enterprises leveraging their unique strengths to penetrate niche markets. Companies such as Huawei, Baidu, and Tencent are collaborating with multiple airports to jointly develop smart airport solutions, capitalising on their technological advantages. Moreover, intelligent digital twin technology provides airports with real-time operational situational awareness and panoramic visualisation capabilities. This enables bidirectional interaction between physical and virtual airports, facilitating optimised resource allocation and comprehensive coordination among operational entities. Future developments in smart airports will continue to deliver more convenient, efficient, and secure aviation services to passengers while revolutionising airport operational management.

9.4.1.2 Personalised services

Personalised services are a key means of enhancing passenger satisfaction and loyalty. Since 2017, CCCSS has continuously advanced its 'Smart Airport' initiative, focusing on informatised flight coordination support, optimising passenger departure procedures, and increasing the deployment of self-service terminals. This aims to improve flight punctuality and reduce delays, while leveraging new technologies such as AI intelligence, facial recognition, and big data analytics to deliver high-quality personalised services. MMMLL has launched its 4S commercial service brand initiative, centred on Service, Satisfaction, Sincerity and Speed. Through satisfaction analysis models and a one-stop customer service centre, it has elevated passenger contentment. Furthermore, the airport's VIP management system project employs technological empowerment to create efficient, convenient VIP travel experiences tailored to individual passenger characteristics and travel requirements. China's airport enterprises are enhancing personalised passenger experiences through technological innovation and service upgrades, thereby better meeting the individualised demands of modern travellers.

9.4.1.3 Cross-sector collaboration and integration

Cross-sector collaboration and integration represent another significant trend driving the development of airport enterprises. SJJJ Group, Huawei Technologies Co., Ltd., and Shanghai Fudian Intelligent Technology Co., Ltd. jointly held the unveiling ceremony for the 'Airport Brain and Artificial Intelligence Joint Innovation Laboratory' to advance smart airport construction. SSZZZ signed a strategic cooperation framework agreement with Huawei, with both parties to conduct research, development, and application promotion in areas such as the Internet of Things and airport cloud services. Several airports have also commenced partnerships with technology firms and innovative enterprises to jointly propel digital transformation. Germany's FFLLK collaborated with SAP to develop digital passenger services, delivering more personalised offerings. Cross-sector partnerships not only enhance airport service quality and operational efficiency but also open new markets and collaborative avenues. The development trajectory of China's airport enterprises primarily manifests in three aspects: intelligent advancement, personalised services, and cross-sector collaboration and integration. Through these measures, airport operators can not only elevate passenger experience and service standards but also secure advantageous positions in the fiercely competitive market landscape.

9.4.2 Policy Recommendations

9.4.2.1 Improve the legal and regulatory framework

To enhance the competitiveness of airport enterprises, it is imperative to refine the legal and regulatory framework. This entails revising and improving existing regulations to enhance their relevance and foresight, ensuring they remain contemporary. Concurrently, innovative approaches to industry oversight should be pursued, advancing intelligent regulatory practices and strengthening research into foreign-related legal systems. These measures will safeguard the scientific rigour and authoritative standing of the legal and regulatory framework.

9.4.2.2 Strengthen policy support measures

Increasing policy support is a crucial means of enhancing the competitiveness of airport enterprises. The government should expand fiscal subsidies for small and medium-sized airports, particularly providing greater support for the construction and operation of airports in plateau and remote regions. Concurrently, airports should be granted tax incentives equivalent to those afforded to industrial enterprises to alleviate corporate burdens. Banking institutions should be encouraged to enhance credit support for hub airports, ensuring adequate and stable funding. Furthermore, governments should integrate airport development into local socio-economic priorities, promptly disburse route operation subsidies, and coordinate solutions for land acquisition and taxation issues arising from airport construction and expansion.

9.4.2.3 Optimise airport planning and layout

Optimising airport planning and layout is pivotal to enhancing the competitiveness of airport enterprises. By taking the airport master plan as the focal point for coordinating internal and external arrangements, functional layouts should be refined, resource allocation optimised, and synergistic integration between airports, cities, and other transport modes promoted. Concurrently, regional airport distribution must align with local economic geography and socio-economic development levels to facilitate the optimal allocation of aviation resources within the region and foster coordinated socio-economic development. Through refining airport layout structures and expanding the number and scale of airports, market demands and major national strategic deployments can be further met.

9.4.2.4 Enhancing the competitiveness of airport enterprises

Enhancing the competitiveness of airport enterprises is central to achieving long-term development. By strengthening corporate governance, we aim to elevate the enterprise's leadership, competitiveness and influence, thereby securing its market position. Concurrently, the introduction of advanced air traffic technologies and training facilities will ensure safety and service quality. Efforts are being made to transform the airport into a multimodal transport hub, improving transport efficiency and service standards. Furthermore, the implementation of green airport policies will drive sustainable development.

References

1. Shin Wan Hongyuan Research, Yan Hai, Luo Shi. Special Report on the Air Transport Industry: Global Civil Aviation Enters a New Normal, Focusing on Long-Term Industry Shifts [EB/OL]. https://new.qq.com/rain/a/20230717A011OR00, 17 July 2023.
2. Wang Di. Civil Aviation Industry Advances Towards Quality Enhancement and Efficiency Improvement Phase [EB/OL]. http://www.news.cn/fortune/20240112/29cc94c76e5a4eac8747aa1e4c4cec44/c.html, 12 January 2024.
3. Liu Jie. Multiple Indicators Hit New Highs: Civil Aviation Industry to Enter New Cycle of Sustained Rapid and Healthy Development in 2024 [EB/OL]. https://news.cctv.com/2024/02/29/ARTIa5NAF0orZV1XK7DerQ6F240229.shtml, 29 February 2024.
4. Wang Xiao. Analysis of the Current Status and Prospects of China's Civil Aviation Industry [J]. State-owned Assets Management, 2010, (I): 63–65.
5. Ma Wenhua. China to Become Aviation Powerhouse by 2035 [EB/OL]. https://www.gov.cn/zhengce/202403/content_6935404.htm?ddtab=true, 2024-03-01.

6. Xinhua News Agency. Xinhua Finance | Civil Aviation Industry Returns to Natural Growth; 2024 Airline Profits on the Horizon [EB/OL]. https://news.qq.com/rain/a/20240117A05CIM00.2024, 2024-01-17.
7. Zhu Ziyue. Century-Long Commercial Warfare in European and American Aviation: Implications for China's Participation in International Competition [EB/OL]. http://att.caacnews.com.cn/mhfzzcgjyxb/mhfzzcgjyxb5th/202111/t20211101_59843.html#,2021-11-01.
8. Cui Guoqiang, Zhang Jun. Building a Super Carrier: China's Civil Aviation Contributed Over 25% to Global Growth Last Year [EB/OL]. https://www.thepaper.cn/newsDetail_forward_2726050, 2018-12-11.
9. Li Yansheng, Guo Xiaoting. From 37th to 2nd: China Emerges as a Global Aviation Powerhouse [EB/OL]. https://www.gov.cn/gzdt/2008-12/04/content_1168516.htm, 2008-12-04.
10. Banliang Finance. Air China's 2023 Revenue Reaches 141.1 Billion Yuan, Up 166.74% Year-on-Year_Tencent News [EB/OL]. https://new.qq.com/rain/a/20240329A06LGT00, 2024-03-29.
11. Xiaolang Financial News. China Eastern Airlines reports first-half revenue of RMB 64.199 billion, up 29.89% year-on-year; net profit of -RMB 2.768 billion, up 55.70% year-on-year; sales expenses increase by 35.26% year-on-year | China Eastern Airlines_Sina Finance_Sina.com [EB/OL]. https://finance.sina.com.cn/stock/aiassist/yjbg/2024-08-31/doc-incmnhnt9855144.shtml, 31 August 2024.
12. Zhang Sunmingshuo. China Eastern Airlines reduces losses by RMB 29 billion in 2023, with international route revenue recovering to 50% | Financial Report Analysis [EB/OL]. https://finance.ifeng.com/c/8YLyhgcfw1E, 2024-03-29.
13. China Southern Airlines. China Southern Airlines Co., Ltd. 2023 Annual Report Summary [N]. 2023.
14. BT Finance. China Southern Airlines I-Q1 2024 Report: Comprehensive Financial Data Analysis [EB/OL]. https://xueqiu.com/2509537588/293493003, 2024-06-13.

Chapter 10
Comprehensive Study on a Pipeline Facilities Operation Enterprise

Malin Song and Xinmin Zhang

Abstract This paper conducts a study on GUWAJ, a pipeline facilities operation enterprise. Firstly, it elaborates on the enterprise's development history from its establishment background to its gradual growth and expansion, and presents its development trajectory. Then, it analyzes the enterprise's business model in detail, and explores its position and advantages in market competition as well as the competitive situation with other related enterprises. Secondly, the paper discusses the challenges faced by the enterprise in the development process, such as technological upgrading and changes in market demand, as well as the opportunities brought by favorable policies and energy structure adjustment. It conducts an in-depth analysis of the enterprise's development strategies, such as expanding business areas and optimizing pipeline network layout, and finally puts forward relevant development suggestions. The purpose of this paper is to provide relevant personnel in the field of pipeline facilities operation with comprehensive information to understand GUWAJ, offer references for enterprise development decisions and industry research, and promote the further development of the pipeline operation industry.

10.1 Introduction

With the rapid development of the economy, energy demand has grown swiftly, and the stable supply and efficient transportation of energy have become crucial pillars for sustained economic development. Pipeline transportation, boasting prominent advantages such as safety, efficiency, environmental friendliness, and cost-effectiveness, is gaining increasing importance in energy transportation. Alongside the optimization of the energy structure, the consumption proportion of clean energy like natural gas

M. Song
Anhui University of Finance and Economics, Bengbu City, Anhui Province, China

X. Zhang (✉)
Dongbei University of Finance and Economics, Dalian City,, Liaoning Province, China
e-mail: 15148635083@163.com

J. Wang and M. Song (eds.), *The Green Path of Chinese Transportation Enterprises*, Resources, Climate and Sustainable Development,
https://doi.org/10.1007/978-981-95-5289-4_10

has gradually risen, bringing significant development opportunities for the construction and operation of oil and gas pipeline facilities. To enhance the level of fair competition and efficiency in the energy market, China has actively carried out energy system reforms. Against this backdrop, GUWAJ came into being. The establishment of GUWAJ represents a major reform in China's oil and gas pipeline operation mechanism. Its goal is to build a national oil and gas pipeline platform that is fair, open, and interconnected, thereby realizing the optimal allocation and efficient utilization of pipeline resources. Conducting research on GUWAJ's development is of great significance. It not only enables in-depth insight into the challenges faced by pipeline facilities operation enterprises in the new market environment and policy framework as well as their countermeasures to achieve steady development, but also provides valuable experience references for other enterprises in the industry to promote the progress of the entire sector. Meanwhile, it holds undeniable practical value for safeguarding national energy security, accelerating energy transition, and improving the energy market structure.

This study aims to conduct an in-depth analysis of GUWAJ's development status. Specifically, by comprehensively examining its operation model, market strategy, technological innovation, and measures to respond to changes in the policy environment, it will reveal GUWAJ's successful experiences and potential problems. It also explores how GUWAJ can find its position amid energy structure adjustment and market reform to achieve efficient operation and sustainable development. Additionally, this study is expected to provide a reference for government departments to formulate more scientific and reasonable energy policies and regulatory measures, so as to promote the healthy and orderly development of China's pipeline facilities operation industry.

10.2 Overview of GUWAJ

10.2.1 Company Profile

GUWAJ was established on December 9, 2019. As a key state-owned backbone enterprise supervised by the State-owned Assets Supervision and Administration Commission of the State Council (SASAC), it undertakes the important mission of safeguarding national energy security and promoting the high-quality development of the energy industry. The company focuses on the investment, construction, and operation of infrastructure such as oil and gas trunk pipelines, gas storage, and peak-shaving facilities, with the goal of building a national oil and gas pipeline platform that is fair, open, and interconnected. Its business scope covers the planning, construction, and operation of oil and gas pipelines; investment, construction, and operation of gas storage facilities; and operation and management of LNG receiving terminals. By integrating resources and optimizing layout, GUWAJ has improved the allocation efficiency of oil and gas resources, providing strong support for China's stable energy supply and economic development.

GUWAJ boasts a well-structured organizational framework, which comprises nine functional departments, four headquarters (including Production and Operation) and three business divisions (including Crude Oil). As directly affiliated second-tier units at the bureau level, these four headquarters and three business divisions serve as a critical link between the company's decision-making and operational levels. Specifically, its internal functional departments include the General Office, Human Resources Department, Planning and Plan Department (Innovation and Strategy Department), Strategy Execution Department (Production and Operation Department), HSE and Enterprise Management Department, Audit and Compliance Department (Inspection Office), Finance Department, and Party and Mass Work Department, each with clearly defined mandates: the Production and Operation Command Headquarters oversees the scheduling of oil and gas pipeline production and operation; the Science and Technology R&D Headquarters focuses on relevant technological research and development; the Engineering Construction Headquarters undertakes pipeline network construction tasks; the Pipeline Integrity Headquarters ensures pipeline safety and integrity; and the Crude Oil Division, Natural Gas Division, and Refined Oil Division are responsible for business operation and management within their respective domains. This specialized organizational setup is strategically designed to facilitate the investment, construction, and operation of infrastructure such as oil and gas trunk pipelines. It further aims to drive the formation of the "X + 1 + X" oil and gas market system—characterized by multi-subject and multi-channel supply in the upstream, efficient collection and transportation via the mid-stream pipeline network, and full competition in the downstream—thereby safeguarding national energy security and advancing the high-quality development of the energy industry [1].

In terms of oil and gas pipelines, GUWAJ is responsible for investing in, constructing, and operating domestic oil and gas trunk pipelines, including crude oil, refined oil, and natural gas pipelines. It has built a nationwide oil and gas transmission network to ensure the efficient allocation of oil and gas resources. Gas storage business is also one of its key focuses: by investing in, constructing, and operating gas storage facilities, it enhances oil and gas reserve capacity to cope with seasonal demand fluctuations and emergencies, thus safeguarding the stability of energy supply. In addition, the operation of LNG receiving terminals constitutes an important part of its business. GUWAJ undertakes the reception of imported liquefied natural gas, as well as its storage, gasification, and transmission to meet domestic natural gas demand. Meanwhile, the company actively engages in relevant technological research, development, and innovation to improve the safety, efficiency, and intelligence of pipeline transportation. In the field of market transactions, it promotes the optimal allocation and fair trade of oil and gas resources to drive the healthy development of the oil and gas market. Furthermore, it continuously expands cooperation with relevant domestic and foreign enterprises to enhance its influence and competitiveness in the international energy sector [2].

10.2.2 Development History of GUWAJ

Start-up Stage (2019–2020): On December 9, 2019, GUWAJ was formally established, with the State-owned Assets Supervision and Administration Commission of the State Council (SASAC) performing the duties of an investor on behalf of the State Council. Since its establishment, the company has been faced with the key task of integrating oil and gas trunk pipelines and related infrastructure assets. On September 30, 2020, GUWAJ held a signing ceremony for the delivery of oil and gas pipeline assets and handover of operations, and jointly signed asset transaction and capital increase and share expansion agreements with ZSSYY, ZSSHH, ZHHAA, etc. After fully taking over oil and gas pipeline infrastructure assets and personnel, GUWAJ officially launched grid-connected operations, marking a major achievement in the market-oriented reform of China's oil and gas pipeline operation mechanism.

Growth Stage (2021–2022): GUWAJ moved forward steadily, continuing to focus on business expansion and resource integration. On March 31, 2021, the group officially took over the equities of TIARA and YYETR previously owned by Kunlun Energy under ZSSYY, successfully realizing the interconnected operation of China's main oil and gas pipelines. The company also signed cooperation agreements with multiple local governments and enterprises, including the People's Government of Guangdong Province, the People's Government of Gansu Province, Angang Group, and China Aerospace Science and Technology Corporation. In-depth cooperation was carried out in pipeline network construction, operation, technical cooperation and other fields, further promoting the interconnection and optimized layout of the national oil and gas pipeline network.

Maturity Stage (2023–Present): GUWAJ has maintained steady development and achieved a number of important outcomes. On February 15, 2023, Hong Kong & China Gas Group signed a market cooperation agreement with GUWAJ. On May 22, it reached a strategic cooperation agreement with HHTKK. On July 20, KKENN signed a long-term and medium-term use agreement for liquefied natural gas receiving terminals with a receiving terminal enterprise under GUWAJ. On November 10, the Tianjin LNG outbound pipeline was successfully put into operation on the first attempt, realizing connection with multiple natural gas trunk pipelines. On November 13, China's first hydrogen-blended natural gas pipeline release jet fire test and confined space leakage combustion and explosion test were successfully carried out. These achievements have further enhanced the company's capabilities in the construction and operation of oil and gas infrastructure, playing an important role in safeguarding national energy security and promoting the development of the energy industry [3].

10.2.3 Current Status of GUWAJ

As of the end of 2021, the total length of GUWAJ's in-service oil and gas pipelines reached 96,400 km, covering 30 provinces (autonomous regions, municipalities directly under the Central Government) and the Hong Kong Special Administrative Region of China [4]. For instance, the West-to-East Gas Pipeline System has a huge annual gas transmission capacity; its cumulative gas transmission volume exceeded 100 billion cubic meters in 2023, strongly guaranteeing energy supply in eastern China. GUWAJ has 8 underground gas storage facilities and 7 LNG receiving terminals in operation and management, with an injection and production volume of 4.24 billion cubic meters. This has further consolidated and strengthened the pattern of a "national integrated pipeline network" and significantly enhanced the country's capacity to safeguard energy security. During the peak gas consumption period in winter, for example, these gas storage facilities and receiving terminals can timely play a peak-shaving role to ensure stable gas supply for residents and enterprises.

Since its establishment, GUWAJ has achieved remarkable results in patent research and development. As shown in the Table 10.1, from 2020 to 2024, its patents cover multiple key technical fields, including pipeline strain testing, third-party damage simulation, magnetic flux leakage internal detection data alignment, pipeline inventory calculation, compressor dry gas seal gas filtration, and retractable direct pipe-laying devices. These patented technologies reflect GUWAJ's high emphasis on the safe and efficient operation of pipelines, as well as its active investment in technological innovation.

Since its establishment, GUWAJ has achieved remarkable results in patent research and development. As shown in the Table 10.1, from 2020 to 2024, its patents cover multiple key technical fields, including pipeline strain testing, third-party damage simulation, magnetic flux leakage internal detection data alignment, pipeline inventory calculation, compressor dry gas seal gas filtration, and retractable direct pipe-laying devices. These patented technologies reflect GUWAJ's high emphasis on the safe and efficient operation of pipelines, as well as its active investment in technological innovation.

In addition, GUWAJ has actively promoted market opening by signing strategic cooperation agreements with provinces such as Guangdong, Hunan, Fujian, Heilongjiang, and Gansu. It has accelerated the construction of the "national integrated pipeline network", attracted numerous upstream and downstream enterprises to participate in cooperation, and promoted the diversified development of the oil and gas market.

In terms of social responsibility, GUWAJ has always kept its mission in mind. For example, in 2024, it successfully completed the natural gas supply guarantee task for the fourth consecutive year. The 18 gas storage facilities connected to the pipeline network injected 15.3 billion cubic meters of gas, with a completion rate of 107%, making every effort to ensure residents' living warmth [5]. At the same time, the company also attaches importance to environmental protection and takes diversified measures to protect biodiversity, realizing the harmonious coexistence of project

Table 10.1 GUWAJ's individual patents

Patent name	Application date	technical brief
Strain testing method and testing device for large-scale high-pressure pipelines under explosion impact	January 11, 2021	Relating to the technical field of strain testing methods and testing devices for high-pressure pipelines, it includes a pipeline fixing assembly, an initiation control assembly, etc. It can conduct strain testing under explosive impact, providing methods for detecting pipeline strain data and its variation laws in explosive environments, and offering data support for blasting operations near pipelines
Method and test device for simulating third-party damage to large-diameter high-pressure pipelines	January 11, 2021	Relating to the technical field of simulation methods and devices for third-party damage to high-pressure pipelines
Natural gas pipeline inventory calculation method, device, equipment and storage medium	June 19, 2023	By dividing the natural gas pipeline into multiple micro-segments, acquiring relevant data of each segment and performing calculations, the inventory of the natural gas pipeline is finally obtained. This method can improve calculation accuracy
A dry gas seal filtration device for compressors	May 31, 2024	Comprising a primary filtration mechanism, a secondary filtration mechanism, an electric heater, etc., which are sequentially connected through pipelines. It can achieve clean and impurity-free gas, thereby extending the service life of dry gas seals
Retractable direct pipe-laying device	May 30, 2024	Including a tunnel boring machine, an outer casing, a cutter head, etc. When the tunnel boring machine needs to be retracted, the connecting structure is disassembled, and the pipe pushing machine can pull the tunnel boring machine and cutter head to retract through the towing part, effectively preventing surface settlement and collapse

construction and operation with the ecological environment. For instance, during pipeline construction, it adopts green construction technologies to reduce the impact on the ecological environment. Furthermore, it actively participates in emergency rescue and disaster relief, poverty alleviation, rural revitalization and other work, giving back to society through practical actions.

10.3 Analysis of the Business Model and Market Compet(I) Analysis of the Business Model

The pipeline infrastructure construction model of GUWAJ encompasses investment, project management, and technology application dimensions. In terms of investment, it expands funding channels through collaborative partnerships with governments at all levels, state-owned enterprises, and social capital, ensuring sufficient financial support for large-scale pipeline projects. It also optimizes capital allocation via financial innovation tools to enhance capital utilization efficiency. For project management, standardized processes are established for every stage from planning and design to construction and acceptance, with real-time monitoring of progress, quality, and costs enabled by advanced project management software. In technology application, GUWAJ introduces internationally advanced construction technologies and adapts them to domestic conditions—for instance, applying high-grade steel pipes and efficient welding processes in long-distance pipeline construction to boost capacity and safety—while prioritizing energy-saving and environmental technologies to minimize resource consumption and ecological impact during construction.

GUWAJ maintains significant advantages in its integrated management model, covering safety, quality, cost, and personnel management. For safety, it deploys advanced detection technologies such as distributed optical fiber monitoring systems and intelligent inspection robots to real-time track pipeline parameters and surrounding geological activities, complemented by regular safety training and emergency drills for employees. In quality management, strict protocols govern material procurement (e.g., X80-grade steel pipes), construction supervision, acceptance testing, and daily maintenance to ensure long-term stability. Cost management relies on refined budgeting, energy-saving measures like optimized pumping station operations, and efficient resource allocation to reduce waste. Additionally, it invests in employee training and career planning, supported by performance appraisal systems tied to safety and energy efficiency, while leveraging information technology to centralize operational data analysis, driving data-informed decision-making and management optimization.

GUWAJ's profit model is built on diversified and sustainable revenue streams. Primarily, it charges transportation fees to oil and gas enterprises based on factors such as throughput, distance, and transportation conditions, generating stable income by providing efficient, reliable transportation services. Secondly, its gas storage facilities cater to domestic demand for seasonal peak shaving and emergency reserves, with revenue derived from rental fees paid by enterprises utilizing these facilities. Finally, LNG receiving stations contribute to profits by offering unloading, storage, and gasification services for imported LNG, with corresponding service fees. This multi-faceted profit structure provides financial backing for the company's sustainable development and ongoing infrastructure investments.

10.3.1 Market Competition Analysis of China National Pipe Network Group

10.3.1.1 Competitor Analysis

In the field of pipeline facility operation, although GUWAJ has become the core entity for domestic oil and gas pipeline transportation since its establishment, traditional energy enterprises such as ZSSYY, ZSSHH, and ZHHAA still exert competitive pressure on it to a certain extent. Before GUWAJ's establishment, the "Big Three Oil Companies" each owned a huge pipeline asset and operation system. Although some of their assets have been injected into GUWAJ, they have profound accumulation and advantages in upstream and downstream businesses. For example, ZSSYY occupies an important position in the field of oil and gas exploration and development, and its strong resource control capability gives it considerable say in the source of oil and gas supply. ZSSHH has strong strength in the refining and chemical industry and boasts an extensive downstream sales network. ZHHAA, on the other hand, has unique advantages in offshore oil and gas development. In terms of data, taking 2023 as an example, ZSSYY's oil and gas output remained at a relatively high level, with crude oil output reaching approximately 105 million tons and natural gas output about 133.19 billion cubic meters [6]. The transportation demand for these oil and gas resources enables the "Big Three Oil Companies" to still have a certain bargaining power in their cooperation with GUWAJ. At the same time, the "Big Three Oil Companies" are also continuously expanding their own pipeline construction and operation capabilities, and may compete with GUWAJ in some specific regions and business fields. For instance, in terms of refined oil pipeline transportation in certain areas, ZSSHH may compete with GUWAJ for customer resources and transportation volume by virtue of its local market share and resource advantages (Fig. 10.1).

With the rapid development of new energy and the diversification of the energy market, GUWAJ is also facing competition from emerging energy enterprises and other transportation methods. In the field of new energy, the popularization of electric vehicles has had a certain impact on the demand for refined oil pipelines. In

Fig. 10.1 Market share

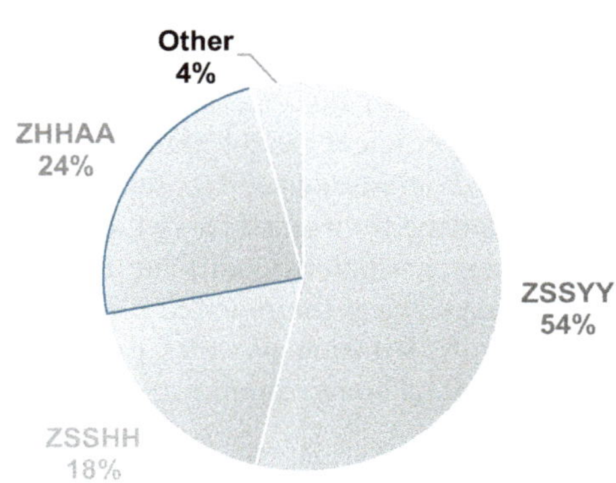

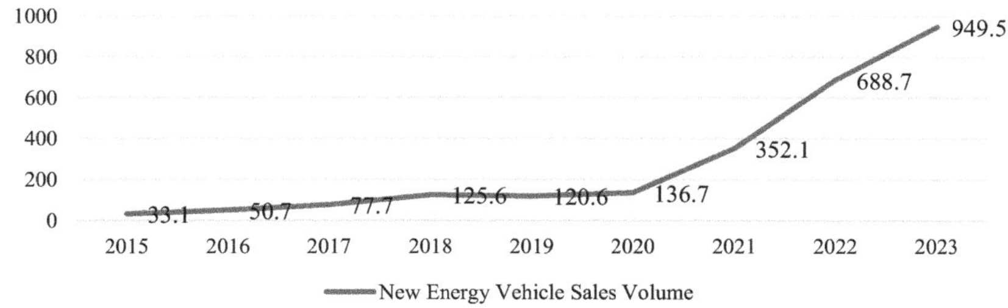

Fig. 10.2 New energy vehicle sales volume

recent years, the sales volume of electric vehicles in China has maintained continuous growth. As shown in the above figure, the sales volume of electric vehicles reached 9.495 million units in 2023, a year-on-year increase of 37.9% [7]. This has put downward pressure on the traditional demand for refined oil transportation. At the same time, in the natural gas sector, the rise of distributed energy suppliers and some small gas enterprises has also posed challenges to GUWAJ's market share. Through flexible operation models and localized services, these enterprises have certain competitiveness in some niche markets. In addition, other transportation methods such as liquefied natural gas (LNG) shipping and railway transportation also compete with pipeline transportation to a certain extent. For example, in some coastal areas, LNG shipping has the advantages of low transportation costs and high flexibility, which may divert part of the natural gas market demand. GUWAJ needs to continuously innovate and optimize its own operation model, and improve service quality and efficiency to cope with the challenges from these emerging competitors (Fig. 10.2).

10.3.1.2 Analysis of Competitive Advantages

GUWAJ owns a vast and widely covered oil and gas pipeline network, which enables it to occupy a dominant position in the field of pipeline facility operation. Up to now, the total length of GUWAJ's in-service oil and gas pipelines has reached tens of thousands of kilometers, connecting major domestic oil and gas resource producing areas and consumer markets. For instance, the West–East Gas Pipeline Project is an important pipeline project under GUWAJ, with a huge annual gas transmission volume. In 2023, the cumulative gas transmission volume of the West–East Gas Pipeline exceeded 100 billion cubic meters, effectively ensuring energy supply in eastern China. In terms of infrastructure construction, GUWAJ has advanced gas storage and peak-shaving facilities such as gas storage tanks and LNG receiving stations. In 2024, 18 gas storage tanks connected to the pipeline network completed gas injection of 15.3 billion cubic meters, with a completion rate of 107%. These facilities play a key role in coping with gas consumption peaks and ensuring supply stability. Compared with traditional energy enterprises, GUWAJ focuses on pipeline facility operation, enabling it to integrate resources more efficiently and improve the utilization efficiency of infrastructure. Taking ZSSYY as an example, although it has strong strength in the field of oil and gas exploration and development, GUWAJ has more advantages in the professionalism and centralization of pipeline facility operation [8]. GUWAJ places technological innovation in an important position and actively adopts advanced technologies and management concepts in pipeline facility operation. In terms of pipeline monitoring technology, the use of advanced intelligent sensors and data transmission systems allows real-time awareness of pipeline operation status and timely identification of potential safety hazards. For example, distributed optical fiber sensing technology can accurately detect parameters such as temperature, pressure and strain of the pipeline. In terms of intelligent operation, an efficient pipeline management system has been developed, which uses big data analysis and artificial intelligence algorithms to conduct in-depth mining of pipeline operation data, optimize pipeline operation plans, and improve transmission efficiency.

10.4 Development Challenges and Opportunities of China National Pipe Network Group

10.4.1 Development Challenges

First, GUWAJ undertakes massive tasks related to the construction and maintenance of oil and gas pipeline networks. As China's energy demand continues to grow, expanding pipeline coverage and improving transportation capacity necessitates enormous capital investment—for example, building long-distance pipelines incurs high costs spanning pre-construction planning, land expropriation, construction, and equipment commissioning. Maintenance is equally arduous: pipelines traverse complex terrains like mountains, rivers, and deserts, exposing them to natural erosion and geological disasters, which demands continuous investment in manpower and materials for inspection, repair, and renewal. Annual maintenance costs account for a significant share of operating expenses, and the urgent need for intelligent upgrading of pipeline facilities, driven by technological advancement, further intensifies pressure on construction and maintenance costs.

Against the backdrop of market opening, GUWAJ faces a complex competitive landscape. On one hand, while committed to fair pipeline network access, balancing the interests of diverse market entities and ensuring equitable allocation of transportation capacity remains practically challenging—enterprises of different scales may encounter varying access costs and conditions, potentially triggering disputes. On the other hand, the rapid development of new energy and diversification of the energy market bring competition from alternative transportation methods: the popularization of electric vehicles may reduce demand for refined oil pipelines, while renewable energy growth could reshape consumption structures and lower reliance on traditional oil and gas pipelines. Additionally, fluctuations in the international energy market, such as changes in global oil prices, impact domestic supply–demand dynamics, thereby affecting pipeline utilization and economic benefits.

GUWAJ also grapples with hurdles in technological innovation and safety assurance. To enhance operational efficiency and safety, continuous improvement of pipeline intelligence is essential, but innovation entails high risks and investments—developing new monitoring technologies and intelligent management systems requires substantial financial and talent support, and translating R&D achievements into practice demands time and testing, with potential issues like immature or unstable technology. For safety, transporting flammable and explosive oil and gas means leaks or accidents could have severe consequences, requiring sustained high safety awareness, robust management, and risk control. This involves investing heavily in safety facilities and improving employees' safety literacy and emergency response capabilities, yet the expanding pipeline network and increasingly complex operating environment continue to raise the difficulty of safety assurance.

10.4.2 Development Opportunities

With the sustained development of China's economy and the continuous optimization of its energy consumption structure, the demand for oil and gas resources has maintained a steady growth trend. As China's largest professional energy infrastructure operator, GUWAJ plays a pivotal role in ensuring energy supply. The continuous expansion of industrial production, accelerated urbanization, and improved residents' living standards have collectively driven up oil and gas consumption: the manufacturing sector's demand for energy like natural gas keeps rising to support production, while urban households increasingly use natural gas for cooking and heating. Additionally, the state's active promotion of energy transition has elevated natural gas' share in the energy structure as a clean energy source, creating broad market space for GUWAJ. Relevant statistics show China's natural gas consumption has grown annually in recent years, with expectations of sustained high growth—allowing GUWAJ to expand its business, strengthen pipeline construction and operation, and enhance transportation capacity to meet rising demand and solidify its market position.

GUWAJ's commitment to promoting oil and gas pipeline interconnection and building a unified national market unlocks multifaceted development opportunities. Pipeline interconnection optimizes resource allocation, improves energy transportation efficiency, and reduces costs. Meanwhile, the unified national market has attracted more market entities to the oil and gas industry, boosting market vitality. Intensified competition pushes GUWAJ to enhance service quality and operational efficiency to better satisfy customer needs. Moreover, the fair opening of pipeline networks generates new business growth: beyond serving traditional oil and gas enterprises, GUWAJ can attract emerging energy firms and distributed energy suppliers to connect to its network, expanding its customer base and diversifying its business. For example, small natural gas producers and distributed energy projects can leverage the national pipeline platform to deliver energy more efficiently, achieving mutual benefit [9].

In the era of rapid technological advancement, scientific and technological innovation provides strong impetus for GUWAJ's development. GUWAJ actively pursues R&D and innovation, making breakthroughs in pipeline construction, operation management, and safety assurance—such as increasing investment in intelligent pipeline technologies, using the Internet of Things, big data, and artificial intelligence to monitor and control pipeline operations in real-time, thereby enhancing intelligence, reducing risks, and cutting costs. Furthermore, the rise of emerging industries like hydrogen energy and carbon capture, utilization, and storage (CCUS) allows GUWAJ to leverage its infrastructure and technological advantages to expand into new areas: it can explore hydrogen pipeline transportation technologies and business models to lay the groundwork for the hydrogen industry, and participate in carbon dioxide transportation and storage projects to support carbon emission reduction goals. Collaborating in-depth with universities and research institutions also enables GUWAJ to drive energy sector innovation, foster new growth points, and lead the industry's transformation and upgrading.

10.5 Development Strategies and Suggestions for China National Pipe Network Group

10.5.1 Development Strategies

Guided by its corporate mission of serving national strategies, meeting people's needs, and boosting industry development, GUWAJ fully leverages the advantages of unified planning, construction, and operation to accelerate the layout of various pipeline network structures and expedite the construction of key projects. For example, the China-Russia East-Route Natural Gas Pipeline was completed and put into operation ahead of schedule, and the Sichuan-East Gas Pipeline II project broke ground. The total length of pipelines built has reached nearly 12,000 km, significantly enhancing natural gas transmission capacity. It has actively promoted the market-oriented integration of provincial pipeline networks, and the "national unified pipeline network" with rational layout, extensive coverage, and complete functions has become increasingly improved, basically eliminating the hard bottlenecks in oil and gas infrastructure for supply guarantee. Through the construction of this network, GUWAJ has realized the efficient allocation and utilization of oil and gas resources, and enhanced the safety and stability of energy supply. GUWAJ firmly implements the Digital China strategy and the strategy of building a strong country through science and technology, taking digital transformation as a major strategic task of the company, with a focus on advancing digital transformation and achieving self-reliance and self-improvement in science and technology. On one hand, it has improved information management capabilities and accelerated its transition to a digital company: established a dedicated team for process reform, carried out sorting and restructuring of business processes, and promoted the launch of application scenarios for "Industrial Internet + Work Safety". On the other hand, it has innovated science and technology management mechanisms by establishing a "gate-style" management system for scientific research projects as well as mechanisms for "soliciting proposals to tackle key problems" and "competitive selection". It has increased investment in scientific and technological research and development, and strengthened research on key core technologies. For instance, in fields such as pipeline network reliability and simulation optimization, efficient utilization of underground space, and hydrogen/carbon dioxide pipeline transportation, it is building first-class basic research and development platforms, cultivating leaders and innovation teams in the scientific and technological field, and focusing on the localization of key equipment, so as to lead the transformation and upgrading of the oil and gas pipeline network industry through scientific and technological innovation.

10.5.2 Development Suggestions

GUWAJ should actively expand into the international market and strengthen cooperation with international energy enterprises and pipeline operators. For example, it can participate in the construction and operation of pipeline projects in countries along the "Belt and Road" Initiative, and learn advanced international pipeline technologies and management experience. Through international cooperation, it can enhance its international competitiveness and influence, and realize the global allocation of resources.

GUWAJ must firmly implement the "dual carbon" strategy and actively practice the "Scientific and Technological Innovation Support Action" for carbon peaking. It should increase investment in the R&D and promotion of green and low-carbon technologies, strengthen methane emission control, implement natural gas venting recovery, scientifically develop projects such as centralized photovoltaic power generation and cold energy utilization, accelerate the green electrification of self-used electricity, and further reduce natural gas consumption per unit. At the same time, it should speed up the deployment of emerging energy storage businesses, advance technical research on hydrogen blending in natural gas pipelines, high-pressure large-diameter pure hydrogen transportation, and carbon dioxide pipeline transportation, expand new core businesses of the pipeline network, and facilitate the implementation of industry standards and technological achievements. While ensuring energy supply, it should attach importance to environmental protection and sustainable development.

References

1. Qi Yu.Organizational Structure of PipeChina Determined-Securities Times. December 9, 2019. https://baijiahao.baidu.com/s?id=1652416062400936119.2019.12.9.
2. PipeChina National Petroleum and Natural Gas Pipeline Network Group Co., Ltd. https://www.pipechina.com.cn/.
3. Shiyilu.Special Topic on Oil and Gas Pipelines:A Detailed Account of PipeChina's Past, Present and Future. December 10, 2019. https://www.vzkoo.com/document/5684390255a16373099d9c0b84b61a95.html.2019.12.10.
4. An Dongping. PipeChina Releases 2021 Annual Social Responsibility Report. https://cpnn.com.cn/news/yq/202208/t20220830_1547915.html.
5. PipeChina National Petroleum and Natural Gas Pipeline Network Group Co., Ltd. PipeChina Achieves a Good Start in Production and Operation in the First Quarter. http://www.sasac.gov.cn/n2588025/n2588124/c30541328/content.html.2024.4.16.
6. Wang Jingjing. PetroChina's Net Profit Reached 161.1 Billion Yuan Last Year, Up 8.3% Year-on-Year, Hitting a Record High. March 25, 2024. https://www.thepaper.cn/newsDetail_forward_26806935.
7. Zhou Xiaoxiao. National Development and Reform Commission: New Energy Vehicle Sales Reached 9.495 Million Units in 2023;Actively Expand New Energy Vehicle Consumption in 2024. January 18, 2024. https://m.21jingji.com/article/20240118/herald/ba7cf1e865f9b6e4e7300ecf9170837f.html.
8. National Energy Administration. PipeChina Unveiled; Oil and Gas Market-Oriented Reform Enters a Critical Stage. December 10, 2019. https://www.nea.gov.cn/2019-12/10/c_138620254.htm.
9. Cao Miao, Li Yuan. PipeChina: Laying a Solid Foundation to Provide Strong Support for the High-Quality Development of Enterprises. July 30, 2024. http://finance.people.com.cn/n1/2024/0730/c436861-40288842.html.

Chapter 11
Analysis of Express and Postal Enterprises

Malin Song and Yuguang Liu

Abstract The logistics industry is a fundamental sector vital to people's livelihoods, and express and postal enterprises serve as crucial pillars within this industry. These enterprises not only form a critical link between production and consumption by handling cargo transportation but also drive efficient supply chain operations and foster coordinated development between urban and rural economies. As the logistics industry continues to expand, the performance of major express delivery companies has gradually diverged. Through unique strategic positioning and market strategies, these companies have demonstrated diversified competitive advantages in areas such as revenue scale, parcel volume, and specialized services, resulting in significant variations among them.

In terms of revenue, SHFE and JIDOW are industry leaders, each achieving annual revenues in the hundreds of billions and leading in R&D investment. In parcel volume, ZTO stands out notably, exceeding 30 billion parcels in 2023 and maintaining its leading position. YTO, STO, YDA, and DBA have also showcased their strengths in specific domains and specialized services. Among them, YTO surpassed YDA in parcel volume in 2023 to become the second-largest logistics company in China by business volume. STO achieved a 35.22% year-on-year growth in parcel volume in 2023, making it the fastest-growing company in this regard. YDA boasts an efficient sorting system and an extensive service network, while DBA is renowned for its expertise in large-item delivery.

This article elaborates on various express delivery companies across four dimensions: company profiles, development history, current status, and future prospects and recommendations.

M. Song
Anhui University of Finance and Economics, Bengbu, China

Y. Liu (✉)
Dongbei University of Finance and Economics, Dalian, China
e-mail: ygliu0101@126.com

J. Wang and M. Song (eds.), *The Green Path of Chinese Transportation Enterprises*, Resources, Climate and Sustainable Development,
https://doi.org/10.1007/978-981-95-5289-4_11

11.1 General Landscape of the Express Delivery Sector

11.1.1 Rapid Industry Growth Fueled by Continual Business Scale Expansion

The year 2023 marked a period of vigorous growth for China's express delivery industry. The annual parcel volume reached 132.07 billion pieces, representing a growth rate of 19.4%. This significantly exceeded the global average growth expectation of 6%, continuing the industry's multi-year expansion and solidifying its leading position in the global market by size (see Fig. 11.1). According to the State Post Bureau, in the first half of 2024, the national express delivery volume reached 80.16 billion parcels, a year-on-year increase of 23.1%, while revenue generated was 653 billion yuan, up 15.1% year-on-year. Despite already operating on a high annual base of hundreds of billions of parcels, the express market maintained double-digit growth, underscoring the sector's robust vitality and development potential. 2023 was also a lucrative year for many major domestic express delivery giants, with all companies achieving varying degrees of growth (see Table 11.1). Concurrently, industry concentration continued to intensify. The CR8 concentration index for express delivery and parcel service brands reached 84.0 in 2023, reflecting a further strengthening of the dominant position held by leading market players.

Against the backdrop of rapid growth in national express delivery volume, regional development has shown significant disparities. Eastern coastal areas, given their economic maturity, demonstrated stable market growth, while central and western

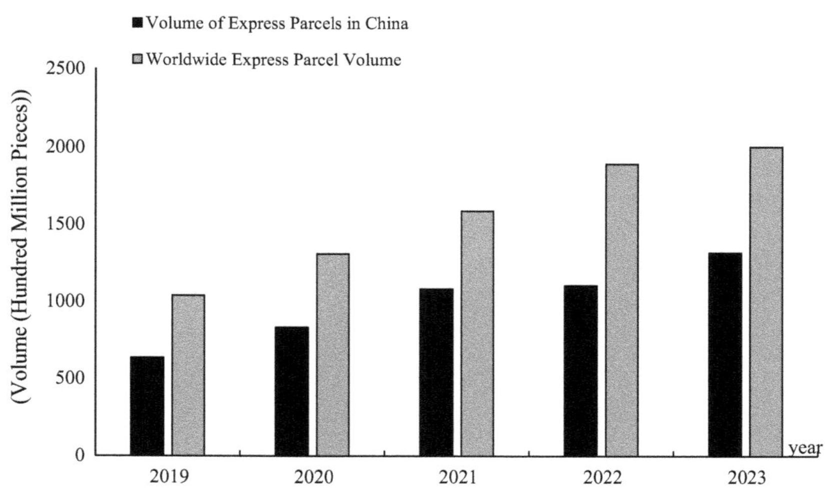

Fig. 11.1 Trend of express parcel volume in China and worldwide, 2019–2023. *Data source* State Post Bureau and BNBA (Pitney Bowes), with the 2023 global parcel volume being an estimated value [1]

Table 11.1 Parcel volume completed by selected enterprises in 2023

Company	SHFE	ZTO	YTO	STO	YDA	JTSD
Parcel volume (hundred million pieces)	119.7	302	212.04	175.07	188.54	188.1
YoY growth (%)	7.45	23.83	21.31	35.22	7.07	29

Source Compiled from respective annual reports

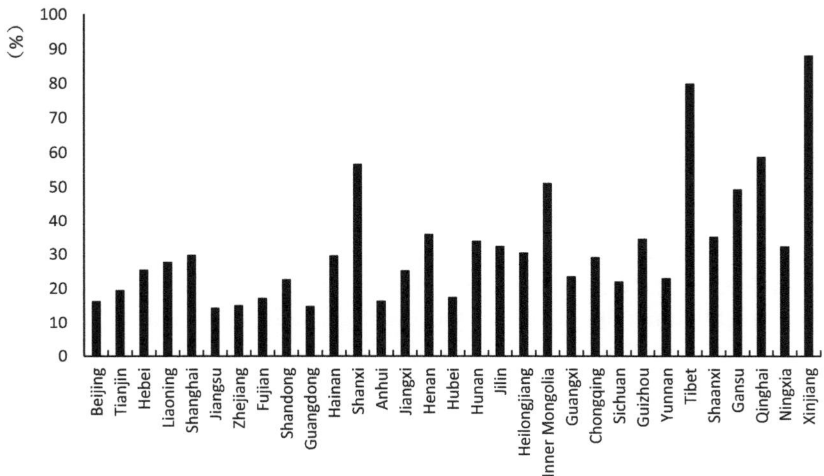

Fig. 11.2 Year-on-year growth of express delivery volume by province in 2023. *Data source* State Post Bureau [2]

regions—benefiting from the expansion of e-commerce into lower-tier markets and improved logistics infrastructure—experienced even higher growth rates (see Fig. 11.2). With the advancement of the "Express Delivery to Villages" initiative, provinces such as Gansu and Guizhou saw year-on-year growth in express parcel volume exceeding 20% in 2023. Guangdong Province led the country with 34.57 billion parcels delivered, followed by Zhejiang and Jiangsu, which handled 26.32 billion and 9.95 billion parcels, respectively. This regionally coordinated development has played a positive role in expanding the overall scale of the industry.

11.1.2 High-Quality Development: Reducing Costs and Increasing Efficiency

Between 2013 and 2023, the average unit price in China's express delivery industry demonstrated a trend of gradual decline amid overall stability. By 2023, the actual

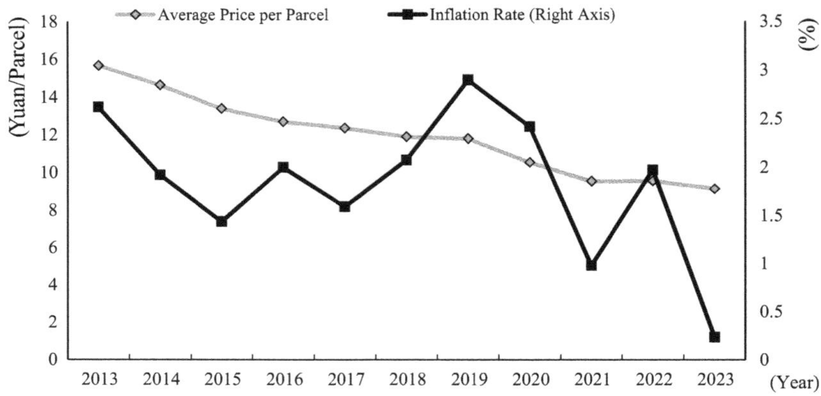

Fig. 11.3 Trend of China's express delivery unit price and inflation rate, 2013–2023. *Data source* Company annual reports and World Bank [3]

price was only 48.98% of the hypothetical level had it increased in line with inflation (see Fig. 11.3). Nonetheless, the decrease in price did not compromise service quality—highlighting the industry's success in reducing costs and improving efficiency through continuous operational optimization. According to the State Post Bureau, the public satisfaction score for express delivery services reached 84.3 points in 2023, an increase of 0.9 points from 2022, reflecting positive progress in the industry's high-quality development. The rapid adoption of technology has been central to boosting efficiency and cutting costs. For instance, the extensive use of automated sorting systems has significantly raised operational efficiency while reducing labor and time requirements. A case in point is YDA: driven by the implementation of new intelligent sorting systems and IT-based management platforms, its core operating cost per parcel in 2023 fell by 11% compared to 2022. Moreover, the large-scale use of new energy vehicles has not only lowered transportation costs but also reduced carbon emissions, aligning with the sustainable development goals of green logistics. The growth in parcel volume has also played a key role in enabling economies of scale. In 2023, the national express delivery volume reached 132.07 billion parcels, a year-on-year increase of 19.4%. This surge helped spread fixed costs across more units and strengthened companies' bargaining power, contributing to a further reduction in cost per parcel.

11.1.3 Intensive Policy Benefits

In October 2014, the State Council issued the *Medium- and Long-Term Plan for the Development of the Logistics Industry (2014–2020)*, which set the goal of establishing a modern logistics service system with rational layout and advanced technology by 2020, providing strategic guidance for the transformation and upgrading of the logistics sector. In October 2015, the State Council released the Several Opinions on Promoting the Development of the Express Delivery Industry, outlining the overall requirements, key tasks, and policy measures to facilitate the growth of the express industry. By August 2017, the General Office of the State Council published the Opinions on Further Reducing Logistics Costs and Increasing Efficiency to Support the Real Economy, proposing measures such as streamlining branch structures of express enterprises, reducing tax burdens, and deepening the integration of logistics with industries to further lower costs and enhance operational efficiency. In April 2018, the State Post Bureau introduced the Work Plan for the Construction of a Credit System in the Express Delivery Industry, aiming to improve credit management regulations, establish comprehensive credit records, gather credit information in full, and strengthen industry credit oversight. In April 2020, the State Post Bureau and the Ministry of Industry and Information Technology jointly issued the Opinions on Promoting In-Depth Integration and Development of the Express Delivery and Manufacturing Industries, emphasizing strengthened supply chain services to drive convergence between logistics and manufacturing. In March 2023, the Ministry of Transport, the National Railway Administration, the Civil Aviation Administration, and other departments jointly released the *Five-Year Action Plan for Accelerating the Building of China's Strength in Transportation (2023–2027)*, highlighting the acceleration of this initiative as a key measure for advancing transportation modernization. This series of intensive policy releases has provided strong support for the development of express delivery enterprises, fueled sustained industry growth, and opened broad prospects for its future expansion.

11.2 The Leader in Premium Express Services: Built on Efficacy and Quality

11.2.1 Introduction

On March 26, 1993, SHFE Holding Co., Ltd. (hereinafter referred to as "SHFE") was officially established in Shunde, Guangdong, and later relocated its headquarters to Shenzhen. After three decades of development, SHFE has grown from a regional courier company into China and Asia's largest comprehensive logistics service provider, as well as the world's fourth-largest logistics enterprise. In 2023, SHFE was ranked 377th on the Fortune Global 500 list, climbing 64 places from the previous year. This marked its second consecutive year on the list, making it the first and only Chinese private express delivery enterprise to enter the Global 500. The company has also been included for two consecutive years in Fortune magazine's "China ESG Influence List" and has received numerous awards, including the 12th China Charity Award for Donating Enterprises, the 2023 "Green Point Award for Sustainable Practices" from Green Point China, the Third Postal Industry Science and Technology Award, and the 2023 "SSL Gold Chain Award - Outstanding

Supply Chain Innovation Award." These accolades reflect SHFE's significant role in promoting the development of the express delivery industry and contributing to economic and social development. "Integrity and Responsibility, Empowering Employees, Serving Customers, Innovation and Inclusiveness, and Pursuit of Excellence" form the core values of SHFE, while "Equality and Respect, Integrity, and Pragmatism" represent its corporate spirit. These principles are deeply embedded in SHFE's culture and development strategy, driving continuous breakthroughs in the express logistics sector.

11.2.2 Development History

SHFE was established in 1993 in Shunde, Guangdong, and pioneered the collection and delivery commission-based operation model between 1996 and 1997. Starting from Shunde, the company rapidly expanded into the Yangtze River Delta, Central China, North China, and Southwest regions. During this process, SHFE adopted a low-price strategy to attract a large number of small and medium-sized customers by offering rates below market prices, quickly capturing market share. To further expand its business, SHFE once utilized a franchise model to achieve rapid growth. However, as the number of service points and employees increased, the drawbacks of franchising became apparent: franchisees operated independently, competed with each other for customers, and even started their own businesses, effectively becoming SHFE's competitors. In response to this situation, Wang Wei began implementing measures to consolidate control in 1999, gradually transitioning from a franchise model to a direct operation model. By 2002, with the establishment of its Shenzhen headquarters, SHFE had successfully completed this large-scale effort to recentralize management [4].

In 2003, amid the SARS outbreak, SHFE seized the opportunity presented by a significant drop in air freight prices by signing a charter agreement with YZJK. By 2009, it became the first express delivery company in China to own its own dedicated cargo aircraft, greatly enhancing delivery speed. In 2005, SHFE independently developed infrared scanners and launched its first-generation handheld scanner. The establishment of SHFE Technology in 2009 marked the company's deeper commitment to logistics technology, a strategic move that significantly accelerated its transformation from a traditional courier company into a technology-driven integrated logistics service provider. In 2013, SHFE introduced external investors for the first time, with three companies—ZSJU, YHKO, and ZXZBE—acquiring a total of no more than 25% of its shares. This transition made SHFE a company with a diversified ownership structure. That same year, SHFE began diversifying its logistics services by establishing a food supply chain division and launching the "SHFE Cold Chain" brand, creating integrated supply chain solutions for the pharmaceutical and fresh food industries. In 2015, SHFE collaborated with STO, ZTO, YDA, and PLSI to introduce the Fengchao Smart Lockers. The year 2017 was exceptional for SHFE: on February 24, the company was listed on the Shenzhen Stock Exchange, formally entering the capital markets and marking a new height in its business development and brand influence. In 2019, SHFE further strengthened its integrated logistics service network by acquiring DHL's supply chain operations in Greater China.

SHFE established its Singapore branch in 2010, marking the beginning of its global expansion. Today, SHFE's international express, freight forwarding, and supply chain services cover 97 countries and regions, while its international parcel delivery network reaches 202 countries and regions. The company has become a significant participant in global logistics and supply chain services. After three decades of dedicated development, SHFE has emerged as one of the benchmark enterprises in China's logistics industry, making substantial contributions to economic and social progress.

11.2.3 Development Landscape

Since commencing operations in 1993, SHFE has established a leading position in China's express delivery industry over the past three decades, recognized for its exceptional logistics services and innovative spirit. From 2016 to 2023, the company achieved remarkable performance: the compound annual growth rate (CAGR) of parcel volume reached 24.51%, while the CAGR of operating revenue grew by 23.86% (see Fig. 11.4).

As shown in Fig. 11.4, SHFE's business maintained steady growth over the past eight years. However, operating revenue declined in 2023 due to decreased demand and falling prices in international air and sea freight, which affected the company's global freight and forwarding services. Specifically, revenue from its supply chain and international segment amounted to RMB 59.98 billion, a year-on-year decrease of 31.7%. Fortunately, strong domestic market demand helped offset some of these losses. Revenue from the express logistics segment reached RMB 191.1 billion, a increase of 9.7% compared to the previous year, partially compensating for the weak performance in international markets.

SHFE possesses strong innovation capabilities, driven by its high-quality R&D team and substantial investment in research and development (see Fig. 11.5). The

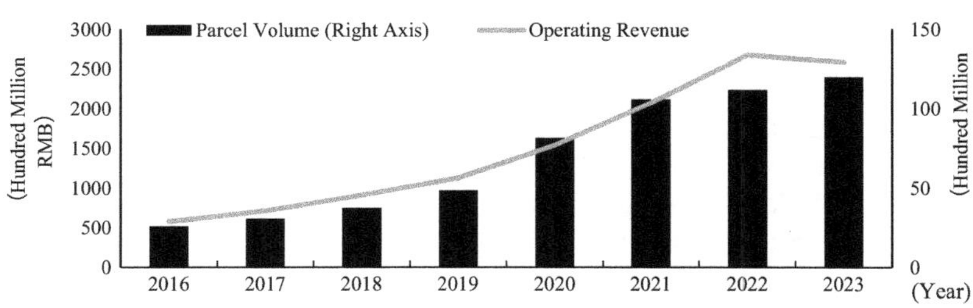

Fig. 11.4 Trend of SHFE's parcel volume and operating revenue, 2016–2023. *Data source* Compiled from company annual reports

company has always emphasized the application of science and technology in production and daily operations, maintaining an industry-leading position in the automation and intelligence of express logistics. In 2023, SHFE deployed more than 630 sets of automated equipment, invested RMB 6.944 billion in the automation upgrade project for express transportation equipment, and allocated RMB 3.003 billion to the development of digital-intelligent supply chain system solutions. These investments have enabled SHFE to significantly enhance operational efficiency, improve customer experience, and achieve cost savings. Additionally, SHFE holds 4093 patents in effect or pending, along with 2490 software copyrights, demonstrating its leading role in technological innovation. Beyond R&D, SHFE continuously improves its infrastructure and operational management. As of the end of 2023, the company operated 86 self-owned freight aircraft, managed over 100,000 line-haul and distribution vehicles worldwide, and maintained 239 express transfer centers, 157 freight transit hubs, and 1900 warehouses. In 2021, SHFE officially released its Carbon Target White Paper, committing to build a sustainable, green, and low-carbon supply chain. Through its self-developed digital platform—the Fenghe Sustainability Platform—the company monitors, integrates, and analyzes carbon data across all operational stages. In 2023, SHFE developed the industry's first waybill-level carbon calculation model, enabling precise carbon footprint measurement and making emissions management more transparent and controllable. SHFE also designed a smart packing algorithm for a motorcycle parts company, simulating packaging processes to optimize material usage and transportation space, thereby enhancing both economic and environmental benefits for the client. The company has deployed photovoltaic power generation systems across multiple industrial parks, generating 18.39 million kWh of solar electricity and reducing greenhouse gas emissions by approximately 10,000 tons.

11.2.4 Outlook and Recommendations

11.2.4.1 Maintain the Brand Image of "Speed"

After three decades of development, "speed" has become a deeply ingrained hallmark of SHFE and one of its most representative and proud advantages. Since the State Post Bureau initiated express delivery service satisfaction surveys in 2009, SHFE has ranked first for 15 consecutive years. Its 72-h on-time delivery rate (exceeding 90%) has led the industry for nine years in a row, earning the trust of a broad consumer base. The company consistently upholds the philosophy that "talent is the primary productive force," placing strong emphasis on employee growth and development. It has established a comprehensive training system covering the entire employee lifecycle and continuously adapts its training programs and management mechanisms to build a high-quality workforce. SHFE recognizes that serving employees well is the foundation for serving customers well. Accordingly, in 2023, the company introduced an innovative incentive mechanism called "Fengdou" (literally "Wind

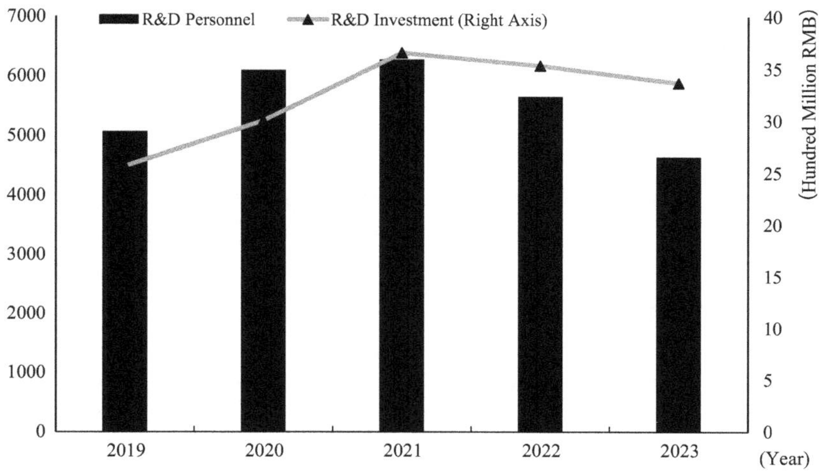

Fig. 11.5 SHFE's R&D personnel and investment, 2019–2023. *Data source* Compiled from company annual reports

Table 11.2 Trend in the number of monthly billed customers of SHFE, 2019–2023

Year	2019	2020	2021	2022	2023
Monthly billed customers (ten thousands)	140	150	164	181	195

Source Compiled from company annual reports

Bean") to encourage self-management among frontline couriers and continuously improve service quality. This strong brand image has allowed SHFE to gain extensive customer trust, with the number of its monthly billed corporate customers growing steadily (see Table 11.2). By deeply understanding the diverse needs of industry clients, SHFE continues to drive healthy and sustainable business growth.

11.2.4.2 Focus on Overseas Market Development

From the perspective of SHFE's revenue distribution, mainland China remains its primary market. However, internet penetration in the country has already reached a considerably high level. According to the Statistical Report on China's Internet Development released by the China Internet Network Information Center (CNNIC), as of December 2023, the number of internet users in China had reached 1.092 billion, with an internet penetration rate of 77.5%. The number of online shopping users also grew to 915 million, solidifying China's position as the world's largest online retail market. Yet, with slowing and even negative population growth, intensifying competition in the express delivery industry, and the domestic market approaching saturation, future growth potential is limited. To address this challenge, SHFE should turn its attention to developing and less-developed regions. These areas are expected to become new growth drivers for the company, supporting its further expansion into the global market.

11.3 E-Commerce-Driven Intelligent Logistics System: Efficient Delivery and Omni-Channel Coverage

11.3.1 Introduction

JIDOW Co., Ltd. (hereinafter referred to as "JIDOW") was established in April 2017, specializing in providing comprehensive supply chain solutions and logistics services for multiple industries. Its operations cover six major sectors: fast-moving consumer goods (FMCG), apparel, home appliances and furniture, 3C products, automotive, and fresh produce. The company operates across China, Europe, Oceania, and numerous other countries and regions. It was listed on the Hong Kong Stock Exchange in May 2021. Guided by the philosophy of "staying customer-centric and creating sustained value through reliable supply chain services," JIDOW is committed to driving efficient global supply chain operations and sustainable development through technology. Its vision is to become the world's most trusted provider of supply chain infrastructure services. Continuously expanding its business footprint, JIDOW has grown into one of the leading comprehensive technology enterprises in China and globally. The company has established six synergistic networks characterized by digitalization, extensive coverage, and flexibility. These include: a digital warehousing network, an integrated transportation network, a last-mile delivery network, a heavy-item logistics network, a cold chain logistics network, and a cross-border logistics network. With nearly 100 bonded warehouses, direct-mail warehouses, and overseas warehouses worldwide, JIDOW enhances its service capabilities through the deployment of advanced automation equipment, upgraded inventory management systems, and optimized operational processes. In 2023, JIDOW reported strong financial performance with net revenue reaching RMB 166.6 billion, a year-on-year increase of 21.27% [5].

11.3.2 Development History

JIDOW was officially established in April 2017, fully opening its services to external clients [6]. That same year, it built the world's first end-to-end unmanned warehouse for B2C operations. In February 2018, the JIND Group raised USD 2.5 billion to develop JIDOW, with investors acquiring a 19% stake in the company, while the JIND Group remained the actual controlling shareholder. In August 2019, JIDOW partnered with seven institutions to co-found the E-commerce Logistics

Packaging Standards Alliance in China and launched a supply chain industrial platform, marking a significant step toward promoting industry standardization. In August 2020, JIDOW acquired KSD, known for its "time-sensitive express services," for RMB 3 billion, strengthening its competitiveness in the high-end express delivery market. The following year, in May 2021, JIDOW was listed on the Hong Kong Stock Exchange, raising approximately RMB 22.9 billion and injecting robust cash flow into the company's development. By July 2022, JIDOW had completed the acquisition of over 50% equity in DBA Holdings. Its warehouse network covered almost all counties and districts across China, with a self-operated fleet of more than 40,000 trucks and around 400 sorting centers operating in mainland China. In August of the same year, JIDOW Airlines officially commenced operations, with three dedicated cargo aircraft entering regular service. This strategic move further solidified JIDOW's competitive advantage in the logistics sector. In April 2023, JIDOW underwent an organizational restructuring, establishing four independent business divisions: Supply Chain, Express Delivery, Freight Transportation, and International Business. This strategic enhancement improved the company's professional operational capabilities and market responsiveness, cementing its position as a leading technology-driven supply chain solutions and logistics service provider in China.

11.3.3 Development Landscape

Over the five-year period from 2019 to 2023, JIDOW achieved leapfrog development in both revenue and business layout, driven by innovation and exceptional execution. The compound annual growth rate (CAGR) of its net revenue reached 35.21% (see Fig. 11.6). Revenue from its core business—integrated supply chain logistics solutions and services—grew steadily, increasing from RMB 71.1 billion in 2021 to RMB 81.5 billion in 2023, demonstrating the company's strong expertise and leading position in the supply chain logistics sector. JIDOW has continued to lead the application of smart logistics technologies, leveraging big data, IoT, and artificial intelligence to optimize operational processes. The company's 41 Asia No.1 Smart Logistics Parks now cover 30 cities across China. With the support of automated equipment and intelligent sorting systems, order processing capacity has been significantly enhanced. The large-scale deployment of Level 4 autonomous delivery vehicles has further reduced operating costs and improved delivery efficiency.

collaboration, the company has closely integrated its national network with regional logistics services to provide customers with highly efficient logistics solutions. In the field of supply chain services, JIDOW has developed diverse and scenario-specific solutions. By serving brands such as WEWO and LXQC, the company's integrated supply chain solutions for automotive after-sales parts have not only shortened delivery cycles but also optimized inventory turnover rates, earning high praise from clients. The strategic acquisition of DBA has been one of the company's key initiatives in recent years. This move not only enhanced its comprehensive logistics service network but also laid the foundation for expanding

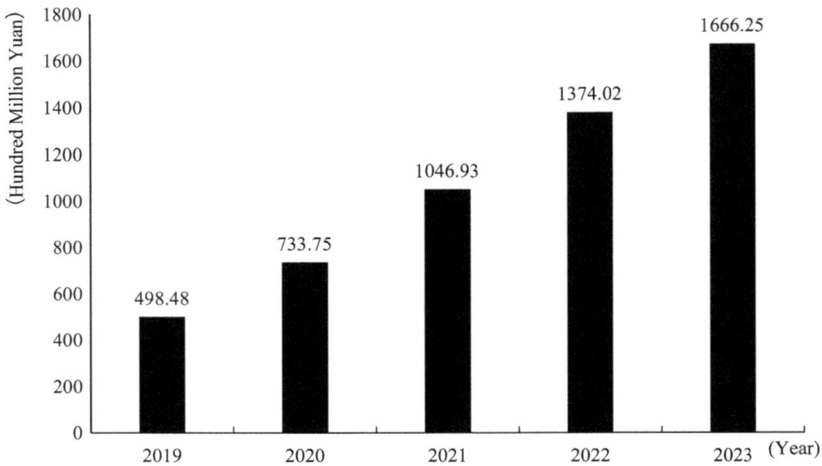

Fig. 11.6 Trend in JIDOW's net revenue, 2019–2023. *Data source* Compiled from company annual reports

its corporate client market. In 2023, leveraging DBA's line-haul transportation capabilities, JIDOW accelerated the growth of its heavy-item logistics business, strengthening its competitiveness in the express and freight transportation markets. While pursuing business growth, the company also emphasizes green logistics initiatives. The promotion of projects such as photovoltaic power generation and eco-friendly packaging across multiple operational centers has set a benchmark for sustainable development in the logistics industry.

11.3.4 Outlook and Recommendations

Despite its promising prospects, JIDOW inevitably faces certain risks: (1) Macroeconomic volatility risk: Domestic macroeconomic growth expectations have moderated, while inflation in developed economies and global geopolitical uncertainties may pose challenges to future performance growth. (2) Policy risk: China has committed to achieving carbon peak by 2030 and carbon neutrality by 2060, which may lead to increased environmental compliance costs and higher operating expenses. In pursuit of its vision to become "the world's most trusted supply chain infrastructure service provider," the company continues to optimize its revenue structure. The proportion of revenue from integrated supply chain solutions has declined for three consecutive years, accounting for 67.91%, 56.33%, and 48.91% respectively. This trend reflects JIDOW's active strategic realignment to better adapt to market changes and customer needs.

11.4 Extensive Network Coverage and Cost-Effectiveness: A Rapidly Growing Mid-Market Logistics Brand

11.4.1 Introduction

Since its establishment on May 8, 2002, ZTO Express (Cayman) Inc. (hereinafter referred to as "ZTO") has grown from a core express delivery business into a comprehensive logistics service provider. Its operations have expanded beyond express delivery to include cross-border logistics, freight transportation, cloud warehousing, cold chain solutions, financial services, and commercial services (Tuxi Life). Guided by its mission to "use our products to create well-being for more people," ZTO is committed to becoming "a respected century-old enterprise." The company was listed on the New York Stock Exchange on October 27, 2016, and on the Main Board of the Hong Kong Stock Exchange on September 29, 2020, making it the first Chinese express delivery enterprise to be publicly traded in both the United States and Hong Kong. ZTO handles the largest parcel volume globally, delivering 30.2 billion parcels in 2023 and capturing a 22.9% market share in China. It has maintained the top position in the industry for eight consecutive years. The company has received several awards, including the "2018–2019 National Civilized Unit in Transportation (Re-evaluation)" from the Ministry of Transport, the "Institutional Donation Award in the First 'Hangzhou Charity Awards'" from the Hangzhou Municipal People's Government, and the "2023 Science and Technology Progress Award" from the China Federation of Logistics & Purchasing. Over the ten years from 2014 to 2023, ZTO achieved steady growth across four key financial metrics: Operating revenue increased from RMB 3.904 billion to RMB 38.419 billion, with a CAGR of 28.93%; Operating costs rose from RMB 2.771 billion to RMB 26.756 billion, with a CAGR of 28.66%; Total assets grew from RMB 4.974 billion to RMB 88.465 billion, with a CAGR of 37.69%; Total liabilities expanded from RMB 1.578 billion to RMB 28.185 billion, with a CAGR of 37.75% (see Fig. 11.7).

11.4.2 Development History

Since its establishment in 2002, ZTO has actively responded to the national logistics development strategy by continuously optimizing its business layout, service structure, and market positioning. Through establishing new branches, strategic mergers and acquisitions, and collaborations with other logistics enterprises, the company has consistently driven its own business transformation and moved toward high-end upgrading of its industrial chain. After more than two decades of development, ZTO has grown into a large comprehensive logistics group with a strong domestic network and international service capabilities [8].

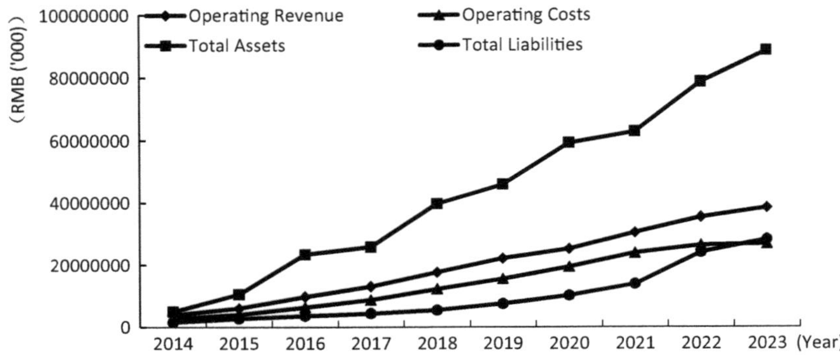

Fig. 11.7 Trend of selected financial indicators of ZTO express, 2014–2023. *Data source* Official website of the U.S. Securities and Exchange Commission (SEC) [7]

11.4.3 Development Landscape

Guided by the development philosophy of "Steady Growth, Strengthening Competitiveness, Long-Term Vision, and Embracing Change," ZTO continues to consolidate and enhance its leading position in the industry while accelerating its transformation into an internationally competitive integrated logistics service provider. Currently, ZTO's service network covers more than 220 countries and regions worldwide, delivering efficient and reliable logistics services to global customers. Figure 11.8 illustrates the structure and evolution of ZTO's business revenue over the past decade. As shown, express delivery revenue has consistently been the core pillar of the company's operating income, accounting for over 90% of the total with a compound annual growth rate (CAGR) of 28.26%. However, as the industry has entered a phase of stock competition, growth in the express segment moderated in 2023. Despite this, annual parcel volume still reached 30.2 billion, a year-on-year increase of 23.77%, demonstrating the company's strong market penetration. Revenue from parts sales achieved notable growth of 43.1% in the fourth quarter of 2023, providing robust momentum for diversifying ZTO's revenue structure. ZTO's strong competitiveness is underpinned by its extensive infrastructure network: More than 31,000 service outlets, 99 sorting centers, 464 automated sorting systems, over 10,000 self-operated line-haul vehicles. This large-scale, stable, and reliable infrastructure provides a solid foundation for logistics services and helps the company maintain its industry leadership amid intense market competition.

ZTO is committed to building a green logistics system, leveraging low-carbon and intelligent approaches to promote energy-saving technologies and environmental measures across transportation, packaging, and sorting processes. The company has continuously increased the proportion of new energy vehicles in its fleet, reducing reliance on traditional fuel-powered trucks and decreasing the carbon footprint of its transportation operations. Efficient line-haul planning, improved economies of scale, and lower fuel prices contributed to a 12.1% reduction in unit transportation

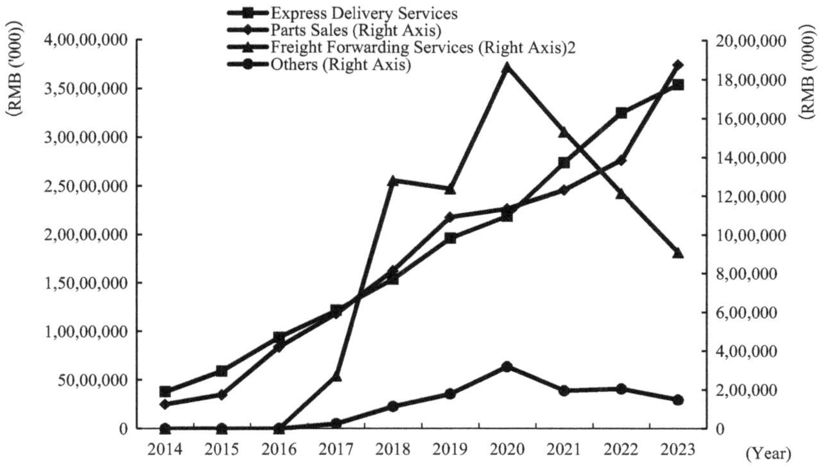

Fig. 11.8 Trend of revenue components of ZTO express, 2014–2023. *Data source* Official website of the U.S. Securities and Exchange Commission (SEC)

costs compared to 2022. ZTO's self-developed Zhongtian System and intelligent routing algorithms dynamically allocate and optimize transportation resources, minimizing inefficient energy consumption during transit. In 2023, the company generated approximately 40,000 MWh of photovoltaic power, a year-on-year increase of 34%, providing stable green energy support for its operations. In terms of packaging, 93.34% of e-commerce parcels no longer undergo secondary packaging, while 91.83% of packaging materials used meet environmental standards. More than 1.465 billion corrugated boxes were recycled and reused, reducing the consumption of single-use woven bags. The company aims to achieve a 20% reduction in greenhouse gas emission intensity per parcel by 2028, driving green development through technology and pioneering new pathways for the industry's ecological transition. Through the "Express Delivery to Villages" initiative, ZTO has improved rural logistics infrastructure and increased service coverage in both urban and rural areas. For many years, ZTO has organized the "Dream Fulfillment 1 + 1" campaign, donating supplies worth millions of RMB to schools in remote mountainous areas of Yunnan, Xinjiang, Shanxi, and other regions. In July 2021, ZTO donated RMB 20 million to the Zhengzhou Red Cross Society to support flood relief efforts in Henan. These active contributions to education in underserved areas and disaster recovery demonstrate the company's strong sense of social responsibility.

11.4.4 Outlook and Recommendations

11.4.4.1 Effective Risk Management and Control

Although the company ranks first globally in parcel volume, it must remain vigilant against risks arising from market competition. The e-commerce market, a major driver of express business growth, is transitioning from explosive expansion to moderate growth, and may even face growth stagnation in the future. Given a largely fixed market demand, competitive pressures in the existing market are likely to intensify. Therefore, ZTO should continue to accelerate its "Going Global" strategy, striving to secure a advantageous competitive position in emerging markets. Furthermore, although intelligent equipment is being widely adopted in the express delivery industry, the sector remains inherently labor-intensive. With the diminishing demographic dividend in China, the risk of rising labor costs is also increasing.

11.4.4.2 Enhance Technological Innovation

Innovation is the core driver of development. ZTO actively promotes an innovation-driven development strategy by establishing a new collaborative market expansion model between its headquarters and branches, effectively enhancing the company's market competitiveness and brand influence. It continues to explore and implement diversified business models, including import and warehousing services, fast-moving consumer goods (FMCG) and retail logistics solutions, to adapt to diverse market demands and sustain business growth. ZTO holds over 200 computer software copyrights and more than 200 patents, demonstrating significant progress and substantial potential in technological innovation. In an era of accelerating logistics technology application, ZTO should further strengthen technology research and development, platform construction, results transformation, and the cultivation of relevant talent.

11.5 Pioneer in National Express Network Expansion: Committed to Innovation and Service Optimization

11.5.1 Introduction

YTO Express Co., Ltd. (hereinafter referred to as "YTO") was established in Shanghai on May 28, 2000. As a comprehensive logistics service provider, it integrates time-sensitive products, digital products, international supply chain solutions, and numerous other products and services. Its business operations cover more than 150 countries and regions worldwide. On October 20, 2016, YTO was listed on the A-share market, becoming the first privately-owned express delivery company to go public in the industry.

11.5.1.1 Corporate Advantages

1. Advantages in Operational Model: After years of practice and optimization, the company's operational model has been increasingly refined. The self-operated model for transfer centers ensures high standards and efficiency in logistics operations, while the franchise model for local outlets fully leverages local market resources and deep customer insights, thereby enhancing service coverage and responsiveness. This superior operational model strengthens YTO's advantages in service quality, customer experience, and overall control over the express delivery network.
2. Extensive Network Coverage Advantage: With over two decades of operations, the company has established a powerful network coverage advantage, which stands as one of its core competitive strengths, delivering an unparalleled service experience to customers. YTO's international service network spans six continents and over 150 countries and regions. The company not only operates its own aviation fleet but has also actively expanded its international route coverage. By combining a global network of overseas hubs and localized professional teams, YTO has strengthened its presence in existing markets while accelerating expansion into new ones. This global network layout has not only enhanced the company's international competitiveness but has also provided customers with more convenient and efficient cross-border logistics solutions.
3. Leadership in Digital Transformation: In the current era of smart logistics driven by cutting-edge technologies such as big data and artificial intelligence, YTO demonstrates distinct advantages in digital transformation. The company consistently prioritizes digitalization as a core strategy for its development and continuously increases investment in research and development to maintain its leading position in the field of intelligent logistics.
4. Advantage of Self-Owned Air Cargo Capacity: In October 2015, YTO Aviation, a wholly-owned subsidiary of YTO, officially commenced operations, marking a significant step in enhancing the company's logistics service capabilities and expanding its international business. The ownership of a dedicated aviation fleet has substantially strengthened YTO's ability to meet diverse market demands, such as fulfilling the delivery requirements for high-value-added products like chilled foods and fresh flowers. Furthermore, the opening of multiple international routes has paved the way for YTO to access global markets.
5. Brand Advantage: Over the past two decades, YTO has successfully earned the satisfaction and trust of numerous customers through its deep commitment to core express delivery services. By continuously innovating and optimizing service channels, the company provides comprehensive and meticulous service experiences through diversified methods and approaches. YTO consistently regards service quality as the core of its corporate strategy and continues to increase investment in brand building, effectively enhancing the brand's market visibility and influence. In 2023, YTO was awarded the honorary title of "AAAAA-Level Logistics Enterprise" and was also listed on multiple prestigious rankings,

including the "2023 Fortune China 500," "2023 China's Top 500 Enterprises," and "2023 Top 100 Chinese Private Enterprises in the Service Industry."

11.5.1.2 Business Layout

YTO's core business encompasses integrated express logistics services and value-added services. By consolidating logistics resources and employing efficient delivery methods, the company captures a wide range of logistics demands. It carefully designs and manages service processes to offer diversified express logistics solutions to the market. YTO also focuses on supply chain management services, leveraging its robust logistics platform to engage in e-commerce logistics cooperation and cross-border e-commerce services. The company's main business segments include: express delivery services, warehousing, distribution and supply chain, solution services, value-added services and benefits, digital products.

11.5.2 Development History

11.5.2.1 Domestic Development History

On May 28, 2000, YTO was officially established in Shanghai. In 2003, YTO became the first company in the industry to propose "year-round operation without breaks," maintaining services on weekends and achieving a turnaround from loss to profitability. In 2005, the company partnered with ALBBA, becoming the first private express delivery enterprise to collaborate with ALBBA. This marked a significant step for YTO in the field of e-commerce logistics. The cooperation not only expanded YTO's business scope but also brought a substantial volume of e-commerce parcels, laying a solid foundation for its rapid development. In 2009, YTO invested heavily in developing its proprietary core business system—the Jingang System—providing crucial support for the company's digital transformation. By the end of 2010, Shanghai YTO Jiaolong Investment Development (Group) Co., Ltd. was established, marking a more substantial move toward corporatization and diversified growth. In August 2014, YTO Aviation, a wholly-owned subsidiary of YTO, received approval for establishment from the Civil Aviation Administration of China. In October of the following year, YTO Aviation officially launched operations, making YTO one of only two private express delivery companies in China at the time to own and operate its own airline [9].

With rising national income and increasing internet penetration, online shopping became a prevailing trend. The company's management accurately assessed the situation and decided to embrace collaboration with ALBBA once again. In May 2015, ALBBA, in partnership with Yunfeng Fund, made a strategic investment in YTO, aiming to drive service upgrades in China's express delivery industry and facilitate development and transformation in areas such as management, service quality, product innovation, international expansion, and rural logistics. In 2016, YTO was listed on the A-share market, becoming the first privately-owned express delivery company in the industry to go public. On May 12 the following year, YTO took the lead in establishing the first national engineering laboratory in the logistics field—the National Engineering Laboratory for Logistics Information Interconnection, Sharing Technology and Applications. In March 2021, YTO signed an agreement to become the official logistics service sponsor of the 19th Asian Games Hangzhou 2022.

11.5.2.2 Overseas Development History

In 2008, Shanghai YTO International Freight Forwarding Co., Ltd. was established, marking the beginning of YTO's expansion into the international business arena. As the company's capabilities grew and international market demand increased, YTO gradually expanded its global operations. By 2016, the company's international business revenue reached RMB 59.94 million. During this period, however, YTO's progress in the international sector was slow due to the presence of established global postal giants, posing significant challenges for the newcomer in the field.

With the advancement of the Belt and Road Initiative, YTO has adhered to the national strategy as its guide, deepening its presence in countries and regions along the Belt and Road to fully tap into their market demand. Through strategic partnerships and equity investments, the company has accelerated its overseas expansion. In 2017, YTO began leveraging key resources such as the China–Europe Railway Express to support Belt and Road construction. It acquired XIADA to further enhance its global network coverage and established a regional distribution center in Malaysia, significantly boosting its cargo handling capacity in Southeast Asia. This move particularly facilitated faster logistics services for cross-border e-commerce between China and Malaysia. In 2018, the company launched international express parcel services in Vietnam, offering comprehensive express and supply chain solutions for clients in China, Vietnam, and beyond. In March 2021, YTO International inaugurated a dedicated cargo flight route from Hangzhou to Kuala Lumpur, providing more than 250 tons of weekly air freight capacity for round-trip operations between the two cities.

11.5.2.3 Development Landscape

Established in 2000 with the mission of "Serving Society, Strengthening the Enterprise, and Contributing to the Nation," YTO has grown into one of China's leading express delivery companies. In 2023, it handled 21.204 billion express parcels, capturing a market share of 16.06%. Over the eight-year period from 2016 to 2023, YTO's operating revenue increased from RMB 16.818 billion to RMB 57.684 billion, achieving a compound annual growth rate (CAGR) of 19.25%. Total assets grew from RMB 11.168 billion to RMB 43.367 billion, with a CAGR of 21.39%, while total liabilities rose from RMB 2.963 billion to RMB 14.02 billion, reflecting a CAGR of 24.86% (see Fig. 11.9).

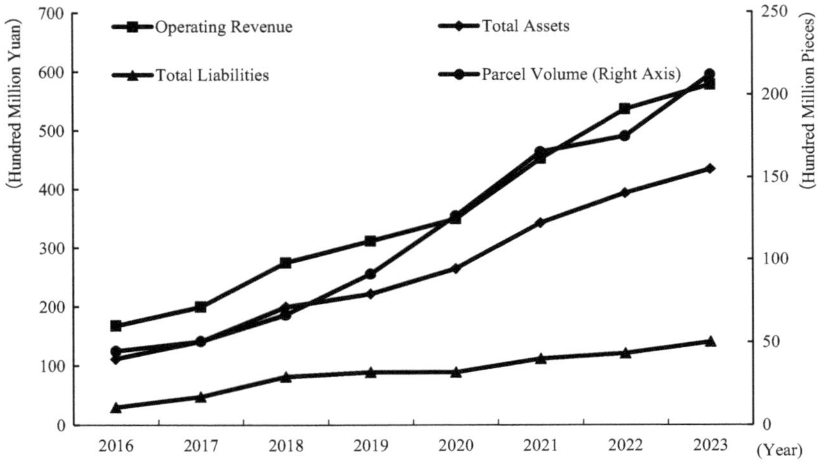

Fig. 11.9 Trend of selected financial indicators of YTO, 2016–2023. *Data source* Compiled from company annual reports

The company not only strives to provide customers with excellent express and value-added services but also continuously improves and strengthens the construction and operation of its infrastructure. Over the six years from 2018 to 2023, YTO consistently invested in its infrastructure, maintaining a high-capacity aviation fleet, self-operated hub transfer centers, and an extensive road transportation network (see Fig. 11.10).

By enhancing vehicle loading capacity, upgrading truck models, promoting direct delivery to stations, and improving smart tools, the company has continuously strengthened its overall network competitiveness. These efforts have allowed YTO to not only maintain but increase both its number of franchisees and its market share over the past five years, despite intensified competition in the domestic express delivery market (see Fig. 11.11). This demonstrates the company's strong market resilience and has delivered exceptional results to investors.

YTO is committed to end-to-end process management to improve its timeliness performance. Through continuous upgrades of intelligent equipment, gradual implementation of coordinated loading and unloading, and strengthened coordination between transfer centers and end-point outlets, YTO has achieved goals such as improving end-point collection and delivery efficiency and reducing delivery time. Compared to the same period last year, the company's overall delivery time in 2023 was reduced by 5.63 h, enhancing parcel timeliness. By optimizing customer service middleware, expanding intelligent customer service functions, increasing robot application scenarios, and promoting the digitalization and standardization of claims processing, YTO significantly improved the customer experience, with the customer complaint rate decreasing by 14 percentage points year-on-year. The company continuously optimizes its transportation capacity structure to ensure rational and efficient resource allocation. In 2023, the per-package transportation cost was RMB 0.46, a

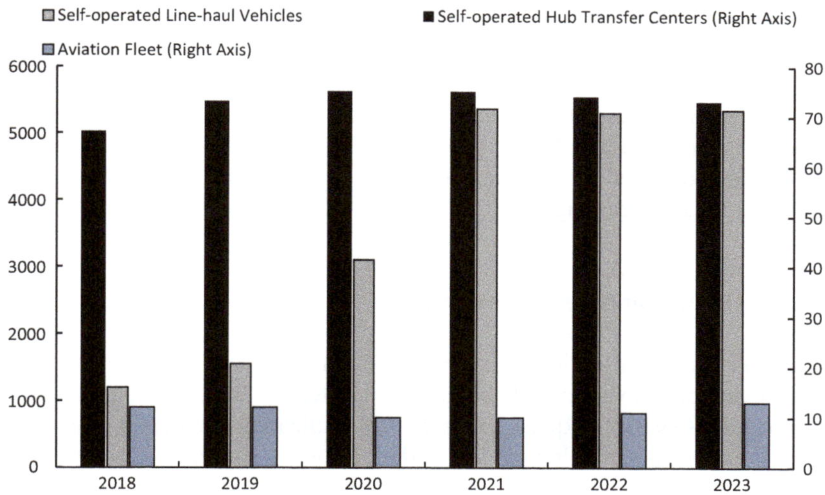

Fig. 11.10 YTO's infrastructure operations, 2018–2023. *Data source* Compiled from company annual reports

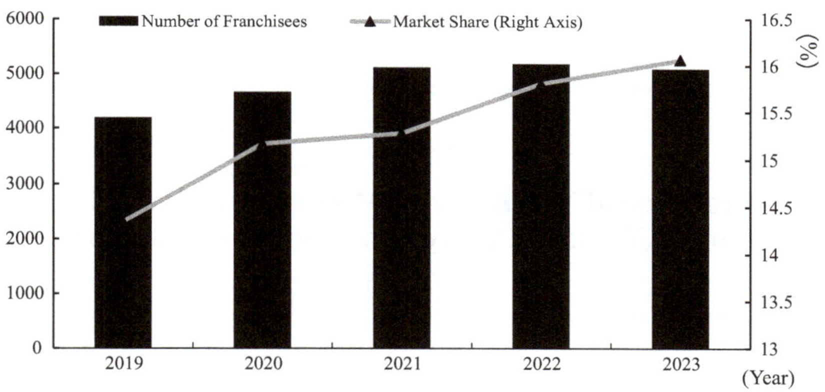

Fig. 11.11 Trend of market share and number of franchisees of YTO, 2019–2023. *Data source* Compiled from company annual reports

year-on-year decrease of 9.67%. YTO also pursues lean production in its transfer centers, where the per-package handling cost reached RMB 0.29, down RMB 0.02 from the previous year. YTO consistently emphasizes high-quality talent cultivation and development, implementing a "Talent Empowers Enterprise" strategy. It has partnered with numerous institutions including Beijing University of Posts and Telecommunications, Nanjing University of Posts and Telecommunications, Zibo Vocational Institute, Shandong Labor Vocational and Technical College, Anhui National Defense Science and Technology Vocational College, Huzhou Vocational and Technical College, Tianhe College of Guangdong Polytechnic Normal University, Heilongjiang Agricultural Engineering Vocational College, Shandong Transport Vocational College, Anshun Vocational and Technical College, Huaxia College of Wuhan University of Technology, and Shandong Light Industry Vocational College to develop talent programs. On May 28, 2023, YTO held the opening ceremony of the "Zhiyuan Academy" at its Shanghai headquarters, aiming to build a knowledge repository and talent training base for the express logistics industry.

11.5.2.4 Outlook and Recommendations

Although key financial indicators such as operating revenue and total assets have increased, the company's net profit has declined instead of rising. According to financial reports, YTO's R&D expenses rose in 2023, indicating that management places high importance on technological innovation. Looking ahead, YTO's investments in R&D areas such as automated sorting, intelligent logistics systems, and green packaging technologies are expected to yield practical results. These advancements will enhance operational efficiency, reduce costs, and ultimately form a competitive advantage. The development of green packaging technology will not only lower the company's operating costs but also position YTO at the forefront of the industry in achieving carbon neutrality goals. As consumers and regulators increasingly prioritize environmental protection, eco-friendly initiatives will become a significant advantage in attracting customers and securing government support.

11.6 Integration of Tradition and Modernity: A Capable Express Company Deeply Rooted in the Local Market

11.6.1 Introduction

STO Express Co., Ltd. (hereinafter referred to as "STO") was established in 1993 and is a comprehensive logistics service provider focused on delivering fast and reliable express services. Under the leadership of its founder, Nie Tengfei, STO has always adhered to a customer-first philosophy, continuously optimized its logistics network, improved service quality, and actively driven progress in product innovation, service upgrades, and market expansion. The company fulfills its mission to "enable customers to enjoy a new lifestyle through express delivery, help employees achieve personal value, and allow the enterprise to shoulder social responsibilities." It rejects narrow zero-sum competition and insists on a path of collaborative win–win partnerships and innovative development. Throughout its growth, STO has remained focused on its core business, continually consolidating and expanding its leading position in China's express delivery industry. At the same time, it has adjusted and optimized its service structure, combining its own strengths with market demand to explore and cultivate new business growth areas. After three decades of dedication, STO has grown into a leading enterprise in China's express logistics sector. It was one of the first privately founded express delivery companies in China, and is now recognized as a 5A-level logistics enterprise, ranked among the Top 500 Chinese Private Enterprises and the Fortune China 500, and has been successfully listed on the A-share market.

11.6.2 Development History

STO was established in 1993 under the original name "Shentong Comprehensive Service Department" and was one of the earliest private enterprises to enter the express delivery business in China. With the rapid development of China's market economy and the rise of e-commerce, STO has grown significantly and has become one of the leading express logistics companies in the country. While consolidating its market position, STO has kept pace with e-commerce trends and leveraged its operational strengths to continuously explore new business areas and growth opportunities, steadily advancing its diversification and international expansion [10].

11.6.3 Development Landscape

Since 2016, STO Express has experienced rapid growth, with its parcel volume increasing from 3.26 billion pieces in 2016 to 17.507 billion pieces in 2023, achieving a compound annual growth rate (CAGR) of 27.14%, surpassing the industry average of 22.85% (see Fig. 11.12).

STO's core businesses include express delivery services, value-added services (such as cash-on-delivery and pre-sale market penetration services), and express packaging supplies (primarily envelopes, document bags, cardboard boxes, and other packaging materials). Furthermore, in recent years, the company has introduced new services such as TEIJA M origin warehouses, grid warehouses, and consolidation warehouses, expanding its service range and enhancing overall competitiveness. STO employs an operational model characterized by "company-operated transfer centers and franchised last-mile outlets." Since transfer centers serve as both geographical hubs and critical nodes in operational processes, maintaining direct control over them helps mitigate risks, while the franchise model for local outlets enables rapid network expansion. A steadily growing self-operated trunk transportation fleet, an increasing number of service outlets, and a relatively stable count of directly managed transfer centers (see Fig. 11.13) constitute the core strengths underpinning STO's efficient logistics services. In recent years, STO has increased investments in operational facilities such as transfer centers. In 2022, the company comprehensively launched a three-year "Ten-Billion Capacity Enhancement Project" aimed at boosting throughput capacity. In 2023, STO carried out 37 capacity expansion projects and 18 renovation

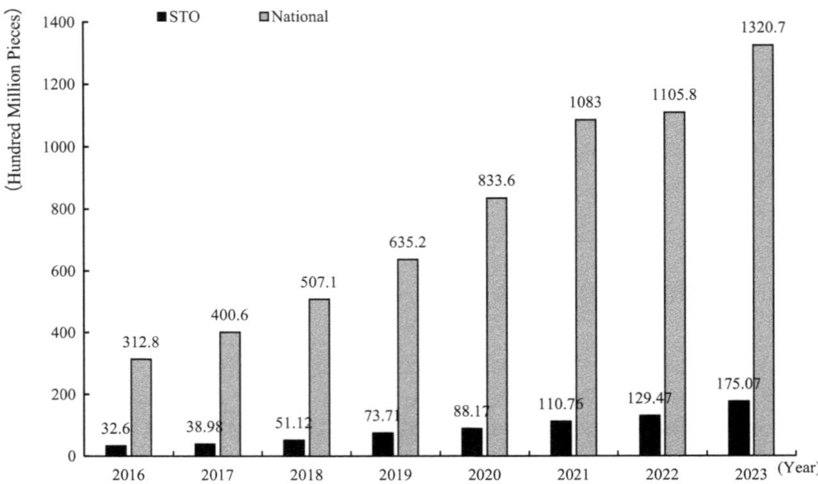

Fig. 11.12 Trend of parcel volume of STO and the national total, 2016–2023. *Data source* Compiled from the State Post Bureau and company annual reports

and optimization initiatives. The company also actively participated in the "5312 Initiative" led by the State Post Bureau and the Ministry of Industry and Information Technology, introducing the first batch of 10 pilot zones and 22 exemplary projects to promote integrated development between the express delivery and manufacturing sectors. The completion and operational launch of these projects have elevated the company's daily processing capacity to over 60 million orders.

Benefiting from the company's continuous investment and optimization in service quality, operational efficiency, and market expansion, STO Express achieved a market share of 13.26% in 2023, an increase of 1.55 percentage points compared to the previous year. The company reported operating revenue of RMB 40.924 billion, a year-on-year growth of 21.54%, and total assets reached RMB 22.75 billion, demonstrating strong performance across key financial indicators (see Fig. 11.14).

STO was the first privately-owned express delivery company in China to establish an information technology team and has consistently pursued technological empowerment to drive corporate development. In 2023, the company launched its self-developed instant logistics platform, Shendongdong, which now covers 113 cities, significantly enhancing delivery efficiency and user experience. By integrating big data and artificial intelligence technologies, the platform enables intelligent order dispatch and route optimization, further strengthening STO's competitiveness in the instant delivery sector. STO continues to optimize existing equipment and deploy new intelligent systems. By the end of 2023, the company had implemented 240 cross-belt automated sorting systems, raising the level of automation in its logistics operations and providing strong support for the industry's digital and intelligent transformation. The application of such smart equipment has not only improved operational efficiency but also reduced labor costs and error rates, allowing STO

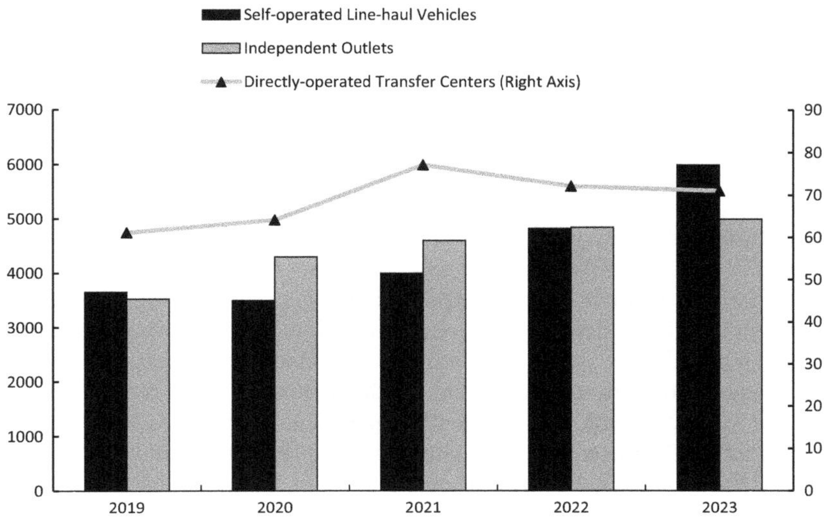

Fig. 11.13 STO's infrastructure operations, 2019–2023. *Data source* Compiled from company annual reports

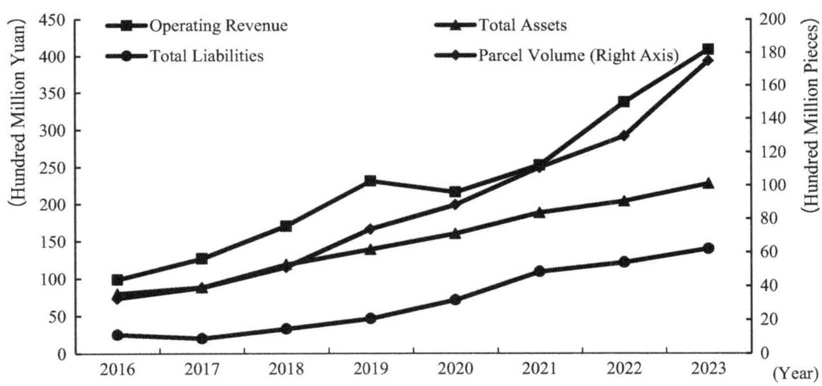

Fig. 11.14 Trend of selected performance indicators of STO, 2016–2023. *Data source* Compiled from company annual reports

to maintain a technological leading position in the logistics industry. To further enhance customer experience, STO intensified employee training and restructured management models in 2023. The Tengyun Training Program, launched early in the year, aimed to improve the capabilities of front-line managers. Through systematic training, it helped employees adapt to new technologies and equipment while strengthening their management skills and service awareness, leading to an overall elevation in team competency. In 2022, STO introduced an employee stock ownership plan, closely aligning individual interests with the company's overall performance.

This initiative has further stimulated employee motivation and innovation, not only fostering a greater sense of belonging among staff but also providing a stable talent foundation for the company's long-term development.

STO's core competitiveness is primarily reflected in five key areas: Efficient Transfer Operation System and Granular End-Point Network Coverage Achieved through increased investment in transfer center assets, enhanced hub throughput capacity, and implementation of flattened network management reforms. Standardized Express Operation System and High-Quality Service Experience Enabled by standardized operational processes, a unified customer service system, and intelligent engineering initiatives. Strong Digital and Intelligent Capabilities with a Refined Management System Supported by a top-down digital ecosystem, product development, and intelligent operation platforms that enable real-time, accurate, and complete parcel status monitoring, achieving automation, intelligence, and informatization in logistics. Stable and Efficient Management Team and Cohesive Organizational Culture Characterized by a culture of "simplicity, trust, and commitment," reinforced through internal communication campaigns such as "Shen Story," "Shen Outlet," and "Shen Benchmark" to promote corporate culture. Long-Term Development Philosophy and Recognized Brand Image Guided by the principle of long-termism, STO ensures customer satisfaction through premium services, empowers employees to realize personal value, and fulfills corporate social responsibilities.

11.6.4 Outlook and Recommendations

11.6.4.1 Adhere to Digital and Intelligent Transformation

STO was the first privately-owned express delivery enterprise to establish an information technology team, and its forward-looking strategic layout laid a solid technical foundation for subsequent development. It was this early technological investment that enabled STO to respond rapidly to market changes in the express logistics industry, improving efficiency and optimizing customer experience through automated and intelligent logistics systems. As markets and technologies continued to evolve, STO consistently increased its R&D efforts in information technology, introducing innovative logistics solutions including intelligent sorting systems, big data analytics, and cloud computing platforms. These initiatives significantly enhanced the company's competitiveness in the industry. In addition to boosting internal operational efficiency, STO also focused on using technological means to improve customer experience. For example, STO's intelligent systems provide real-time logistics tracking, ensuring that customers are always informed about the status of their parcels. At the same time, the company uses big data analytics to predict market demand trends, optimize resource allocation, reduce inefficient transportation and resource waste, and enhance the sustainability of logistics processes. Driven by the wave of digital transformation, STO has strengthened monitoring and quality control over express delivery processes through a series of technological innovations, ensuring service stability and reliability. These technologies have not only supported the company's sustained business growth but have also set a benchmark for the digital development of the entire express logistics industry. Looking ahead, STO will continue to leverage its strong technological capabilities to drive industry innovation and provide customers with more efficient and convenient express services.

11.6.4.2 Explore International Markets

Driven by the Belt and Road Initiative, STO has accelerated the development of its international business, expanding into more countries and regions along the routes. By establishing overseas warehouses, optimizing cross-border logistics channels, and deepening cooperation with local enterprises and e-commerce platforms, the company can enhance its global service network and strengthen its international competitiveness. Furthermore, through collaboration with local governments and business associations, STO can better integrate into local markets and reduce operational risks when entering new regions. Against the backdrop of an increasingly saturated domestic market, STO needs to respond flexibly to market changes, further intensify its "going global" efforts, and enhance market resilience through diversified business strategies.

11.7 Balanced National Distribution Development: Dedicated to Cost-Effectiveness and Service Improvement

11.7.1 Introduction

YDA Holdings Co., Ltd. (hereinafter referred to as "YDA") was established on August 8, 1999, and is headquartered in Shanghai. The company went public on December 23, 2016. With steady business development, YDA holds a significant position in the express delivery industry thanks to its high-quality services. In terms of delivery timeliness, from 2017 to 2021, YDA ranked among the top two in end-to-end delivery time among comparable franchise-based peers (ranking first for three consecutive years from 2017 to 2019), demonstrating its highly efficient service capabilities. Additionally, according to data from the State Post Bureau, YDA performed excellently in express service satisfaction, with a comprehensive index for complaint handling reaching 98.68 points, ranking second among comparable peers. This reflects the company's exceptional level of customer service quality. Guided by its mission to "Deliver Love, Warmth, and Convenience," YDA strives to achieve its vision of "becoming a respected, trustworthy, and service-driven first-class express enterprise." For over two decades, YDA has focused on express delivery as its core business while actively expanding into services such as supply chain, international logistics, and cold chain transportation. By optimizing its product structure and enhancing service quality, YDA is committed to becoming a comprehensive express logistics service provider that satisfies its customers.

11.7.2 Development History

Since setting sail in August 1999, YDA has kept pace with market trends by optimizing its business layout, service structure, and operational systems. Through strategic establishment of subsidiaries and targeted acquisitions, YDA has actively promoted business diversification and high-end upgrades of its industrial chain, gradually evolving into a leading comprehensive express logistics service group. In August 2003, YDA established a subsidiary—Shenzhen YDA Express Co., Ltd.—a key move in its strategic presence in South China that enhanced both its market influence and logistics distribution efficiency in the region. As Shenzhen is at the forefront of China's reform and opening-up, boasting strong economic vitality and innovation capabilities, YDA leveraged the city's locational advantages and open business environment to drive business development and brand building in Southern China. In 2007, the establishment of Beijing YDA Express Co., Ltd. strengthened the company's footprint in North China. Capitalizing on Beijing's geographical and market advantages as a core city, YDA enhanced its market penetration, operational efficiency, and service quality in the region. In 2011, YDA Freight made an appearance at the China E-Commerce Expo, showcasing advanced logistics technology and service concepts. This attracted increased attention from enterprises and consumers, further raising brand awareness. In 2017, YDA achieved 100% self-operation of all transfer centers, strengthening control over core resources and trunk networks. This move helped optimize parcel processing workflows and improve efficiency. The company has consistently pursued an international strategy, initiating cross-border logistics services as early as 2013 through an international department in active response to the "Going Global" strategy. Today, its international business covers 41 countries and regions [11].

11.7.3 Development Landscape

Over the eight-year period from 2016 to 2023, YDA's annual parcel volume increased from 3.214 billion to 18.854 billion, achieving a compound annual growth rate (CAGR) of 28.76%, with a current market share of 14.28%. During the same period, its operating revenue grew from RMB 7.35 billion to RMB 44.983 billion, reflecting a CAGR of 29.54% (see Fig. 11.15). In 2023, although the company's operating revenue experienced a decline, its operating costs decreased by 5.61 percentage

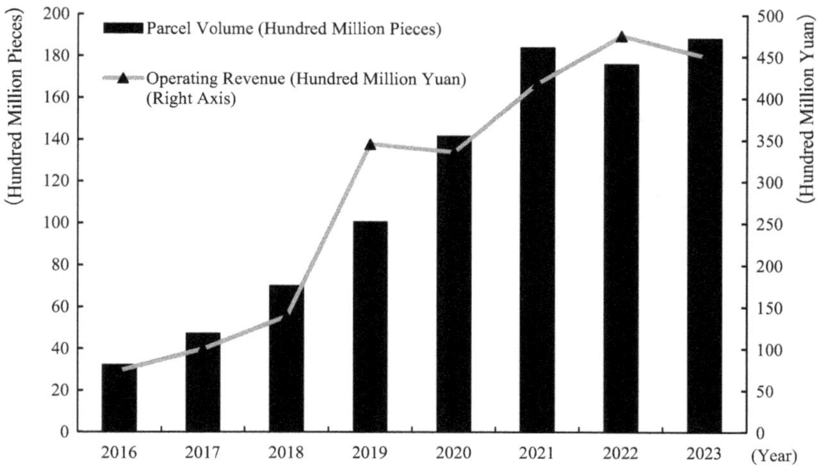

Fig. 11.15 Trend of parcel volume and operating revenue of YDA, 2016–2023. *Data source* Compiled from company annual reports

points compared to the previous year. Net profit reached RMB 1.625 billion, a year-on-year increase of 9.58%, demonstrating strong capabilities in resource integration and cost control. During the peak business period from November to December, YDA's monthly parcel volume growth outpaced the industry average, highlighting its comprehensive advantages in service capacity and market expansion.

YDA adopts a business model of "self-operated hub transfer centers and franchised last-mile networks." By maintaining self-management of transfer centers, the company ensures better control over critical resources, trunk networks, and service quality. Meanwhile, pickup and delivery operations are handled by franchised partners with express delivery qualifications, enabling extensive service coverage and improved efficiency. Currently, YDA has 4851 primary franchisees nationwide, with a franchise ratio of 100%. The company operates over 93,000 end-point service outlets, with its network covering 2837 county-level units. Notably, 42 additional county-level units were added in western regions to expand service coverage. YDA's delivery service now reaches all 31 provincial-level administrative regions in China. Except for Golog Prefecture and Sansha City, all prefecture-level cities are included in its service scope. The company's products and services include express delivery products, value-added services, YDA Supply Chain, and YDA International.

YDA has deeply implemented the "9218 Initiative," actively promoting new energy vehicles, photovoltaic power generation, and green packaging. The company currently operates 1320 new energy vehicles, advancing the development of low-carbon transportation. YDA has deployed photovoltaic power generation equipment across multiple operational centers to supply energy for daily operations. In 2023, the company reduced its greenhouse gas emissions by 6.28% year-on-year and decreased energy consumption by 6.27%, effectively lowering carbon emission intensity. YDA has installed approximately 20,000 green recycling devices nationwide and introduced 20 million reusable transit bags, significantly reducing the use of non-degradable plastics. Through initiatives like "Xinjiang Products to National Markets," YDA delivers high-quality agricultural products from Xinjiang to markets across the country, boosting agricultural sales and injecting vitality into rural revitalization. In recognition of its outstanding performance in expanding rural service networks and supporting the distribution of agricultural products, YDA was awarded the "Rural Revitalization Contribution Award."

11.7.4 Outlook and Recommendations

With the revision and improvement of laws and regulations, China's express delivery industry is entering a new phase of development, where competition in service quality will become the dominant trend, and high-quality development is a common goal for major express giants. YDA should continue to focus on its core business, leveraging technological innovation as its primary driver and serving the real economy as its central mission. Key priorities for the company include: Enhancing the quality and volume of its main business, Increasing operational saturation margins, Maintaining service leadership, and Continuing to implement product stratification to meet diverse customer needs. Looking ahead, YDA will use innovation as an engine to optimize operational processes through technological empowerment, advance digital transformation, and explore smart logistics solutions. It will accelerate its presence in areas such as smart warehousing, automated sorting, and intelligent distribution, continually improving service precision and response speed to provide customers with more efficient, green, and intelligent express services.

11.8 Specialized Provider of Heavy Freight and Bulk Logistics Solutions

11.8.1 Introduction

DBA Logistics Co., Ltd. (hereinafter referred to as "DBA") was established in 1996 and listed on the Shanghai Stock Exchange in 2018. It is a comprehensive logistics service provider integrating small parcel delivery, heavy freight delivery, logistics services, solutions, and value-added services. Guided by the mission of "Smooth Flow of Goods, Full Realization of Talent," DBA demonstrates not only its commitment to logistics efficiency but also its emphasis on unlocking and respecting employees' potential. With core values centered on "customer focus, empowering the driven, achieving efficiency through teamwork, and persistent self-reflection," the company highlights its dedication to clients, employees, collaboration, and continuous improvement. DBA focuses on its express freight business, consolidating its competitive position in the logistics market. It continually innovates and optimizes service processes, actively adjusts its business structure, and leverages its extensive experience in logistics along with market demands to identify new growth opportunities. After more than two decades of operation, DBA has significantly enhanced its brand influence, service capabilities, and market competitiveness, achieving annual revenue of RMB 36.279 billion and establishing itself as a leading player in the logistics industry. Furthermore, DBA's international operations cover 220 countries and regions, including the United States, Canada, and Japan. Its well-established global logistics network further strengthens its competitiveness in the international market. Adhering to a "Zero Damage" philosophy, DBA enforces strict cargo protection and packaging standards, leading to a year-on-year reduction in cargo damage rates and continuous improvement in customer satisfaction. The company has been honored with awards such as the "2021 Outstanding Logistics Service Provider in China," the "2022 Award for Ensuring Smooth Flow and Supply Chain Resilience," and a place among the "Top 500 Chinese Private Enterprises." In 2023, DBA maintained an average comprehensive ranking of first place in the industry according to the CANI Index.

11.8.2 Development History

In 1996, DBA's predecessor, "Cui's Freight Company," was established, marking the first step in DBA's journey amid the growth wave of China's logistics industry. In the following years, DBA continued to innovate and develop, gradually emerging as a notable player in the sector. In 2000, the company launched its second-generation information management system, becoming one of the first enterprises to provide real-time cargo tracking and inquiry services. The early 2000s were a period full of opportunities and challenges for the express logistics industry. With the rapid rise of e-commerce, logistics demand surged, and numerous logistics companies sprang up. Seizing the opportunity, DBA actively embraced business innovation while continuously optimizing its logistics network and improving service quality. Since 2001, DBA ventured into the road freight transportation sector. In 2004, it innovatively introduced the "Truck Flight" service, which established the company's strong position in the domestic less-than-truckload (LTL) market. In November 2013, DBA, which originated in freight services, launched its express delivery business—a critical move in its strategic transformation. In May 2016, the company officially began cross-border operations, and in July of the following year, it launched electronic waybills for individual customers. This initiative not contributed to energy conservation and environmental protection but also improved billing efficiency eightfold. The same year, DBA provided approximately 230,000 direct employment opportunities to society. In 2018, DBA introduced an integrated sorting model for large and small parcels. This sorting system significantly enhanced the company's operational flow efficiency and site handling capacity. In 2019, DBA launched a new freight product called "Heavy Parcel Home Delivery," specializing in the delivery of large items weighing 60–200 kg with door-to-door service. In 2023, DBA implemented the CAST system and a 360-degree data analysis framework to strengthen digital operation management [12].

11.8.3 Development Landscape

DBA focuses on heavy freight transportation as its core business, providing door-to-door precision delivery services covering various sectors such as home appliances, furniture, and building materials. Over the seven-year period from 2017 to 2023, DBA's operating revenue increased from RMB 20.35 billion to RMB 36.279 billion, achieving a compound annual growth rate (CAGR) of 10.12%. Total assets grew from RMB 6.521 billion to RMB 17.577 billion, with a CAGR of 17.97%, while total liabilities rose from RMB 3.45 billion to RMB 9.877 billion, reflecting a CAGR of 19.16% (see Fig. 11.16). In 2023, the freight business contributed RMB 32.28 billion, accounting for 88.98% of total revenue, with a year-on-year increase of 18.19%, maintaining its position as the company's core operation.

As shown in Fig. 11.16, DBA's operating revenue has demonstrated steady growth over the past seven years without significant fluctuations due to the pandemic, highlighting the company's exceptional risk resilience, strong market adaptability, and robust operational capabilities. This stable performance can be attributed to its forward-looking strategic planning and effective operational management in multiple areas. These include: An expanding logistics network, reflected in DBA's well-developed infrastructure with 9194 service points, a 93.94% township coverage rate, and 190 distribution centers. A high-quality talent pool, with 23.7% of employees

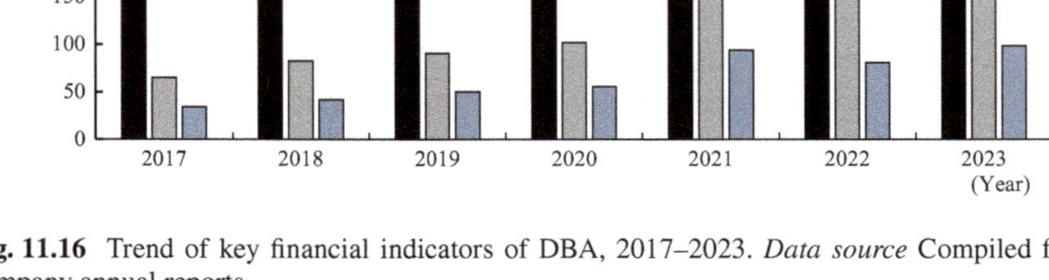

Fig. 11.16 Trend of key financial indicators of DBA, 2017–2023. *Data source* Compiled from company annual reports

holding a bachelor's degree or higher. The company offers a dual career advancement system encompassing both management and professional development paths. Continuously improved governance mechanisms, supervision systems, integrity culture, and anti-corruption education frameworks, leading to more efficient whistle-blowing channels. In 2023, the company received 1936 reports related to corruption and misconduct, reflecting not only DBA's zero-tolerance policy toward unethical behavior but also the trust employees and the broader community place in its integrity governance system. Driven by the Belt and Road Initiative, DBA has launched 1338 international routes, including 616 air freight routes and 584 sea freight routes, covering markets in Europe, America, and Southeast Asia, thereby supporting the rapid development of cross-border e-commerce. In 2023, the company leveraged its logistics network to help promote specialty agricultural products such as Maoming lychees and Yangshan peaches to national markets. For lychee transportation, DBA provided cold-chain pickup and custom packaging services, while for peaches, it adopted an air-land multimodal transport solution. These measures ensured delivery quality while reducing costs for farmers, contributing to increased agricultural income.

11.8.4 Outlook and Recommendations

In recent years, DBA has adopted a differentiated competition strategy, focusing on the mid-to-high-end heavy-item delivery market to meet customers' specialized demands for large parcel logistics. The company's overall business volume remains relatively low, with suboptimal capacity utilization and untapped economies of scale, indicating significant potential for future growth. To further enhance its market competitiveness, DBA should strengthen the optimization of its business structure while expanding its market scale. This will help improve capacity utilization and ensure the full realization of economies of scale. DBA's relatively low investment in R&D may reduce short-term capital expenditure and positively impact its financial performance. In the long run, however, this conservative approach may constrain the company's competitiveness in technological innovation and product development, thereby weakening its market position. Against the backdrop of increasing digitalization and intelligence in the logistics industry, insufficient R&D investment could hinder DBA's ability to keep pace with industry advancements. The company should more firmly embrace a development philosophy driven by technological innovation and increase investment in core technologies such as digitalization, intelligence, and automation. In the face of intensifying homogeneous competition, DBA needs to innovate in business models and service offerings, building new competitive advantages through high-quality services. Looking ahead, DBA should actively participate in promoting green logistics by deploying new energy vehicle fleets, promoting eco-friendly packaging, and adopting other sustainability measures. These efforts will align with societal demands for sustainable development, enhance the company's brand image, and position DBA as a more responsible and innovative logistics service brand.

11.9 State-Owned Enterprise Advantages: Comprehensive Coverage and Policy Support

11.9.1 Introduction

In June 2010, with the approval of the State Council, YZJ, as the main initiator, established YZW Express Logistics Co., Ltd. (hereinafter referred to as "YZW"). YZW is the most long-standing and extensively networked comprehensive express logistics service provider in China. Guided by the service philosophy of "Treasure Every Moment, Strive for Every Step," the company is committed to delivering meticulous and attentive service at every stage of the express logistics process, striving to build an efficient, secure, and green logistics system. After more than 40 years of operation, YZW has grown into an enterprise handling over 10 billion parcels annually with annual revenue exceeding RMB 100 billion.

11.9.2 Development History

The development of YZW traces back to its roots as the core force of the national postal service, providing essential mail delivery services. On July 15, 1980, it launched international express mail service (EMS), marking the beginning of China's express delivery industry. In December 1985, China Express Service Corporation was established, becoming the country's first professional express delivery enterprise. In January 1994, YZJ introduced online mail tracking and inquiry, ushering in the information era [13]. It subsequently issued barcode standards for containers, letter boxes, and mail packaging, facilitating standardized processing and creating conditions for logistics informatization. In 1999, YZJ initiated the establishment of the Universal Postal Union EMS (Global Express Mail Service) cooperation body and has served as a council member ever since. That same year, it implemented a profit-oriented financial management method called the "Balance Package Responsibility System" to turn losses into gains [14]. In 2003, YZJ launched a logistics distribution network and began offering China Post Express Delivery and cold chain logistics services. By 2006, YZJ's business revenue reached RMB 73.05 billion, forming a diversified operational structure encompassing five major sectors: Delivery Services: Continued consolidation of traditional letter and parcel delivery while actively expanding express logistics in response to market demand, initially building a logistics network covering both urban and rural areas. Financial Services: Through the establishment and development of the Postal Savings Bank, it gradually became a key pillar of the group, offering a wide range of savings, loan, and wealth management services, particularly excelling in rural financial services. Newspaper and Periodical Distribution: Maintained its position as the largest distribution channel in China, providing stable services for domestic and international publications. Philatelic Business: Leveraged the cultural and collectible value of stamps to continually attract philatelic enthusiasts and enhance cultural influence through commemorative stamp issuances. Information Services: Combined communication and modern technology to provide various modern services, including email and electronic payments.

Since 2008, the company has ventured into supply chain financial services and introduced milk run services, which not only effectively reduced transportation costs but also significantly enhanced customer satisfaction. In June 2010, YZW Express Logistics completed its joint-stock reform and officially established "YZW Express Logistics Co., Ltd." In March 2014, the company launched the EMS "Express Fresh – Direct from Origin" platform. Leveraging the advantages of its self-operated postal aviation network, it coordinated operations across its entire network to provide strong support for the development of specialty fresh agricultural products. In 2018, YZW actively responded to the central government's call to "win the Poverty Alleviation Campaign." It assisted 67 specialty agricultural projects across 18 provinces and 57 national-level poverty-stricken counties, generating agricultural sales exceeding RMB 10 million. In 2021, YZW's EMS annual parcel volume surpassed 10 billion pieces, marking a significant milestone in the domestic logistics industry. Amid increasingly fierce market competition, YZW continues to optimize its service system, expand network coverage, improve transportation efficiency, and strengthen last-mile delivery to consolidate its market position. YZW is actively promoting the international development of China's express postal services. In February 2023, it signed an agreement with Xiamen Airlines to purchase nine aircraft, and in June the following year, it launched its first self-operated intercontinental flight route, enhancing its air freight capabilities in Belt and Road countries and RCEP regions [15].

11.9.3 Development Landscape

In 2023, YZW's annual parcel volume exceeded 15 billion, marking a significant breakthrough in China's logistics industry. Its express service revenue increased by 9.8% year-on-year, international business revenue grew by 20.9%, and logistics revenue rose by 6.4 percentage points compared to the previous year. Currently, YZW's aviation fleet operates over 40 transport aircraft and has launched the intercontinental route "Nanjing–Luxembourg." Future plans include opening routes to North America and Oceania, building a comprehensive air logistics network covering East Asia, North America, Europe, and Southeast Asia. The company actively participates in the "5312 Project" jointly organized by the State Post Bureau and the Ministry of Industry and Information Technology, with nine of its projects successfully selected for the program, aimed at promoting deep integration between the express delivery and manufacturing industries. YZW also contributed to the construction of the Xiong'an New Area by launching a direct postal route from "Beijing West to Xiong'an," reducing mail delivery times. It has vigorously advanced the development of rural logistics channels, achieving full coverage of express delivery services in villages. YZW's business includes express services, logistics services, and industry solutions. Specific offerings include but are not limited to Domestic Express Mail, Medium-Speed Express, International Parcels, supply chain consulting, supply chain finance, and customized solutions. This diversified service portfolio enables YZW to meet the personalized needs of clients across various industries. The company continuously optimizes its smart logistics systems, leveraging technologies such as big data and artificial intelligence to enhance operational efficiency and provide end-to-end visible logistics tracking. YZW has not only achieved remarkable progress in smart logistics but also remains committed to green logistics initiatives. In 2018, it released the "Green Postal Construction Action Three-Year Plan Outline (2018–2020)," dedicated to reducing packaging usage and promoting biodegradable and recyclable materials to drive high-quality green development. During the "14th Five-Year Plan" period, YZW continues to enhance green transportation, actively responds to national calls for reducing plastic pollution, adopts new energy vehicles, and establishes green logistics centers, further advancing the environmental transformation of the logistics industry.

11.9.4 Outlook and Recommendations

Over the seven-year period from 2016 to 2022, global parcel volume increased from 64 to 161 billion pieces annually, while the domestic Chinese market grew from 31.28 billion to 110.58 billion pieces. It is not an overstatement to say that the strong growth of the global logistics market has been largely attributed to the rapid expansion of the Chinese market. Considering the significant growth in China's parcel volume continued into the first half of 2024, short-term domestic parcel volume is likely to maintain positive growth. However, as internet penetration approaches its ceiling, birth rates decline, and macro-economic policies aimed at promoting the real economy may alter consumption habits, the growth potential of the domestic market is limited. In contrast, the international market shows greater potential. For example, India's parcel volume grew by 18% in 2022 due to rapid e-commerce development, while Sweden saw a decline of 11%, Canada 9%, and Germany 7%. These figures highlight regional disparities in logistics demand worldwide. Such differences are reflected not only in growth rates but also in the demand for and quality of logistics services. For instance, some emerging markets in Asia seek more efficient logistics solutions to support their rapidly growing e-commerce sectors, while more mature markets in Europe place greater emphasis on sustainability and environmental considerations. As the "national champion" of China's logistics industry, YZW should focus on expanding into overseas markets and consistently implement its "Going Global" development strategy [16]. The company can leverage its global supply chain network and accumulated domestic experience to strengthen international resource integration and collaboration capabilities. By enhancing cross-border and in-country logistics services, YZW can build a global service system that allows consumers in more countries to access high-quality Chinese products and services, thereby setting a new benchmark for the development of the global logistics industry.

References

1. https://www.pitneybowes.com/us
2. https://www.spb.gov.cn
3. https://databank.worldbank.org/source/world-development-indicators20(29).pdf
4. https://www.sf-express.om/hn/s
5. https://www.jdl.om
6. https://corporate.jd.com
7. https://www.se.gov
8. https://www.zto.om
9. https://wap.yto.net.n
10. https://www.sto.n/p
11. https://www.yundaex.om/n/index.php
12. https://www.deppon.om
13. Shixiong Wu. China Post is rapidly advancing in mechanization, automation, and informatization [J]. Journal of Nanjing University of Posts and Telecommunications (Social Science), 2000 (04).
14. Lifeng Wu. Strategies for Strengthening the Financial Management of China Post [J]. Journal Of Xi'an Institute of Posts and Telecommunications, 2000 (04).
15. https://www.ems.om.n
16. Juncheng Ye. Research on the Competitive Strategy of China Post [D]. Southwestern University of Finance and Economics, 2009.
17. https://www.ls.n/detail/557615

Chapter 12
Automotive Manufacturers: Competitive Landscape and Sustainable Development Strategies

Malin Song and Mengjie Zhu

Abstract Guided by national policies and driven by market demand, China's automotive manufacturing sector has achieved remarkable progress, particularly in new energy vehicles and intelligent connected vehicle technologies. With breakthroughs in Level 3/4 autonomous driving technologies and the strengthening trend toward multi-center industrial clustering, China's automobile production and sales exceeded 30 millison units in 2023, maintaining its position as the world's largest market for the 15th consecutive year. The market penetration rate of new energy vehicles surpassed 50%, signifying market maturity and deep technological integration. Domestic brands like BYD and L2 have demonstrated formidable competitiveness globally through technological innovation and international market expansion. However, the industry still faces severe challenges, including chip shortages, supply chain instability, overcapacity, and international economic uncertainties, which exert significant pressure on sustainable corporate development. This paper reveals the competitive strengths and weaknesses of Chinese automotive manufacturers in the global market by thoroughly analyzing their technological development, market performance, policy support, and industry structure. The article further proposes strategies to strengthen independent R&D, optimize supply chain management, rationally plan production capacity layouts, and enhance technological capabilities and global market share through international cooperation. Research findings indicate that future technological innovation and globalization strategies will be key to maintaining sustainable competitive advantages for China's automotive manufacturing sector.

M. Song · M. Zhu (✉)
Anhui University of Finance and Economics, Bengbu, China

J. Wang and M. Song (eds.), *The Green Path of Chinese Transportation Enterprises*, Resources, Climate and Sustainable Development,
https://doi.org/10.1007/978-981-95-5289-4_12

12.1 Current Development Status of China's Automobile Manufacturers

12.1.1 Policy Framework

The policy evolution for China's automotive enterprises reflects a shift from initial industrial sector management to comprehensive support and structural adjustment, particularly emphasizing and promoting emerging technologies such as new energy vehicles and intelligent connected vehicles. In 1994, China introduced the "Automotive Industry Development Policy" (1994), marking the birth of the nation's first formal automotive industry policy. This policy charted the industry's direction and, for the first time, advocated encouraging automobile consumption and private car ownership. It also established specific requirements for the localization of joint venture products, laying the foundation for the strategy of introducing technology, digesting and absorbing it, and pursuing independent development. By 2004, the government introduced the "Automotive Industry Development Policy" (2004), which for the first time integrated automotive industrial policy with automotive consumption policy. It emphasized the significance of the automotive industry as a vast industrial chain for economic growth, social development, and consumer habits. Subsequently, the Chinese government issued multiple automotive industry policies to drive continuous improvements in corporate R&D and innovation capabilities. Domestic brands gradually gained market recognition, accelerating industrial restructuring and consolidation, enhancing industry concentration, and expanding economies of scale for enterprises. During this period, major enterprises such as YIIQQ, SSQQQ, CCCAA, BBQQQ, and DDFFF all surpassed 500,000 units in sales in 2004, with FAW achieving sales of 1,007,500 units. In 2009, the government implemented the "Automotive Industry Adjustment and Revitalization Plan (2009)." Guided by this plan, small-displacement vehicles significantly increased their market share due to advantages in safety, environmental protection, and energy efficiency. Concurrently, the pace of mergers and acquisitions accelerated, further enhancing industry concentration, while the market share of domestic brands also rose markedly. By 2020, the State Council issued the New Energy Vehicle Industry Development Plan (2021–2035), which outlined the development trends, overall deployment, fundamental principles, and vision for the new energy vehicle industry. This plan aims to promote the high-quality and sustainable development of new energy vehicles, accelerating China's progress toward becoming a leading automotive power.

Over the past five years, with strong support from national macro policies, China's automotive manufacturers have demonstrated robust development momentum. Particularly in the field of intelligent connected new energy vehicles, policies such as the "Notice on Launching the Pilot Program for Market Access and On-Road Operation of Intelligent and Connected Vehicles (ICVs)" have provided substantial support for Level 3/4 autonomous driving technologies, accelerating their industrialization. By September 2023, XINWD had accumulated over 70 million kilometers of safe operational mileage for its Level 4 autonomous driving technology. Concurrently, policies encourage enterprises to increase R&D investment, enhance technological innovation capabilities, develop products with independent intellectual property rights, and implement brand strategies. For instance, BYMOTOR launched the first product under its premium brand, the Yangwang U8 Luxury Edition, priced at 1.098 million yuan, signaling its strategic expansion into the high-end electric luxury vehicle segment. The release of China's Top Ten Automotive Technology Trends for 2024 further underscores the state's support and guidance for advancing technologies like Level 3 autonomous driving and NOV (Navigation-based On-Vehicle) city pilot assistance. Driven by these policies, China's automakers have achieved significant progress in technological innovation and market expansion, demonstrating robust development momentum and promising future prospects.

12.1.2 Market Scale

12.1.2.1 Steady Market Growth

Despite complex global economic conditions, China's automotive market has maintained steady growth over the past decade, with continuous expansion in scale. According to data from the China Association of Automobile Manufacturers, both production and sales exceeded 30 million units in 2023, reaching 30.161 million units and 30.094 million units respectively, securing China's position as the world's largest automotive market for the 15th consecutive year. Furthermore, 2024 sales are projected to reach 32 million units, marking a year-on-year increase of approximately 7% and setting another historical record. Among these, BYD, leveraging its robust performance in the new energy vehicle sector and extensive domestic and international market presence, achieved sales of 3.0244 million units in 2023—a 61.9% year-on-year surge—solidifying its position as China's top-selling automaker. New energy vehicles have emerged as a pivotal driver of the automotive industry's development. In 2023, China's new energy vehicle stock reached 4.92 million units, growing 29.13% year-on-year, with annual additions exceeding one million units for three consecutive years. By 2024, the penetration rate of new energy vehicles is projected to reach 50.8% for the first time, marking their comprehensive adoption in the Chinese market. PwC forecasts that new energy vehicle sales will reach 11.5 million units in 2024, representing approximately 20% growth (Fig. 12.1).

12.1.2.2 Regional Industrial Concentration

The regional distribution of Chinese automotive enterprises exhibits a pronounced concentration trend, primarily in coastal areas, particularly Guangdong, Shandong,

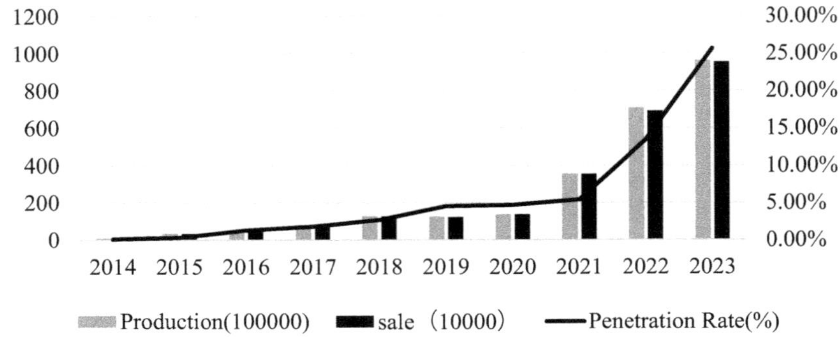

Fig. 12.1 Sales Volume and Penetration Rate of New Energy Vehicles in China, 2014–2023. *Data Source* China Association of Automobile Manufacturers, provided by CEIC Database

Zhejiang, and Jiangsu. Guangdong Province leads the nation in the number of enterprises across the automotive supply chain, hosting numerous representative companies in the automotive electronics sector such as XINWD, CHWEZ, and DESAX, which operate across both upstream and midstream segments. Additionally, Guangdong is home to significant vehicle manufacturing enterprises including GUQII, XIAPE, and YILOX. Within the vehicle manufacturing sector, enterprises are highly concentrated in Guangdong, Shanghai, and Jilin, with Guangdong and Jilin ranking notably high in national automotive production volumes. Notably, the Northeast automotive cluster centered around Changchun has gathered manufacturers like BIYSD, HUCBA, and HAFJT, leveraging a robust foundation in traditional automotive manufacturing. Regarding the new energy vehicle supply chain, core technologies are primarily concentrated in the battery sector, while the southeastern coastal regions boast a more comprehensive industrial chain covering additional segments. Jiangsu Province hosts representative companies like Nanjing, Huatian, and Mada; Anhui Province includes Qirui, Wela, and Jihu; while Guangdong Province features GAC, BYD, and Xapeng. China's automotive industry distribution exhibits a pattern of coastal concentration and gradual inland development, with Guangdong Province demonstrating particularly prominent performance in the new energy vehicle sector.

12.1.3 Corporate Structure

12.1.3.1 Structural Adjustment and Transformation

China's automotive manufacturing sector is undergoing profound structural adjustment and transformation, encompassing comprehensive changes in production methods, market positioning, and corporate strategies. Mass customization has emerged as a key direction for this transformation, enabling enterprises to offer personalized products and services at near-mass-production costs, thereby enhancing market competitiveness. For instance, SSQQQ's "Spider Smart Selection" platform offers users up to 15 billion vehicle configuration options, digitally connecting the entire value chain from model definition to pricing. Simultaneously, networked production models are becoming increasingly feasible through rapid advancements in information technology. Factories equipped with smart manufacturing capabilities achieve efficient management and optimized resource allocation via networked production, significantly reducing production costs. Take LATQC as an example: it has built a flexible production model using 5G and industrial internet technologies, successfully achieving networked production. Furthermore, smart manufacturing is a critical component of the automotive industry's transformation. By integrating technologies like big data, cloud computing, and the Internet of Things (IoT), companies can achieve real-time monitoring and management of production processes, further improving production efficiency and product quality. For instance, SSQQQ's MEB smart factory integrates digital twins with expert knowledge modules through 5G technology, while GUQII Honda's Huangpu plant collaborates with HUWII to build a wireless, paperless smart factory. These intelligent manufacturing technologies not only boost production efficiency but also help enterprises respond to rapidly changing market demands. At the WOD Automotive Manufacturing Digital Transformation Summit, industry participants extensively discussed the application of technologies like artificial intelligence, machine learning, and big data analytics. This marks China's automotive manufacturing sector advancing into a new developmental phase characterized by mass customization, networked production, and intelligent manufacturing.

12.1.3.2 Competitive Landscape and Brand Emergence

China's automotive manufacturing sector has achieved significant progress in structural adjustment and transformation in recent years, particularly in terms of competitive landscape and brand emergence. According to multiple data sources, Chinese automotive brands have continuously increased their market share and influence in terms of sales volume and brand competitiveness. In terms of sales volume, Chinese brands' market share has climbed steadily, reaching 14.596 million units in 2023 with a 56% market share—a 6.1 percentage point increase from the previous year. Notably, Chinese brands captured 80.6% of the new energy passenger vehicle market. From a competitiveness perspective, the rise of domestic brands like BYD, JILI, and CATL has further squeezed the market space for other brands. BYD surpassed Tesla in 2022 to become the world's top-selling new energy vehicle manufacturer. JILI's cumulative sales exceeded 1.5 million units from January to November 2023, with November sales reaching 200,079 units—setting a new monthly sales record. CCAAA ranks among the top three domestic brands. These Figures demonstrate that the rise of domestic brands has not only expanded market share but also strengthened brand competitiveness. This surge has also driven companies to continuously optimize product portfolios, enhance quality, and boost innovation capabilities. In 2023, JILIL launched its Galaxy series, establishing separate product lines for hybrid and

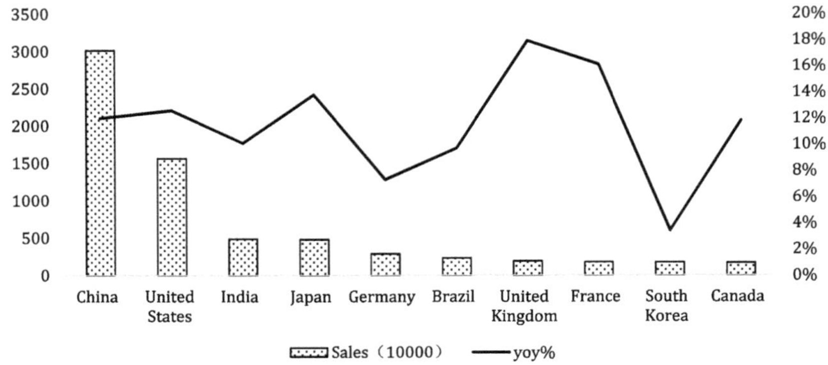

Fig. 12.2 Top 10 Countries by Global Automotive Sales Volume and Year-on-Year Growth Rate in 2023. *Data Source* MarkLines (Nikkei Group), company disclosures, and GlobalData

pure-electric vehicles positioned in the mid-to-high-end segment, signaling its full-scale push into the new energy vehicle sector. BIYSD has maintained its momentum in the new energy vehicle sector, launching multiple popular models. In 2023, China's automotive market achieved "three global leaderships": China has led the world in automobile production and sales for 15 consecutive years, maintained global leadership in new energy vehicle production and sales for nine consecutive years, and ranked first globally in automobile exports (Fig. 12.2).

12.1.4 Technological Development

12.1.4.1 Technological Innovation

Technological innovation among China's automotive manufacturers primarily manifests in new energy, intelligent connectivity, and vehicle integration. Significant achievements have been made in new energy vehicle technology, with companies like BYD and WELAQ establishing leading positions in this field. The deep development of intelligent connectivity in China has accelerated the penetration of autonomous driving and intelligent connectivity technologies within the automotive industry. Similarly, the 2024 China Automotive Technology Trends Report highlights that multiple technologies—including Level 3 autonomous driving, NOV Urban Pilot Assist, electric drive systems exceeding 20,000 rpm, lithium-rich manganese cathode materials, active suspension systems, domestically produced EHB, and multi-material multi-point integrated molding technology—are poised for breakthrough progress. China's automotive manufacturers have achieved remarkable technological innovations.

12.1.4.2 Supply Chain Restructuring

As automotive technology undergoes profound transformation, the pace of supply chain restructuring accelerates. The widespread adoption of smart manufacturing and industrial internet has effectively reduced production waste, enhanced efficiency, and elevated overall industrial chain performance. Take DDFFF as an example: by integrating high-efficiency combustion systems, electrified accessories, and intelligent control technologies, it significantly boosts hybrid system thermal efficiency. This enables the 1.5 T hybrid-specific engine to achieve thermal efficiency exceeding 45%, reaching internationally leading levels. Companies like WECHG have employed aluminum alloy heat-treatment-free integrated die-casting technology, simultaneously reducing costs and achieving bodyweight reduction, thereby enhancing vehicle range and overall performance. As the industry advances toward high-end and intelligent development, technological breakthroughs further promote coordinated development across upstream and downstream supply chains. Notably, in central computing electronic/electrical architecture, enterprises such as LIPPA and XIAPE have successfully established a "central computing + regional control" framework, driving deep integration of software and hardware. Simultaneously, digital technology applications have enhanced supply chain management efficiency. Real-time data analytics optimize inventory and logistics scheduling, strengthening the industry chain's adaptability to market fluctuations and resilience against risks. Cross-industry integration between automotive manufacturing and emerging sectors like internet, big data, and artificial intelligence has spurred the emergence of new business and service models, further refining the industrial ecosystem. Collectively, these transformations drive the optimization and upgrading of the automotive industry chain, laying a solid foundation for its sustainable development.

12.1.4.3 Digital Transformation

Digital transformation spans multiple stages including R&D, production, management, and marketing. It plays a pivotal role in shifting value within the industrial chain of Chinese automotive manufacturers, injecting new momentum into industry revenue generation. Digital R&D achieves "software-defined vehicles" through hardware-software integration, shortening new vehicle iteration cycles; Digital production utilizes IoT and AI technologies to optimize manufacturing processes; digital management maximizes data asset utilization through data management platforms; digital marketing achieves low-cost, high-efficiency user acquisition through online-offline synergy. Digital technologies not only enhance operational efficiency but also optimize consumers' vehicle purchasing and usage experiences. CRRC Sifang has successfully applied digital twin technology in intercity and suburban trains. Its independently developed CINOVA 2.0 intelligent intercity/suburban train incorporates "digital twin" technology, enabling comprehensive self-sensing of train status, self-diagnosis of faults, and autonomous maintenance decisions. This enhances train intelligence while reducing lifecycle maintenance costs. Moreover, Chinese consumers demonstrate high enthusiasm for digital service experiences, particularly in in-vehicle digital lifestyles. The proportion of Chinese consumers using navigation apps to plan driving routes leads other countries.

12.1.5 Industry Competitive Landscape

12.1.5.1 Diversified and Tiered Competitive Landscape

The competitive landscape among Chinese automakers is diverse and multi-tiered. Analyzed by sales volume, companies can be categorized into distinct tiers: SSQQQ leads the first tier with annual sales exceeding four million units; YIIQQ, GUQII, DDFFF, and CCAAA form the second tier with annual sales between 2 and four million units; while other manufacturers with annual sales between 1 and two million units constitute the third tier. From a technological perspective, China's automotive industry pursues a diversified development path, with multiple technical routes coexisting, including pure electric, plug-in hybrid, and fuel cell technologies. Market participants in China's automotive sector include traditional state-owned automakers, private manufacturers, numerous joint ventures, and foreign brands, reflecting a highly diversified landscape. Product offerings span the entire spectrum from economy to premium luxury vehicles, encompassing passenger cars, commercial vehicles, and specialty vehicles to meet diverse consumer demands.

12.1.5.2 Intensified Industry Consolidation and Survival of the Fittest

Shifts in the global political landscape and anti-globalization trends have heightened risks for China's industrial and supply chains. Automakers now face challenges in supply chain restructuring, making industrial chain integration an inevitable trend to enhance stability and mitigate risks. Furthermore, the competitive tiers within China's automotive sector are becoming increasingly defined. Significant disparities between tiers are accelerating market consolidation and the survival of the fittest. Amid intensifying technological innovation, resources and technologies are increasingly concentrated among large automakers that already dominate the market. This further elevates their core competitiveness, widening the gap with smaller enterprises. Consequently, it hastens the demise of some smaller players and accelerates the industry's internal consolidation process.

12.1.6 International Competitive Environment

12.1.6.1 Global Leadership in Production and Sales for Multiple Consecutive Years

China has become the world's largest automotive market, with production and sales volumes ranking first globally for consecutive years. In 2023, China's automotive production reached 30.161 million units, while sales totaled 30.094 million units, maintaining its global leadership. New energy vehicle (NEV) production and sales demonstrated particularly outstanding performance, securing the top global position for the ninth consecutive year. NEV sales in 2023 reached 6.887 million units, representing a year-on-year increase of 93.4%. SSQQQ ranked first with sales of 5.0209 million units. Although this represented a 5.31% year-on-year decline, it maintained its leading position in China's automotive market. Among them, BIYSD Holdings achieved sales of 3.0244 million units, a 61.8% year-on-year increase, ranking second. Moreover, its passenger vehicle sales reached 3,012,906 units, establishing it as a significant force in China's new energy vehicle market. CCAAA ranked third with sales exceeding 2.55 million units. CCAA achieved significant growth in both domestic new energy vehicle sales and overseas market share, with domestic new energy vehicle sales surpassing 470,000 units and overseas sales exceeding 230,000 units—both reaching record highs. These Figures demonstrate the robust vitality of China's automotive market and the continuous enhancement of Chinese automotive brands in global competition.

12.1.6.2 Export Volume Shows Significant Growth

China's automobile exports have experienced remarkable growth in recent years. In 2023, total vehicle exports reached 4.91 million units, marking a 57.9% year-on-year increase. Among these, new energy vehicle exports amounted to 1.203 million units, surging by 77.6% compared to the previous year. Chinese automakers have progressively strengthened their competitiveness in international markets, particularly in the new energy vehicle sector. SSQQQ led the pack with 1.099 million units exported, maintaining its position as China's top automotive exporter for consecutive years while expanding its overseas market footprint and brand influence. QIRUQ ranked second with 925,000 units exported, demonstrating exceptional performance in overseas markets—particularly in Russia where its sales saw significant year-on-year growth, establishing it as a key player in the local automotive market. JILIL secured third place with 408,000 units exported. Through continuous technological innovation and brand development, JILIL has enhanced its products' competitiveness in international markets, driving sustained export growth (Fig. 12.3).

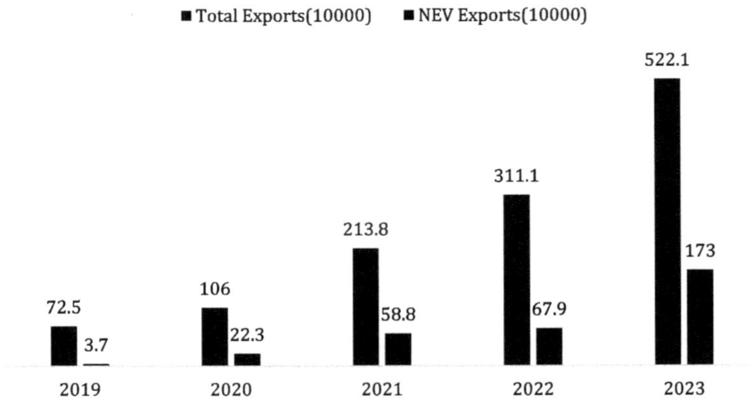

Fig. 12.3 China's Automobile Exports and New Energy Vehicle Exports Over the Past Five Years. *Data Source* Online analysis reports on Chinese automotive manufacturers

12.1.6.3 Globalization of the Industrial Chain

Globalization of the industrial chain is a critical pathway for China's auto parts enterprises to achieve strategic transformation. In the process of globalization, enterprises face challenges such as rising raw material costs, supply chain uncertainties, and the need to reshape R&D models. They must respond through strategies including optimizing customer structures, providing digital transformation support, and building industrial chain ecosystems. In terms of global expansion, Chinese auto parts companies are accelerating international business growth by expanding production capacity in Europe, North America, and Southeast Asia. They are actively pursuing overseas M&A opportunities to achieve goals such as customer resource exchange, acquisition of core technologies, broadening product lines, and increasing overseas revenue scale. Additionally, to enhance global competitiveness, companies are promoting localized supply chain development, strengthening supply chain risk and cost management, and making continuous R&D investments to drive technological innovation and product upgrades. Chinese auto parts companies are progressively elevating their standing within the global automotive supply chain, bolstering their international competitiveness, and pursuing sustainable development within the broader context of globalization. As a leading global supplier of power batteries, NINDD is actively expanding its footprint in Europe, North America, and other regions. By establishing R&D centers and production bases, NINDD is accelerating the pace of its international business expansion. Through technological innovation and product upgrades, NINDD continuously enhances its position within the global automotive supply chain while seeking sustainable development amid globalization. Similarly, WECHJ is actively expanding its global market presence through overseas acquisitions and collaborations, particularly in Southeast Asia. By establishing joint ventures and production bases, it has strengthened its global supply chain layout. WECHJ is also leveraging the opportunity of digital transformation to optimize its customer structure, provide digital transformation support, and build an industrial chain ecosystem.

12.2 Development Trends of China's Automobile Manufacturers

12.2.1 Government Policies

12.2.1.1 High-Quality Development and Industrial Structure Optimization

The Outline of the 14th Five-Year Plan (2021–2025) explicitly states that the government will accelerate the high-quality development of the automotive manufacturing industry. As a strategic and pillar industry in the national economy, the transformation and upgrading of the automotive manufacturing sector is crucial. Therefore, policy support to drive the industry's shift from quantitative expansion to qualitative improvement and optimize industrial structure has become an inevitable future trend. Currently, China's automotive manufacturers are transitioning from reliance on external technologies toward independent innovation. The government is driving technological advancement and industrial upgrading through initiatives like the "Manufacturing Powerhouse" strategy, propelling the sector toward high-quality development. New energy vehicles (NEVs), as a key direction for industrial transformation, have received significant government support and become a core focus of policy initiatives. Simultaneously, promoting the integrated development of NEVs with related industries such as energy, transportation, and information and communications technology (ICT) represents a crucial measure for further optimizing the industrial structure. Take SSQQQ as an example. It ranks first nationally in vehicle production and sales, new energy vehicle sales, and exports, while actively expanding into mobility services, financial services, and innovative technology sectors. By integrating with related industries, SSQQQ not only responds to the national call for independent innovation and high-quality development but also lays a solid foundation for optimizing and upgrading its own industrial structure.

12.2.1.2 Environmental Protection and Sustainable Development

Under the "dual carbon" strategy, driven by environmental pressures and government policies, China's automotive manufacturers are steadily advancing toward environmental protection and sustainable development. Companies actively promote new energy vehicles, reduce pollution emissions during production, and drive the industry's transition toward green and low-carbon operations[2]. By enhancing energy efficiency in factories, implementing energy substitution and carbon reduction upgrades, and improving resource recycling rates, enterprises have effectively lowered energy consumption. Simultaneously, automakers continuously upgrade production processes and equipment to achieve energy savings throughout manufacturing.[3] Against this backdrop, GUQII launched the "GLASS Green Clean Initiative," embedding sustainability principles across product design, R&D, production, and operations. By increasing investments in energy and energy ecosystems, GUQII has established an integrated new energy industrial chain and plans to achieve carbon neutrality across the entire product lifecycle by 2050. Notably, GUQII has obtained carbon neutrality certification from the Guangzhou Carbon Exchange, establishing its first zero-carbon factory and signifying zero carbon emissions throughout its production processes.

12.2.2 Corporate Ecosystem

The ecosystem of China's automotive manufacturers is actively responding to shifting market demands through upstream-downstream coordination, industrial chain clustering, new energy technology advancement, and mergers and acquisitions. These efforts enhance efficiency, drive economic growth, and optimize resource allocation. As the market rapidly evolves, automakers face intense competition and constantly changing consumer needs. To meet these challenges, companies must continuously improve coordination capabilities across the supply chain, build robust collaborative networks, and achieve synergistic effects throughout the industrial chain[4]. Currently, China's automotive industry is at the peak of cluster-based development, having formed six major automotive industrial clusters: Northeast, Central, Western, Pearl River Delta, Yangtze River Delta, and Beijing-Tianjin-Hebei. The Northeast region, anchored by Changchun, hosts enterprises like YIIQQ and HUCBA, possessing a solid foundation in traditional automotive manufacturing. The Western region, centered around Chongqing and led by Changan Automobile, boasts robust heavy industrial capabilities. The Yangtze River Delta region, with Jiangsu, Shanghai, and Zhejiang as its core, features benchmark companies such as JILIL and SSQQQ, whose industrial parks generate annual output exceeding 10 billion yuan. Similarly, new energy vehicle enterprises continue to expand in scale, with emerging corporate groups centered on new energy and intelligent technologies gradually taking shape. Within China's six major automotive clusters, the development of new energy and intelligent vehicles is particularly pronounced. Taking the Yangtze River Delta as an example, the region hosts over 100 industrial parks with annual output values exceeding 10 billion yuan, gathering major enterprises like SAIC Motor, JAC Motors, ZOTY, and Dongfeng's bus, truck, and passenger vehicle divisions. Statistics indicate that among the 30 cities in the Yangtze River Delta cluster, over 14 have planned or established new energy vehicle projects. These encompass more than 20 initiatives with a cumulative planned production capacity exceeding three million vehicles and total planned investments surpassing 100 billion yuan.[5]

12.2.3 Technological Capabilities

China's automotive manufacturers continuously enhance competitiveness, optimize production efficiency, and reduce costs by strengthening R&D, advancing intelligent and automated production, implementing digital management, and pursuing independent technological innovation. Enhanced R&D enables enterprises to better adapt to industry trends and improve product market competitiveness.[7] Advanced manufacturing technologies—including precision, flexibility, integration, networking, globalization, virtualization, intelligence, and clean production—offer new development opportunities for the automotive manufacturing sector[8]. Through persistent technological breakthroughs, YIIQQ has broken the long-standing technical monopoly of international suppliers and mastered core powertrain manufacturing technologies.[9] LIIXI achieved 100% automation across multiple processes in its final assembly workshop at the Changzhou plant, leveraging intelligent systems to automatically deliver thousands of components to workstations. NEEZA implemented digital simulation, production automation, and management visualization to achieve digital lean manufacturing, boosting production capacity by 60%.[10] WELAQ pioneered the industry's first mass production of an independently developed lidar main control chip through technological innovation. This chip reduces power consumption and point cloud processing latency while saving hundreds of yuan per unit. XIAPE employed integrated aluminum die-casting technology for front cabs and rear floors, cutting investment in the welding shop's underbody line by 30%, reducing land use by 40%, and boosting material utilization to over 97%.

12.2.4 International Cooperation

International cooperation serves as a vital pathway for China's automotive manufacturers to advance global integration and enhance worldwide competitiveness. By deeply integrating into the global automotive industry, China's automotive sector has achieved rapid development, gradually establishing itself as a major global manufacturing hub.[12] Concurrently, implementing strategies of internationalization, independence, and autonomous development is key to strengthening the industry's global competitiveness. Strengthening ties with international markets enables China's automotive manufacturing to better navigate global competition.[13] For instance, JILIL expanded its production capacity and influence in Southeast Asia by acquiring BATEQ; GUQII established factories in Europe, enhancing market competitiveness through localized R&D and production; SSQQQ collaborated with DEGDZ to develop new energy vehicles, driving technological advancement and market expansion; CHCHQ built the Tula plant in Russia to serve the local market and extend reach to surrounding regions; QIRUQ achieved sales growth by expanding into the Middle East, Africa, and South America, becoming a representative of China's automotive internationalization.

12.3 Challenges Facing China's Automobile Manufacturers

12.3.1 Imbalance in Automotive Chip Supply and Demand

In recent years, China's automotive industry has significantly enhanced its innovation capabilities. However, deficiencies persist in certain key technologies and core components, with semiconductor shortages posing the most prominent challenge.[13] Chinese automotive semiconductor firms face not only design and manufacturing constraints but also the first-mover advantage of traditional automotive chip industries. They have yet to develop sufficiently competitive technologies, struggling to meet the stringent requirements of automotive semiconductors. Furthermore, frequent geopolitical conflicts and natural disasters like the pandemic have disrupted operations at factories owned by international chip giants, turning chip shortages into a global issue. With the rise of new energy vehicles and the widespread adoption of intelligent applications, demand for automotive chips has surged dramatically. However, chip production capacity expansion has been relatively slow, exacerbating the imbalance between supply and demand. Some leading domestic automakers have initiated independent R&D for automotive chips. For instance, WELAQ has successfully developed a 7-nanometer automotive-grade chip featuring high integration, low power consumption, and strong performance, with its first self-developed chip "Yangjian" nearing mass production. JILIL's Lynk & Co brand debuted the "Longying No.1" 7-nanometer automotive-grade chip, independently developed by Xingqi Technology, in its Lynk & Co 08 model. Despite these advancements, China still lags behind international standards in automotive chips, and current domestic production offers limited relief for the industry's chip demand. From January to September 2023, the China Association of Automobile Manufacturers (CAAM) reported that chip shortages impacted the supply of approximately 600,000 chips, resulting in a production loss of 606,000 vehicles. The CAAM chairman publicly stated, "'Chip shortages and high electricity costs' have severely impacted automotive production capacity."

12.3.2 Supply Chain Instability

The supply chain instability faced by China's automotive manufacturers primarily stems from three issues: increased global supply chain volatility, raw material price fluctuations, and geopolitical and economic risks. Factors such as pandemic-induced supply chain disruptions, raw material price volatility, and international trade frictions have exposed automakers to risks of raw material shortages, rising costs, and production interruptions, significantly heightening global supply chain instability. Second, raw material price volatility has seen substantial cost increases for battery materials like lithium and nickel. Battery manufacturers are passing these cost pressures down to automakers, creating challenges for automakers in managing their new energy vehicle supply chains and reducing costs. Finally, geopolitical shifts and economic fluctuations also impact supply chains. Major automotive manufacturing nations such as the United States, Japan, and Germany face elevated supply chain risks. Trade issues and cyclical global economic adjustments have made supply chain disruptions more frequent and prolonged. Even leading new energy vehicle manufacturers like BIYSD, which have achieved some success in raw material cost management and supply chain optimization, still face pressure from rising raw material prices. Similarly, NINDD, a global leader in power vehicle manufacturing, confronts challenges of rising raw material costs and supply chain management, currently relying on technological innovation and scale expansion to reduce costs.

12.3.3 Low Profitability

Despite significant achievements in market share and total revenue, China's automotive manufacturers maintain relatively low profitability. According to National Bureau of Statistics data, the automotive manufacturing sector achieved cumulative operating revenue of RMB 10.0976 trillion in 2023, representing an 8.7% year-on-year increase. However, total profits amounted to RMB 508.6 billion, a 4.4% year-on-year decline, resulting in an industry sales profit margin of merely 5.0%. Chinese automakers face multiple external and internal challenges to profitability. External factors include rising costs and chip shortages. Increased prices for power battery raw materials and insufficient chip supply have extended delivery cycles, thereby impacting production, sales, and profits[14]. Internal factors manifest as insufficient brand competitiveness and suboptimal capital structures. For instance, CCAA's profitability is constrained by its weaker independent brand competitiveness and reliance on joint ventures. Additionally, price reductions by certain joint-venture brands have significantly compressed profit margins,[15] while JILIL faces intensified competition in the new energy vehicle market and cost control challenge[s.][16]

12.3.4 Overcapacity Issues

China's automotive manufacturers face overcapacity primarily due to multiple factors: a global economic downturn, slowing domestic demand growth, local government incentives distorting market mechanisms, and reckless corporate expansion. From a macroeconomic and policy perspective, external demand growth is constrained by the global economic slowdown and the "manufacturing reshoring" trend in Europe and the US. Simultaneously, the slow pace of domestic economic restructuring and upgrading perpetuates the overcapacity issue[17]. Second, at the corporate level, the rapid expansion of China's automotive industry has driven swift capacity growth, yet market demand has failed to keep pace, resulting in overcapacity [18]. Finally, local governments often adopt preferential policies like low-cost land supply to attract investment and stimulate economic growth. This not only distorts market mechanisms but also exacerbates the overcapacity issue.[19] Due to overcapacity, some automakers have begun reducing or halting production: By 2023, CCAAA's production capacity had been cut by nearly one million vehicles from its peak, and the X model was discontinued. BEJXI sold off the power battery division of one of its regional factories in a bundled deal valued at 1.62 billion yuan. Currently, the automotive industry indeed faces the challenge of overcapacity, with some battery manufacturers already announcing production halts. Examples include SOXIX, MGUXC, HNEGF, and the new energy power battery projects under BNEQC.

12.4 Recommendations for the Development of China's Automobile Manufacturers

12.4.1 Strengthen Independent R&D and Innovation

Chinese automotive manufacturers must enhance their independent R&D and innovation capabilities, particularly in the semiconductor sector, by adopting multiple strategies. First, companies should increase investment in automotive chip development to drive breakthroughs in domestic chip technology and advance industrialization.[20] Examples include BIYSD's breakthroughs in automotive-grade IGBTs and WEEGF's leading position in power and analog chips. Second, companies should strengthen collaborations with universities and research institutions to cultivate specialized talent and advance industry-academia-research synergy, particularly accelerating the R&D and production of automotive-grade chips to reduce reliance on foreign technologies.[11] Beyond independent R&D, enterprises should actively engage in international cooperation and competition, acquiring advanced technologies through mergers and acquisitions—such as WTAKJ's acquisition of ASBAN and WEEGF's acquisition of HAWEK.

12.4.2 Optimizing Supply Chain Management

Under the influence of economic globalization, Chinese automakers must achieve supply chain optimization by actively participating in global industrial chains while strengthening internal supply capabilities. On one hand, their engagement in global industrial chains should not only enhance cooperation with international automakers, diversify supply chains, and improve risk resilience, but also restructure collaboration models with foreign enterprises—including overseas M&A and restructuring—to expand national markets and actively integrate into global industrial systems. On the other hand, they must integrate domestic resources and technologies to support and encourage the supply of domestic automotive components, particularly critical parts like chips and batteries. Companies that have already achieved notable progress include: LILOL, which has enhanced its influence in the global tire supply chain by establishing overseas production bases and actively expanding into international markets; and HYUCC, which holds a high market share in automotive interior/exterior trim and functional components. Its investments in intelligent and new energy transformation are helping to boost its competitiveness within the supply chain.

12.4.3 Rational Capacity Planning

When rationally planning production capacity, Chinese automotive manufacturers must closely align with market demand, technological innovation, and policy direction. Companies should closely monitor market changes and utilize data analysis to forecast demand, avoiding blind expansion and overcapacity. Additionally, companies should actively respond to national policy requirements, such as the measures outlined in the "Work Plan for Stabilizing Growth in the Automotive Industry (2023-2024)" to expand new energy vehicle adoption, stabilize gasoline vehicle consumption, and ensure supply chain stability. Policy guidance should be leveraged to optimize production capacity allocation. As an industry leader, SSQQQ has demonstrated a leading advantage in capacity planning. It has not only optimized its product portfolio by increasing R&D and production of new energy vehicles but also actively expanded into overseas markets, achieving rational capacity allocation and efficient utilization. As a major supplier of power batteries for new energy vehicles, NINDD has successfully met rapidly growing market demand through technological innovation and capacity expansion.

12.5 Conclusion

This paper comprehensively examines the current development status of China's automotive manufacturers and provides an in-depth analysis of future trends. Driven by proactive policy guidance and market demand, China's automotive manufacturing industry has undergone a phase of rapid development, achieving significant accomplishments particularly in new energy vehicles and intelligent connected vehicles. The expansion of market scale, accelerated technological innovation, restructuring of the industrial chain, and advancement of digital transformation have collectively shaped a new competitive landscape for China's automotive manufacturing sector. Despite uncertainties in the global political and economic environment, China's automotive manufacturing industry continues to demonstrate strong resilience and potential. The globalization of the industrial chain, increased export volumes, and deepened international cooperation present new opportunities for the international development of China's automotive manufacturing sector. However, challenges such as the imbalance between supply and demand for automotive chips, supply chain instability, insufficient profitability, and overcapacity also pose obstacles to the sustained and healthy development of China's automotive manufacturing sector. To address these challenges, it is recommended that Chinese automotive manufacturers strengthen independent research and development and innovation, optimize supply chain management, rationally plan production capacity, and actively engage in international cooperation. Through these measures, China's automotive manufacturing industry can further enhance its competitiveness in the global market, achieve sustainable development, and make greater contributions to the advancement of the global automotive industry.

References

1. Zhu Shenglai. Promoting High-Quality Development of the Automobile Manufacturing Industry [J]. Shanghai Automotive, 2021.
2. Chi Hai. Research on Government Regulation Strategies for the Development of China's Automobile Industry [D]. Northeastern University, 2005.
3. Green and Low-Carbon Development Roadmap for the Automotive Industry 1.0.
4. Zhou Wei. Research on the Development Strategy of Automobile Manufacturing Enterprises in China [D]. Central China Normal University, 2018.
5. Economic Information Daily. China's Automotive Industry Clusters Trigger "Chain Reaction" Competitiveness [EB/OL] (2021.6.26) [Accessed: 2024.8.28] http://www.xinhuanet.com/auto/20210626/1844fda2253b44378c8f8ae65e460d59/c.html
6. https://m.thepaper.cn/baijiahao_26772396
7. Mou Hongtao. Development Trends and Key Technology Analysis of Automobile Manufacturing [J]. Times Automotive, 2017.
8. Wang Deping. Brief Discussion on the Characteristics and Development Trends of Advanced Automotive Manufacturing Technology in China [J]. Science & Technology Information, 2009.
9. http://www.xinhuanet.com/politics/2023-07/14/c_1129749148.htm
10. http://www.news.cn/fortune/2023-10/27/c_1129942316.htm
11. Xie Shenkun, Wang Zhezhe, Ren Yan, et al. Challenges and Opportunities of the Automotive "Chip Shortage" [J]. Electronic Product Reliability and Environmental Testing, 2022.
12. Lu Yuduo, Hou Tieshan, Yu Guifen. Problems and Development Strategies of China's Automotive Industry [J]. Automotive Industry Research, 2000(06).
13. Wang Juncheng, Hao Sichun. Research on the Development of Key Chip Technologies for Intelligent Vehicles [J]. Modern Telecommunication Technology, 2017.
14. Jin Yidan. The Automotive Industry Faces Multiple Challenges in Profitability [N]. China Securities Journal, 2021-11-04(A06).
15. Yang Mengru. Analysis of the Profit Quality of Changan Automobile [D]. Shanxi University, 2020.
16. Wang Qian. Profitability Analysis of Geely Automobile [J]. Modern Business, 2022.
17. Yu Shuxiu. Research on the Overcapacity Problem in China's Manufacturing Industry [D]. Northwest University, 2020.
18. Yang Chenchu, Zhang Jun. Analysis of the Causes of Overcapacity in China's Automotive Industry [J]. Finance & Economics, 2016.
19. Jiang Feitao, Geng Qiang, Lü Daguo, et al. Regional Competition, Institutional Distortion, and the Formation Mechanism of Overcapacity [J]. China Industrial Economy, 2012.
20. Li Xianjun, Liu Jianli, Yan Mei. Reshaping Industrial Chain Advantages: Measures Taken by Countries to Address the Automotive Chip Shortage and China's Countermeasures [J]. Contemporary Economic Management, 2022.

Chapter 13
Exploration of Green and Intelligent Transformation in Shipbuilding Enterprises

Malin Song and Xue Xing

Abstract Against the backdrop of sustained growth in global trade and the deepening implementation of the "Belt and Road" Initiative, China's shipbuilding industry is entering a critical period of transformation and upgrading. This paper aims to conduct a comprehensive discussion on the development status, challenges, opportunities, and future trends of China's shipbuilding enterprises. The article first analyzes the positive progress made by China's shipbuilding enterprises in terms of market position, revenue, and profits. Then, it delves into the challenges faced by enterprises, such as rising costs, high-end technological bottlenecks, and intensifying international competition. Subsequently, considering national policy directions, changes in market demand, and technological innovation trends, strategic suggestions are proposed to promote the high-quality development of China's shipbuilding enterprises. Finally, the paper prospects the new positioning and role of China's shipbuilding industry in the global industrial chain.

13.1 Introduction

The shipbuilding industry is one of the key supporting industries of China's national economy and an important driving force for promoting global trade growth and economic progress. In recent years, the deepening global economic integration and the vigorous development of the marine economy have opened up broad development space for the global shipbuilding industry. As a leading global shipbuilding power, China holds a pivotal position in the international shipping industry. Against the backdrop of cyclical fluctuations in the global ship market, rapid advancement of technological innovation, and increasingly complex international competition, Chinese shipbuilding enterprises are facing a series of severe challenges. How to promote the achievement of high-quality development for enterprises has become a

M. Song · X. Xing (✉)
Anhui University of Finance and Economics, Bengbu, China
e-mail: 19956325055@163.com

J. Wang and M. Song (eds.), *The Green Path of Chinese Transportation Enterprises*, Resources, Climate and Sustainable Development,
https://doi.org/10.1007/978-981-95-5289-4_13

core issue urgently needing resolution for Chinese shipbuilders. This paper comprehensively analyzes the development status and challenges faced by China's shipbuilding enterprises from multiple perspectives, considers changes in the domestic and international environment comprehensively, and proposes targeted development strategies. Through in-depth research, this paper aims to provide strategic guidance for the sustainable development of China's shipbuilding enterprises, help them occupy a more favorable position in the global shipbuilding industry competition, and contribute to China's implementation of the strategy to become a maritime power.

13.2 Overall Development Situation of Chinese Shipbuilding Enterprises

The shipbuilding industry is classified within the manufacturing sector under the transportation equipment manufacturing category. It focuses on producing various types of ships, including powered and non-powered vessels, as well as the manufacturing of related equipment such as ship fittings, covering three major categories: water transport, water operations, and underwater operations. Modern ships are diverse and can be classified into steel ships, ocean-going vessels, motor ships, self-propelled ships, civilian ships, etc., based on usage requirements and technical performance.

13.2.1 Analysis of the Shipbuilding Industry Chain

The shipbuilding industry chain is a vast and complex system encompassing upstream, midstream, and downstream segments. The upstream stage mainly involves the supply of raw materials (especially steel), ship supporting equipment supply (such as main engines, cabin facilities, deck machinery, etc.), and ship design. The midstream stage focuses on production manufacturing and ship research and development (R&D), serving as the core of the industry chain; it covers key links such as ship assembly, equipment installation, and system commissioning. The downstream segment covers ship application and service markets, injecting continuous development momentum into the industry chain. Among these, maritime transport, national defense military industry, and offshore engineering constitute the main parts of the application market, especially the maritime transport market. As a long-established and largest application field, it not only plays a key role in global trade but also significantly promotes the sustained development of the shipbuilding industry. The shipbuilding industry chain is a complex system where all links are closely connected, interdependent, and highly synergistic. In the future, Chinese shipbuilding enterprises need to enhance investment in industry chain integration and management, and improve the competitive strength and economic efficiency of the entire industry chain through in-depth collaboration with upstream, midstream, and downstream enterprises, thereby achieving complementary advantages and synergistic development among enterprises.

13.2.2 *Current Development Status of Chinese Shipbuilding Enterprises*

13.2.2.1 Increasingly Prominent Global Market Position

The market share of Chinese shipbuilding enterprises has ranked first globally for many consecutive years. According to the latest data, throughout 2023, the national ship completion volume reached 42.32 million deadweight tons (dwt); orders on hand reached 139.39 million dwt; new orders received reached 71.20 million dwt, with all three major indicators showing an upward trend. As of the first half of 2024, China's ship completion volume was 25.02 million dwt, a year-on-year increase of 18.4%; new orders received increased by 43.9% year-on-year to 54.22 million dwt; orders on hand were 171.55 million dwt, also a year-on-year increase of 38.6%, accounting for 55.0%, 74.7%, and 58.9% of the global total, respectively. Currently, orders from domestic and international clients for many shipyards are continuously rising, with some shipyards' order books scheduled until 2028. According to statistics, China's shipbuilding industry has ranked first globally in order deliveries, orders on hand, and new orders received for 14 consecutive years. This series of brilliant achievements is powerful proof of the comprehensive strength and international status of China's shipbuilding industry. In the future, China's shipbuilding industry will continue to promote the transformation, upgrading, and high-quality development of the shipbuilding industry, contributing greater strength to the prosperity of the global shipbuilding industry (Fig. 13.1).

13.2.2.2 Sustained Growth in Profits and Revenue of Large-Scale Shipbuilding Enterprises Above Designated Size

In the 2023 fiscal year, the main business revenue of China's large-scale shipbuilding enterprises above designated size achieved a year-on-year increase of 20.0%, with the total amount climbing to 623.7 billion RMB, and the total profit increased by as much as 131.7% year-on-year. This significant achievement not only demonstrates the continuous optimization of the internal structure and the obvious improvement of production efficiency within the shipbuilding industry but also profoundly reflects the strong demand for high-quality ship products in both global and domestic markets. According to the latest data, in the first five months of 2024, the total profit of China's shipbuilding industry achieved significant growth. Among them, large-scale shipbuilding industrial enterprises achieved strong growth in operating

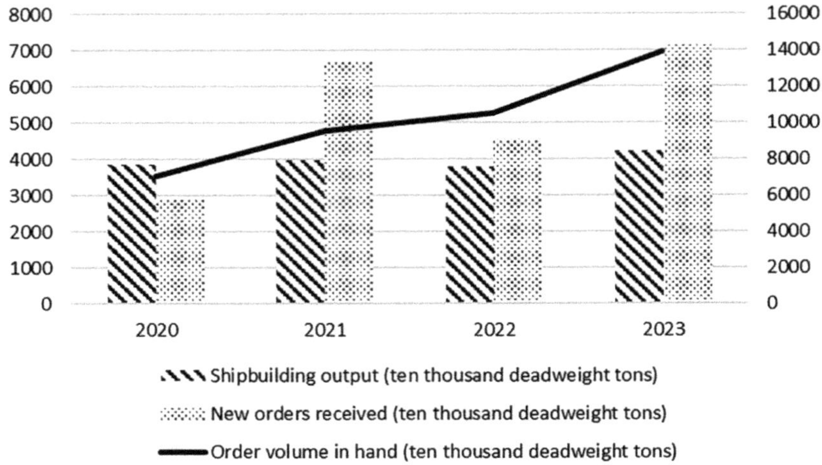

Fig. 13.1 Development of China's three major shipbuilding indicators from 2020–2023. *Data source* Ministry of Industry and Information Technology

revenue, increasing by 26.3% compared to the same period last year, with the total amount climbing to 253.24 billion RMB; the cumulative total profit reached 16.0 billion RMB, surging by 187.5% compared to the same period last year. Behind this series of impressive data lies the deep attention and active investment of China's shipbuilding industry in technological innovation, green transformation, and intelligent manufacturing. (Data source: China Association of the National Shipbuilding Industry).

13.2.2.3 Outstanding Performance of Leading Enterprises

China has numerous and large-scale shipbuilding enterprises. ZGCBJT holds a pivotal position in the shipbuilding industry with its exceptional strength, advanced technology, and broad market layout. According to the latest data, its operating revenue exceeded 70 billion RMB in 2023, ranking first in China's shipbuilding industry. It was followed by ZGZ and ZCF, whose operating revenues reached 46.694 billion RMB and 16.146 billion RMB, respectively, with year-on-year growth of 5.75% and 26.19%. This steady growth trend reflects the continuous efforts of these two major enterprises in technological innovation and market expansion. Although the operating revenues of JLK and YGK are slightly lower compared to other shipbuilding enterprises, they also play important roles in promoting the development of China's shipbuilding industry and have contributed significantly (Fig. 13.2).

(1) ZGCBJT was established in 2019 and is a leader in China's shipbuilding industry. It has vast total assets, a large number of employees, strong production capacity, and market competitiveness, controlling China's largest shipbuilding

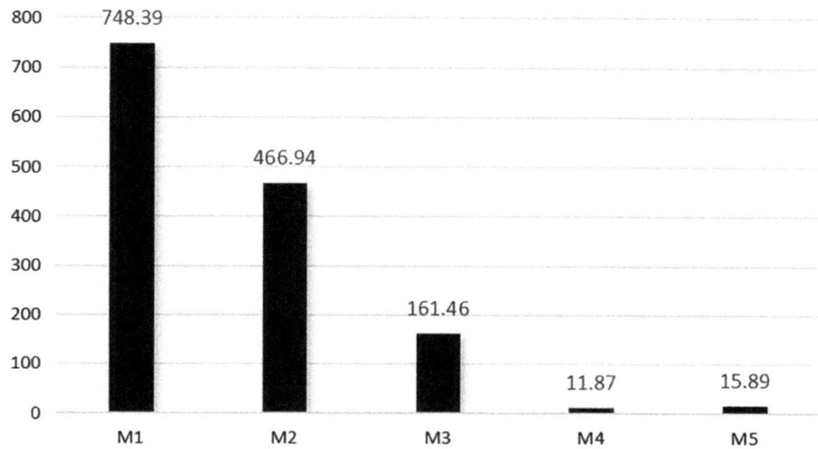

Fig. 13.2 Operating revenue of Chinese shipbuilding enterprises in 2023. *Data source* Annual reports of respective shipbuilding enterprises

and repair base. The group's business scope is extensive, mainly covering the R&D, design, manufacturing, sales, and service support of weapons and equipment, while also involved in multiple industries such as ships and offshore engineering. This diversified business structure gives the company high market adaptability, enabling it to respond quickly to market fluctuations and effectively meet customers' diversified needs. In specific areas such as the global car transport ship market, China's shipbuilding industry continues to lead in market share, with ZGCBJT holding nearly 50% of the global market share in the PCTC (Pure Car and Truck Carrier) sector.

In April 2024, ZGCBJT signed an agreement with an energy company to build 18 of the world's largest 271,000 cubic meter ultra-large LNG carriers, a feat that broke the global record for the scale of a single shipbuilding order. This order was undertaken by Hudong-Zhonghua Shipbuilding, demonstrating China's leading position in the design and manufacturing of large LNG carriers. According to the latest data, in 2023, ZGCBJT's orders on hand reached 139.39 million dwt, a year-on-year increase of 32.0%, accounting for 55.0% of the global total; new orders received reached 71.20 million dwt, a year-on-year increase of 56.4%, accounting for 66.6% of the global total. This indicates that the company will maintain stable business growth and market share for a period of time, further consolidating the group's leading position in the global shipbuilding industry. In the 2023 fiscal year, ZGCBJT's total operating revenue achieved a year-on-year increase of 25.81%, reaching 74.839 billion RMB, showing a continuous growth trend, while net profit attributable to the parent company also reached 2.957 billion RMB, a year-on-year increase of 1614.73%. This astonishing growth rate is mainly due to the rapid growth of the company's business and improved profitability. Entering 2024, this good momentum continued and accelerated. In the first quarter of 2024, operating revenue reached 15.27 billion RMB, a

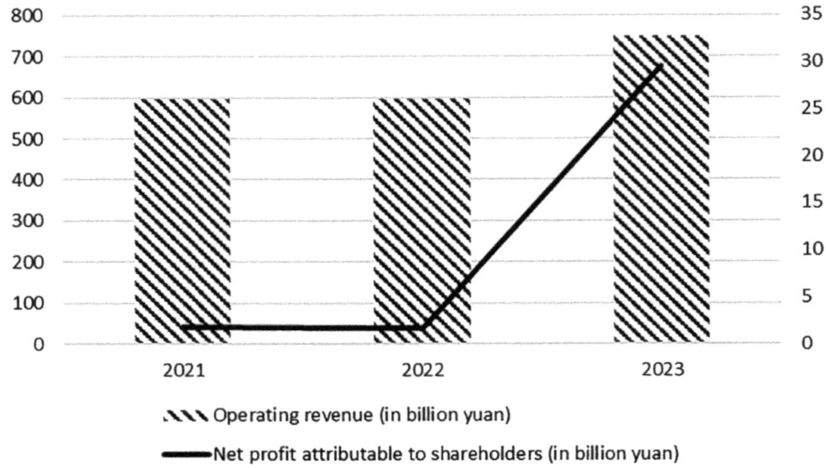

Fig. 13.3 CSSC operating conditions from 2021–2023. *Data source* Company annual report

year-on-year increase of 68.84%; net profit attributable to the parent company also increased by 821.12% year-on-year. The group performs excellently in both shipbuilding and financial performance, not only occupying a leading position in the global market but also achieving sustained and stable growth for the group. In the future, with the further recovery of the global economy and the continuous development of the shipping market, the group will continue to maintain strong growth momentum (Fig. 13.3).

(2) ZGZ was established in 2008 and focuses on the R&D, design, and manufacturing of the entire ship industry chain. Its business scope is wide and deep, mainly covering various equipment and facilities for maritime defense and development, as well as ship repair and modification. Additionally, it is actively involved in multiple emerging business areas promoted by the military-civilian integration strategy. According to relevant data, ZGZ achieved a year-on-year increase of 5.75% in total operating revenue in 2023, reaching 46.694 billion RMB. Specifically, the operating revenue from maritime defense and maritime development equipment was 9.783 billion RMB, accounting for 21% of the company's total revenue; revenue from marine transportation equipment was 14.125 billion RMB, accounting for 31%; revenue from ship supporting and electromechanical equipment was 8.815 billion RMB, accounting for 19%; revenue from deep-sea equipment and ship repair and modification was 8.576 billion RMB, accounting for 19%; revenue from strategic emerging industries related to military-civilian integration and others was 4.375 billion RMB, accounting for 10%.

As an enterprise with significant influence in the domestic shipbuilding field, ZGZ has the following major advantages: First, market advantage. ZGZ plays a key

R&D and supply role for domestic naval vessel equipment, holding a high market share. It has strong R&D capabilities and technical level, occupying an important position in the global shipbuilding market [1]. In recent years, the company has continuously increased its investment in independent R&D in the field of ship equipment, aiming to further expand domestic and international market share by launching high-value-added and highly competitive products. Second, technological advantage. ZGZ has excellent R&D strength and technological advantages, overseeing 28 research institutes specializing in military ships, covering multiple professional fields such as navigation technology, ship electronic systems, and communications. The company continues to invest in and deepen R&D work in several key technology areas, including top-level planning and design of ship equipment, power systems, and advanced manufacturing technology and information integration. It also focuses on perfecting the technological innovation system. The group has established a complete chain from basic research, applied research to achievement transformation, forming an innovation model closely integrating industry, academia, research, and application. By continuously increasing R&D investment, the company has accumulated a large number of cutting-edge technological achievements at the international leading level and made major breakthroughs in several key technology areas, becoming a benchmark and main force leading industry technological innovation.

The company's latest financial report shows that in the first half of 2024, ZGZ demonstrated strong growth momentum and excellent operating performance. Operating revenue reached 22.102 billion RMB, a year-on-year increase of 31.04%. In terms of profitability, the company's net profit reached 539 million RMB, a significant year-on-year increase of 199.44%. The remarkable achievements of the company are not only due to the optimization of the product structure and strong market demand but also closely related to the company's efficient cost control and continuous technological innovation (Fig. 13.4).

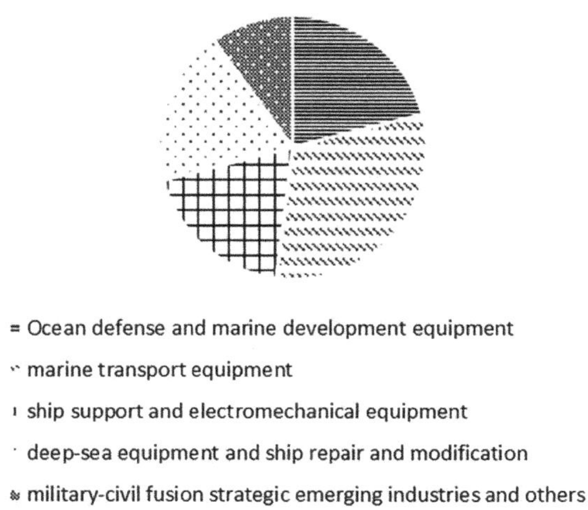

Fig. 13.4 CSIC revenue composition in 2023. *Data source* Company annual report

(3) Since its establishment in 2003, JLK has always placed technological innovation at the core. It is an important R&D and production base for steel-aluminum ships in China. Compared to its peers, JLK demonstrates excellent technical expertise, focusing on the R&D and practice of technologies such as multi-material hull application exploration, comprehensive application of new technologies, green production processes, and clean energy propulsion. These efforts have enabled JLK to make significant progress at the forefront of ship technology, promoting the industry's green transformation and technological innovation. At the same time, the company has a rich product line, able to provide diverse ship products covering various uses and specifications aimed at meeting the needs of different customers. Whether in public service law enforcement, tourism and leisure, or emergency rescue fields, JLK can provide tailored solutions. According to the company's latest financial report, in the 2023 fiscal year, JLK achieved a total operating revenue of 1.187 billion RMB, a year-on-year increase of 74.4%. Entering 2024, it showed strong growth momentum in the first half of the year alone, with operating revenue reaching 766.8 million RMB, also a year-on-year increase of 76.5%. In terms of profitability, the company's net profit attributable to shareholders reached 21 million RMB, a significant increase of 62.21% compared to the same period last year. JLK's remarkable achievements are due both to the continuous optimization of the product structure and sustained strong market demand, as well as the company's efficient cost control strategies and continuous technological innovation efforts.

(4) Established in 2003, YGK is a high-tech enterprise with profound technological accumulation and broad market influence in the field of ship electronics and informatization. Since its establishment, the company has always focused on the R&D, manufacturing, and sales of ship navigation, communication, monitoring, and intelligent systems, committed to tailoring comprehensive, efficient, and reliable solutions for global customers. In terms of technological R&D, YGK has an R&D team composed of industry experts and senior engineers. With their profound professional knowledge and rich practical experience, they continuously promote technological innovation and product upgrading. Meanwhile, a series of ship electronic products independently developed by the company, such as high-precision satellite navigation systems, intelligent communication terminals, and integrated monitoring platforms, are not only performant but also highly customizable, able to meet the diverse needs of different customers and different ship types. In terms of product quality, YGK always adheres to strict quality control standards. Every link from raw material procurement to finished product leaving the factory undergoes meticulous inspection and strict testing to ensure that each product meets international advanced levels. At the same time, the company has also established a perfect after-sales service system, providing customers with timely and professional technical support and maintenance services, winning the trust and praise of numerous customers. Relying on excellent product performance and high-quality service, the company has established a good brand image in the field of ship electronics and informatization. Its products are widely used in merchant ships, warships, yachts, and offshore

engineering equipment, providing solid support for the safe and efficient operation of the global shipping industry. According to the company's latest financial report, the operating revenue in 2023 was 1.589 billion RMB. After entering 2024, affected by multiple factors such as fluctuations in the global economic environment, adjustments in the shipping market, and supply chain uncertainties, the company's operating revenue in the first half of the year was 557 million RMB, showing a certain degree of decline compared to the same period in 2023. This change indicates that although YGK has a solid industry foundation and advanced technological advantages, it still needs to flexibly adjust its strategies when facing rapid changes in the external environment to cope with market uncertainties and challenges.

13.3 Analysis of the Development Environment for Chinese Shipbuilding Enterprises

13.3.1 Analysis of Driving Forces for the Development of Chinese Shipbuilding Enterprises

13.3.1.1 Growth in Market Demand

As global trade continues to grow, the demand for shipping is also expanding. The recovery of the global economy and the increasing frequency of international trade activities are prompting countries to engage in closer exchanges in areas such as goods, services, and technology. Particularly, the international trade of bulk commodities such as oil, natural gas, and iron ore relies heavily on large-scale ship transport, directly driving substantial growth in cargo transport demand, thereby promoting the prosperous development of the shipbuilding industry. Against this background, the deepening implementation of the "Belt and Road" Initiative adds a rich touch to the global trade landscape. This initiative aims to promote connectivity among countries along the route through infrastructure construction, enhanced trade exchanges, and deepened financial cooperation. China, as the initiator and core driving force of the "Belt and Road," has opened up unprecedented market space for Chinese shipbuilding enterprises and also guided the global shipbuilding industry towards a new stage of development.

13.3.1.2 Policy Environment

On the one hand, the state has listed the shipbuilding industry as a key supported area under the strategy of strengthening the country through science and technology, strengthening investment in its technological innovation. To promote industrial progress, the state actively advocates collaborative innovation among enterprises, higher education institutions, and research entities, striving to overcome core technological difficulties, and has implemented a series of talent cultivation and introduction projects aimed at building a professional team of ship technology talents with global competitiveness. These strategies have significantly enhanced the core competitive advantages of the shipbuilding industry. In the context of implementing the strategy of strengthening the country through manufacturing, the state has set clear development blueprints and standards for the shipbuilding industry, clearly pointing out the core position of the shipbuilding industry in the manufacturing system. To this end, the state actively promotes the digital and intelligent transformation of shipbuilding enterprises, advocates the active introduction of cutting-edge manufacturing technologies, and promotes the advancement of China's shipbuilding industry towards a green and sustainable direction, thereby further enhancing the comprehensive strength and international competitive position of China's shipbuilding industry. Furthermore, the government attaches great importance to the development of the shipbuilding industry and has issued a series of policy measures to support the transformation, upgrading, and innovative development of enterprises. Policy documents such as "Made in China 2025" and the "Opinions on Accelerating the Construction of an Ecological Civilization" emphasize actively promoting the green development of the shipbuilding industry, providing more comprehensive and systematic support. The "Shipbuilding Industry Green Development Action Outline (2024–2030)" issued by the Ministry of Industry and Information Technology and others also clearly proposes the overall requirements and development goals for promoting the green development of China's shipbuilding industry, emphasizing the need to promote the implementation of the international ship greenhouse gas emission reduction strategy while accelerating the green transformation of the shipbuilding industry. On the other hand, the IMO, as a specialized agency of the United Nations responsible for maintaining maritime safety and preventing pollution of the marine environment from ships, has issued a series of restrictive policies on ship emissions aimed at promoting the transition of the global shipping industry towards green and low-carbon directions. Facing the severe situation of global climate change, the IMO requires shipping companies to adopt more efficient ship technologies and accelerate the phasing out and updating of old ships. This measure not only aids technological R&D in the new shipbuilding market and the emergence of new ship technologies but also opens up new growth paths and development directions for shipbuilding enterprises.

13.3.2 Challenges Facing the Development of Chinese Shipbuilding Enterprises

13.3.2.1 Cost Challenges

Chinese shipbuilding enterprises face the challenge of rising costs. On the one hand, continuously rising raw material prices have become a significant challenge that cannot be ignored for shipbuilders. As a resource-intensive industry, shipbuilding production is highly dependent on raw material supply. The continuous rise in raw material prices has brought huge cost pressure to shipbuilders. Steel is the most basic and largest-consumption raw material in shipbuilding. However, in recent years, affected by multiple factors such as imbalance in Supply and demand relationship, rising production costs, and tightening environmental policies, steel prices in the global market have shown a continuous upward trend, increasing the production costs of shipbuilding enterprises. At the same time, the prices of auxiliary materials such as paint and cables are also continuing to rise, putting great pressure on shipbuilding enterprises. On the other hand, the continuous rise in labor prices has become another severe challenge faced by shipbuilding enterprises. Shipbuilding, as a typical labor-intensive industry, requires a large amount of labor to participate in its production process, from initial cutting, welding, and assembly to later painting, debugging, and other various links. However, due to continuous global economic development and changes in the labor market, labor costs are generally rising, and shipyards face huge pressure on labor costs.

13.3.2.2 Market Challenges

Currently, another major challenge faced by global shipbuilding enterprises comes from the uncertainty of the international trade environment, which in turn stems from the complex and volatile global economic situation. On the one hand, the rise of trade protectionism has become a significant andcannot be ignored problem. Some countries, out of consideration for protecting local economic interests, implement a series of trade restriction measures such as raising tariffs and setting up non-tariff barriers on products of shipbuilding enterprises, directly hindering their international circulation and trade activities. The strengthening of trade protectionism may lead to reduced export orders and decreased market share for those Chinese shipbuilding enterprises that rely on exports, thereby increasing the difficulty and cost of their entry into the international market. On the other hand, the shipbuilding sector is greatly affected by global macroeconomic fluctuations, and its phased market characteristics are relatively obvious. In situations of global economic prosperity and active international trade, shipbuilding enterprises will usher in a surge in orders during the peak period of shipping demand. Once the global economy experiences recession or fluctuations, shipping demand decreases, and the ship market will quickly cool down, leading to problems such as a sharp drop in orders and overcapacity. The operational risks brought by this phased market characteristic to shipbuilders are equally cannot be underestimated.

13.3.2.3 Technological Challenges

When Chinese shipbuilding enterprises are committed to developing high-tech level ship types, they face shortcomings at the conceptual design level and still rely to a large extent on the mature design schemes of foreign companies. This phenomenon not only leads to enterprises having to pay high design license fees and intellectual property transfer fees but, more importantly, hinders their ability to quickly adjust and achieve customized designs according to their own needs and market changes, thereby weakening the competitiveness of Chinese ship enterprises in the international high-end ship market. In addition, the lack of core equipment supporting capabilities in Chinese shipyards is also a key factor restricting their growth. In shipbuilding, core components such as power systems and electronic information equipment like communication and navigation, whose efficiency and quality directly affect the overall performance and market competitiveness of the ship [2]. However, currently, some Chinese shipbuilding enterprises still have weak supporting capabilities in these areas and still need to rely on importing some equipment from abroad. This status quo not only increases the cost of shipbuilding but also further exacerbates the supply chain risks for enterprises.

13.4 Market Trends and Strategic Recommendations

13.4.1 Future Market Trends of the Shipbuilding Industry

13.4.1.1 Parallel Advancement of Intelligence and Greenification

With the continuous development of information technology and intelligent technology, intelligence and greenification are regarded as the twin drivers promoting the transformation and upgrading of the shipbuilding industry. The application of intelligent and green technologies can not only improve the production efficiency of shipbuilding enterprises and the operational efficiency of ships but also significantly reduce energy consumption in ship transport. Through intelligent monitoring systems, enterprises can monitor ship status in real-time and prevent potential failures [3]; using big data analysis, they can optimize route planning and reduce operating costs; adopting fluid dynamics simulation technology to design hulls with less resistance can significantly reduce energy consumption in ship transport; utilizing renewable energy sources such as solar and wind power as auxiliary or main power sources for ships can significantly reduce energy consumption during ship operation. Greenification and intelligence are inevitable trends for the future development of the shipbuilding industry. In future development, Chinese shipbuilding enterprises must actively respond to the requirements of the national green development strategy and intelligent manufacturing strategy, promoting green production and intelligent transformation. By adopting environmentally friendly materials and energy-saving technologies, and combining advanced technologies such as the Internet of Things, big data, and artificial intelligence, they can further improve enterprise production efficiency and reduce ship energy consumption [4].

13.4.1.2 Active Promotion of Personalized Customization and International Development

Personalized customization is one of the future development trends of the shipbuilding industry. Facing increasingly diversified and personalized customer demands, shipbuilding enterprises are ushering in a new era of personalized customization. Shipbuilding enterprises should pay more attention to listening to customer needs and provide tailored solutions rather than just being satisfied with the production of standardized products. This transformation is reflected not only in the appearance design and internal layout of the ships but also in core areas such as performance parameters and functional configuration. Through close cooperation with customers, shipbuilding enterprises can not only create ship products that meet market demands and have unique competitive advantages but also flexibly meet the personalized needs of customers. Furthermore, the trend of international development in the shipbuilding industry is becoming increasingly evident against the backdrop of globalization and the continuous development of the marine economy. To expand the international market for shipbuilding, Chinese shipbuilders need to strengthen their international strategic layout, which includes strengthening exchanges and cooperation with the international ship market, as well as actively participating in the formulation and promotion of international standards. Through international development, Chinese shipbuilding enterprises will be able to more effectively face the opportunities and challenges of the global market, thereby ensuring the achievement of sustained and stable development for the enterprises.

13.4.2 Strategic Recommendations

13.4.2.1 Strengthen Technological Innovation and R&D

In the context of increasingly fierce global competition, technological innovation has become the core driving force for the continuous growth of enterprises. This is particularly significant for technology-intensive industries like shipbuilding. Chinese shipbuilding enterprises, as a key part of China's manufacturing industry, are facing dual challenges and valuable opportunities from both domestic and international markets. To gain an advantage in the fierce market competition, shipbuilding enterprises urgently need to increase R&D investment to enhance the technical level and added value of shipbuilding [5]. On the one hand, enterprises need to enhance the intensity of investment in scientific research and R&D. During the research process, enterprises must strengthen the protection of intellectual property rights in the shipbuilding field, using methods such as patent layout and technical confidentiality to defend the achievements and advantages of their own technological innovation. On the other hand, enhancing the ability to develop new products and the added value of products is crucial. Chinese shipbuilding enterprises should be guided by market demand, increase efforts in new product development, accelerate the transformation of scientific research achievements into products with market competitive advantages, and meet user demand for high-performance, high-efficiency, and low-energy consumption ships [6]. At the same time, by introducing new processes, enterprises can significantly improve the performance, safety, environmental protection characteristics, and intelligence level of ship products, thereby further enhancing the market competitive advantage of shipbuilding enterprises [5].

13.4.2.2 Optimize Human Resource Allocation

Shipbuilding is a labor-intensive industry. For shipbuilding enterprises to enhance overall operational efficiency and core competitiveness, optimizing human resource allocation is also a key link. By implementing lean production and management, promoting flexible employment systems, and strengthening employee skill training and improvement, shipbuilding enterprises can utilize human resources more efficiently, thereby reducing the risk of rising labor costs. Specifically, it can be divided into the following aspects: First, actively implement lean production and lean management. For shipbuilding enterprises, lean production is a production model that pursues zero waste and continuous improvement. Enterprises can conduct in-depth analysis of the existing production process, identify and eliminate non-value-added activities within it (such as waiting time, over-processing, inventory backlog, etc.), thereby reducing the phenomenon of labor waste. This includes optimizing production layout and introducing high-efficiency production facilities and logistics control systems. Second, a flexible employment system must be implemented in enterprises. Shipbuilding enterprises need to adopt more flexible employment strategies to cope with market demand fluctuations and the uncertainty of production tasks. This includes establishing cooperation with labor dispatch companies for introduction, cooperating with universities to recruit interns, and utilizing online platforms to recruit part-time employees. Furthermore, to achieve effective cost control, the number of employees and working hours can be flexibly adjusted according to seasonal and project needs. Through a flexible employment system, enterprises can reduce the risk of fixed costs while responding more flexibly to market changes. Third, increase the intensity of employee vocational skill training. Employees are one of the most important resources of an enterprise. Improving employees' vocational skill levels is crucial for optimizing human resources. Enterprises need to establish a complete training system for employees, including training content for new employees, while also encouraging employees to actively participate in industry certifications and qualification exams, improving personal professional literacy and competitiveness to enhance employees' shipbuilding skills, achieve the goal of one person for multiple posts and one post with multiple skills, and thereby meet the diverse production needs of shipbuilding.

13.4.2.3 Strengthen Supply Chain Management

Shipbuilding enterprises can strengthen cost control and improve operational efficiency by strengthening supply chain management to cope with the problem of rising raw material prices. First, establish diversified sourcing channels. To expand procurement channels, shipbuilding enterprises should actively search for and evaluate new suppliers. For example, enterprises can reach more potential suppliers by participating in industry exhibitions, utilizing internet platforms, and forming professional procurement teams. They can disperse purchase orders to multiple suppliers to reduce risks brought by a single supplier. At the same time, competition among multiple suppliers also helps enhance the enterprise's bargaining power, thereby obtaining more favorable prices and better services. Second, establishing long-term stable cooperative partnerships with relevant shipbuilding raw material suppliers is also an important reason for cooperation between the two parties. In situations where market prices fluctuate greatly, enterprises and suppliers can jointly negotiate to formulate flexible pricing mechanisms and risk-sharing schemes, as well as sign long-term supply contracts and agree on price adjustment mechanisms. At the same time, clearly define the rights and obligations of both parties in the contract when prices fluctuate, thereby reducing the adverse impact of market price fluctuations on the production costs of enterprise shipbuilding.

13.5 Conclusion

After decades of continuous struggle and determined innovation, Chinese shipbuilding enterprises have successfully joined the broad stage of the global shipbuilding industry and occupy a pivotal leadership position. This outstanding achievement not only demonstrates the strong strength of Chinese manufacturing but also fully reflects the comprehensive competitiveness of China's shipbuilding industry in the global shipbuilding sector. Looking forward, Chinese shipbuilding enterprises need to demonstrate a firmer attitude to cope with the fierce competition in the global ship market. In the subsequent development process, Chinese ship enterprises will uphold the core concepts of innovation-driven and green development, continuously strengthen competitive advantages and sustainable development capabilities, inject more Chinese momentum into the prosperity and progress of the global shipbuilding industry, and thereby occupy more competitive seats in the international market.

References

1. Shaolin Li. The Second Round of Asset Injection of China CSSC Has Started [N]. China Securities Journal, 2011-04-12(A08).
2. Xinhu Wang. Continuous Improvement of Quality Management in Ship Manufacturing Process [J]. Ship Materials and Market, 2024, 32(05): 112–114.
3. Ya Shu. Risk Assessment and Response Strategies for Ship Manufacturing Projects Based on Big Data Analysis [J]. Today's Wealth, 2024, (18): 65–67.
4. Chenshuo Zhang. Research on the Development Strategy of JN Clean Energy Shipbuilding Company [D]. Shandong University of Finance and Economics, 2024.
5. Zhaolin Gu, Dashan Wang. Economic Analysis of New Production Capacity in China's Shipbuilding Industry Under the Dual Circulation Background [J]. Ship Materials and Market, 2024, 32(05): 7–10.
6. Jindan Liang, Zhen Chen, Mengyuan Yang, et al. Application Analysis of Intelligent Manufacturing Technology in Shipbuilding [J]. China Water Transport, 2024, (05): 66–68.

Chapter 14
Exploration of Growth Paths for Rail Transit Vehicle Manufacturing Enterprises

Malin Song and Jialuo Xia

Abstract This article provides an overview of the basic profile, characteristics, classification, current status, and challenges of rail transit vehicle manufacturing companies. These companies specialize in the design, production, assembly, and maintenance of rail transit vehicles. They are characterized by strong technical expertise, a well-developed industrial chain, a high degree of internationalization, and a high degree of safety and customization. Their products generally include subways, intercity trains, and trams. Currently, the industry's market size continues to grow, its industrial chain continues to improve and progress, and it is actively expanding into overseas markets, moving towards internationalization. However, facing challenges such as technological difficulties and insufficient independent innovation, it is necessary to increase R&D investment and enhance the company's technological capabilities. At the same time, the government needs to provide support and policies that respond to the market. Overall, rail transit vehicle manufacturers play a vital role in the development of the rail transit industry, but they still need to continue to work hard on technological innovation and market expansion. Finally, according to statistics, N1 has always been in a leading position among rail vehicle manufacturers. This article provides a detailed introduction to the company's basic profile, business layout, international trade, economic benefits, technological innovation and market position. It can be concluded that the company's core competitiveness is extremely strong in rail transit vehicle manufacturing. In the future, with the continuous development of the global rail transit market and the continuous improvement of N1's own strength, its market position is expected to be further consolidated and improved.

M. Song · J. Xia (✉)
Anhui University of Finance and Economics, Bengbu, China
e-mail: 15255381893@163.com

J. Wang and M. Song (eds.), *The Green Path of Chinese Transportation Enterprises*, Resources, Climate and Sustainable Development,
https://doi.org/10.1007/978-981-95-5289-4_14

14.1 Basic Information of the Enterprise

Rail transit vehicle manufacturers specialize in the design, production, assembly, and maintenance of rail transit vehicles. They typically possess strong technical R&D capabilities, advanced production equipment, and rigorous quality control systems to meet the performance, safety, comfort, and environmental requirements of various rail transit systems (such as subways, light rail, trams, intercity rail, and high-speed rail). These companies typically possess strong technical expertise to meet the complex and ever-changing demands of rail transit systems. For example, the requirements vary from city to city and from system to system, necessitating customized design and production tailored to local requirements. From vehicle design and material selection to manufacturing processes and control systems, every aspect requires high-precision technical support, resulting in exceptionally high vehicle safety performance. Specific manufacturing processes typically involve multiple supply chains, including raw material supply, component manufacturing, and complete vehicle assembly, each of which is closely interconnected, forming a synergistically developed industrial chain. Due to the high level of technical expertise, these companies often exhibit a high degree of internationalization. On the one hand, they need to introduce advanced international production technologies and management expertise to enhance their competitiveness. On the other hand, they also need to continuously expand into international markets and participate in international competition and cooperation. This requires a global perspective, strategic thinking, and the ability to communicate and collaborate across cultures.

Enterprises are categorized by organizational nature into state-owned enterprises, private enterprises, foreign-invested enterprises, and mixed-ownership enterprises. In the rail transit vehicle manufacturing sector, state-owned enterprises typically possess significant scale and strength, undertaking the construction and production of major national rail transit projects. They possess advanced technology and extensive experience, enabling them to provide high-quality rail transit vehicle products and services. Private enterprises typically possess flexible operations and keen market insight, enabling them to promptly adjust their product mix and production strategies based on market demand to meet diverse customer needs. Some foreign-invested enterprises also hold a significant market share in China's rail transit vehicle manufacturing sector. These enterprises typically possess advanced technology and management experience, enabling them to bring international products and services to the Chinese market. The industry chain is divided into upstream, midstream, and downstream enterprises. Upstream enterprises primarily engage in vehicle design and raw material supply, providing technical support and raw material security to downstream manufacturing enterprises. Midstream enterprises, the core link in the entire industry chain, are mostly manufacturers responsible for transforming designs into actual products, producing and testing them, and bear significant responsibility for product quality and performance. Downstream enterprises primarily engage in the operation, maintenance, and upkeep of rail transit vehicles.

14.2 Current Status of Enterprise Development

14.2.1 Market Size Continues to Grow

In recent years, with the acceleration of urbanization and the continuous growth of the population, the demand for urban rail transit has continued to grow, driving the rapid development of the rail transit vehicle manufacturing industry. According to information obtained from relevant statistical bureaus, the total market value of my country's rail transit vehicle industry has increased from 653.7 billion yuan in 2018 to 967.3 billion yuan in 2022, with an average annual compound growth rate of 10.3%. The number of rail transit vehicles has also increased from 14,400 in 2013 to 57,300 in 2021. It is expected that in the next few years, with the continuous improvement and expansion of urban rail transit networks, the market size of the rail transit vehicle manufacturing industry will continue to grow, gradually moving towards internationalization [1] (Fig. 14.1).

The expansion of the market has led to a continuous improvement in the regional layout of manufacturing companies. According to Qianzhan.com, first- and second-tier companies are concentrated in first-tier cities such as Beijing and Shanghai, while third-tier companies are located in Shanxi, Henan, Jiangsu, Zhejiang, Guangdong, and Sichuan. These cities have relatively complete industrial chains and supporting systems, providing strong support for the development of these companies. Furthermore, with the advancement of the Belt and Road Initiative and the continuous development of international trade, rail transit vehicle manufacturers are also actively expanding into overseas markets and participating in international competition and cooperation.

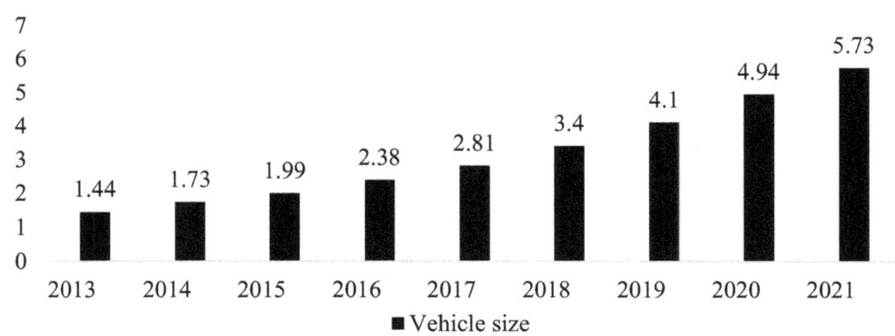

Fig. 14.1 Rail vehicle scale from 2013 to 2021. *Data source* National Bureau of Statistics

14.2.2 The Competitive Landscape is Becoming Increasingly Fierce

Competition in the rail transit vehicle manufacturing industry is fierce. As rail transit vehicle manufacturing is a technology-intensive industry, technological advancements are rapidly evolving, and this rapid iteration further intensifies competition [2]. Despite the large number of companies in the industry, market share is relatively concentrated, resulting in many companies receiving only a small share of market profits. Companies can be divided into different competitive tiers based on revenue. N1, the industry leader, boasts rail transit manufacturing revenue exceeding 100 billion yuan, placing it in the first tier. Other companies, such as N2, N3, N4, and N5, also possess strong competitive strength, but still lag significantly behind N1. This is likely due to their core strengths: companies in other tiers typically possess strong advantages in technology research and development, production, manufacturing, and marketing. However, within the industry chain, these companies have collectively driven the rapid development of the rail transit vehicle manufacturing industry.

14.2.3 Continuous Updates of Technology and Policies

Technological innovation is an inexhaustible driving force for the sustainable development of rail transit vehicle manufacturers. With the rapid advancement of science and technology and the diversification of market demands, companies are striving to increase R&D investment, strengthen technological innovation, and upgrade products [3]. For example, N1 has achieved significant breakthroughs in independent R&D, successfully developing a series of Chinese-standard subway trains and intelligent urban rail trains with fully independent intellectual property rights. These trains have reached internationally leading levels in key component independence and system integration. In today's big data era, companies have made significant progress in intelligent, green, and lightweighting. In terms of intelligence, companies are leveraging advanced technologies such as the Internet of Things, big data, and artificial intelligence to achieve intelligent dispatching, monitoring, and maintenance of rail transit vehicles. In terms of greening, companies are reducing energy consumption and emissions of rail transit vehicles by adopting new energy and energy-saving technologies. In terms of lightweighting, they are reducing the weight of rail transit vehicles and improving operating efficiency through the use of new materials and optimized structural designs.

The policy environment has a significant impact on the development of rail transit vehicle manufacturing companies. In recent years, the Chinese government has attached great importance to the construction and development of urban rail transit and has introduced a series of policy measures to support the development of the rail transit vehicle manufacturing industry. For example, documents such as the "Notice on Supporting the Construction of New Generation Artificial Intelligence Demonstration Application Scenarios" and the "Opinions on Accelerating the Construction of a Strong Transportation Nation through Technological Innovation" propose accelerating the development of the rail transit manufacturing industry and promoting independent research and development and industrialization of rail transit vehicle manufacturing. These policy measures provide a favorable development environment and opportunities for rail transit vehicle manufacturers. During this period, companies should actively seize these policy dividends, optimize their capital structures, and focus on developing diversified financing channels, such as attracting strategic investors and issuing corporate bonds, to diversify financial risks and ensure the long-term stability of their projects. Conversely, when policies tighten and financing channels are limited, companies need to strengthen capital management and be prudent in their budgets. They should ensure the security of project funding chains by optimizing cost structures and improving operational efficiency, and adjust project schedules and scales appropriately to cope with market uncertainties.

14.3 Difficulties in Enterprise Development

14.3.1 Technical Level

The current rail transit vehicle manufacturing market tends to favor supporting product manufacturers with simultaneous design and development capabilities. This places high demands on companies' capabilities in computer-aided design, mold manufacturing, and analytical testing. However, the development of these technologies requires significant R&D investment, as well as high rigid requirements for production and testing equipment, requiring substantial capital investment. Companies seeking to enter the market are often initially constrained by technical and financial constraints, making it difficult to meet the requirements of vehicle manufacturers. For companies to truly enter the market, the process of introducing, digesting, and learning advanced manufacturing processes and technologies, developing independent R&D and innovation capabilities, and gaining technical recognition from vehicle manufacturers requires long-term accumulation and extensive industry experience, making it difficult for new entrants to achieve this in a short period of time [4].

While Chinese rail transit vehicle manufacturers have made some progress in technological innovation in recent years, their R&D investment remains relatively insufficient compared to their international counterparts. This is primarily reflected in the low ratio of R&D expenditure to operating revenue and limited dedicated funds for technological innovation. According to relevant statistics, some leading international companies, such as Germany's Siemens and the United States' General Electric, generally have a high R&D expenditure-to-operating-revenue ratio, investing heavily in the development of new products and technologies. However, this ratio is relatively low for Chinese rail transit vehicle manufacturers, limiting their investment and development in technological innovation. While the number of patents has been increasing year by year, there remains a significant gap in both the quantity and quality of patents compared to their international counterparts. In particular, Chinese companies have a relatively weak foundation in the R&D of key systems and core components, lacking core technologies with independent intellectual property rights. This results in relatively weak competitiveness in the international market, making it difficult for them to compete with leading international companies.

14.3.2 Management Level

Some manufacturing companies may be overly focused on short-term profits, neglecting long-term development planning. This can lead to a lack of resilience in the face of market changes and technological innovations. Furthermore, they fail to clearly define their market position in their development strategies, resulting in severe product homogeneity and a lack of competitive advantage. This can damage a company's brand image over time, leading to consumer doubts about its products and services, and ultimately impacting its profits and long-term development. Failure to promptly adjust a company's strategic focus and flexibly adapt to specific market changes can lead to losses in operating income.

Rail transit vehicles have extremely high safety and reliability requirements, but some companies may have redundant processes in their production processes, leading to low production efficiency, increased costs, and subpar products. This can also lead to quality control loopholes, impacting product quality and brand image. Secondly, supply chain management is a crucial aspect for manufacturers. Some companies may face issues such as poor supplier selection, inventory backlogs, and high logistics costs, resulting in unsatisfactory operational results and a loss of profits. Furthermore, the vehicle manufacturing industry primarily involves infrastructure transfers, and companies may not fully consider details when compiling asset inventories, resulting in poor quality. During the billing process, asset ledger management is not standardized, and there is a mismatch between physical assets and ledger records, hindering the effective utilization and management of assets. Finally, companies may experience inefficient asset management due to a lack of data interaction between management systems, large inventory counts, and a lack of standardized and efficient methods.

14.3.3 Market Level

With the acceleration of urbanization and the rise of emerging markets, market demand for rail transit vehicles is influenced by a variety of factors, including urban development plans, government policies, and the economic environment. These fluctuations can lead to fluctuations in market demand, profoundly impacting companies' operational strategies, market layout, and long-term development. Furthermore, developing and entering emerging markets requires time and resources, carries certain risks, and is prone to input-output imbalances. Companies also need to pay attention to market demand in these specific regions and develop corresponding market strategies, which in turn requires a certain level of market adaptability. With market expansion, competition is inevitable. The rail transit vehicle manufacturing industry is a highly competitive market, with numerous domestic and international companies vying for market share. Domestic companies not only face competitive pressure from international giants but also face fierce competition from their domestic peers. Consequently, some companies, in an effort to seize market share, may resort to low-price strategies or compromise product quality, leading to a vicious cycle of price wars and low-quality competition. This not only damages companies' brand image and long-term profits but also impacts the healthy development of the entire industry.

The market is increasingly demanding companies' ability to diversify their production. With the continuous development and improvement of rail transit systems, customer demands for vehicles are also becoming more diverse. Different customers have different requirements for vehicle performance, functionality, and appearance, which poses greater challenges to companies' R&D and production capabilities. To meet customer needs, companies need to provide customized products and services. However, customized production can increase production costs and complexity, requiring companies to make corresponding adjustments in production management and production [5].

In some regions, the rail transit vehicle manufacturing industry chain is underdeveloped, leading to difficulties for companies in sourcing raw materials and sourcing parts during production. This increases operating costs and reduces production efficiency. Rail transit vehicle manufacturing involves multiple industrial links, including raw material supply, parts manufacturing, vehicle assembly, and operations and maintenance. Problems in any link of the supply chain can significantly impact a company's production and operations. Therefore, insufficient coordination between upstream and downstream industries can affect the efficiency and competitiveness of the entire industry chain.

14.4 Development Strategies

14.4.1 Obtaining Government Support

14.4.1.1 Issue Relevant Policies

The government can introduce special support policies for enterprises, such as the "Chengdu City Several Policies to Promote the High-Quality Development of the Rail Transit Equipment Manufacturing Industry (Trial)." This policy covers multiple aspects of enterprise development and can serve as a reference for the government's introduction of relevant policies. Effective policies provide enterprises with a favorable development environment and market opportunities, helping to promote the sustained and healthy development of the rail transit vehicle manufacturing industry. The government can directly invest in relevant enterprises through three aspects: enterprise qualifications, product verification and trial, and procurement. Enterprise qualifications provide a one-time reward to enterprises that obtain qualifications to manufacture EMUs, high-altitude locomotives, and freight vehicles. For example, Chengdu City offers rewards of 5 million yuan, 3 million yuan, and 1 million yuan, respectively, to enterprises that obtain these qualifications. Product verification and trial rewards encourage industry chain enterprises to conduct trial verification of key products that have not yet been used in the local rail transit sector. Direct purchasers of products that are used on operating lines will receive a one-time reward. Procurement rewards provide enterprises that purchase qualified key products and production equipment with rewards of varying amounts based on the purchase amount, encouraging them to actively promote related products and equipment into the supply chain.

14.4.1.2 Organizational Cooperation and Process Simplification

The government can strengthen the promotion of cooperation between domestic rail transit vehicle manufacturers and international certification agencies to obtain internationally recognized product certification and quality management system certification; continuously connect with and mutually recognize international standards to enhance the competitiveness of enterprises in the international market; organize or participate in exhibitions, forums and other activities in the international rail transit field to build a platform for enterprises to communicate and cooperate with international peers. Through these platforms, enterprises can understand the latest developments and trends in the international market and find partners and business opportunities.

The government should further streamline the approval process for rail transit vehicle manufacturers, reducing unnecessary steps and time, improving approval efficiency, and facilitating the handling of various procedures by enterprises through reforms such as "one-stop service" and "one-stop service." Regarding the service system, the government should strengthen the construction of government service windows, improve the professional capabilities and service levels of staff, and ensure that enterprises are provided with efficient, convenient, and high-quality services. At the same time, a sound enterprise complaint handling mechanism should be established to promptly resolve issues encountered by enterprises in the process of government services.

14.4.2 Deepening Enterprise Reform

14.4.2.1 Increase Investment in Technology and Talent

Enterprises should continue to increase their investment in technological research and development, especially in key components, control systems, intelligent technologies, etc., to enhance the core competitiveness and market share of their products; build a technology innovation system with themselves as the main body, market-oriented, and deep integration of industry, academia and research, strengthen cooperation with universities and research institutes, and jointly overcome technical difficulties. At the same time, use advanced technologies such as big data, cloud computing, and artificial intelligence to promote the development of rail transit vehicle manufacturing towards intelligence and digitalization, and improve production efficiency and product quality.

Enterprises should establish and improve talent development systems in the field of rail transit vehicle manufacturing. They should strengthen cooperation with universities and vocational colleges to cultivate high-quality technical and management personnel. They should increase efforts to attract high-end talent, attract outstanding domestic and foreign talents to join the rail transit vehicle manufacturing industry, and enhance the innovation capabilities and core competitiveness of enterprises. They should also establish and improve talent incentive mechanisms to stimulate the enthusiasm and creativity of talents and provide a strong talent guarantee for the sustainable development of enterprises.

14.4.2.2 Understand Market Trends and Formulate New Plans

With the continuous expansion and upgrading of the domestic rail transit network, companies should actively seize market opportunities, strengthen cooperation with local governments, and participate in the construction and operation of rail transit projects. They should also participate in international competition and, through international cooperation platforms such as the Belt and Road Initiative, promote rail transit products to the global market and enhance their international influence and competitiveness. As companies develop, they need to cultivate their brand image, focusing on brand building and market promotion to enhance brand awareness and reputation, and build a rail transit vehicle manufacturing brand with international influence [6].

Enterprises should closely monitor national and local government policy guidance and industry trends, and, based on their own actual conditions and market environment, formulate clear development strategies and plans, identifying development goals, paths, and measures. They should actively participate in the formulation and revision of relevant standards for rail transit vehicle manufacturing, enhancing their voice and influence in the industry. Finally, they should strictly comply with national laws, regulations, and industry standards to ensure compliance and sustainable development.

14.5 Representative Enterprises

N1 Co., Ltd. ("N1") is a leading company in rail vehicle manufacturing. N1 holds a significant position in the industry, both in China and globally, demonstrating exceptional business development, operating performance, technological innovation and R&D, market position, and brand influence. With the continued development of the global rail market and the deepening implementation of China's Belt and Road Initiative, N1 is expected to maintain its leading position and achieve even more robust growth.

14.5.1 Basic Profile of the Enterprise

N1, approved by the State Council and licensed by the State-owned Assets Supervision and Administration Commission of the State Council, was formed through the merger of N6 Co., Ltd. and N7 Co., Ltd., based on the principle of reciprocity, to become a publicly listed company with both A -shares and H-shares. With the approval of the China Securities Regulatory Commission, it was successfully listed on the Shanghai Stock Exchange and the Hong Kong Stock Exchange on June 8, 2015. Headquartered in Beijing, N1 comprises 46 wholly-owned and majority-owned subsidiaries, employs over 170,000 people, and fully inherits the entire business scope and asset base of N6 Co., Ltd. and N7 Co., Ltd. As a leading global supplier of rail transit equipment, N1 holds a leading position in scale, boasts a diverse product range, and world-class technology. This demonstrates its significant influence and strong competitiveness within the industry, contributing significantly to and playing a leading role in the development of the global rail transit industry (Fig. 14.2).

The company's core business encompasses the research and development, design, manufacturing, maintenance, sales, leasing, and technical services of railway locomotives, EMUs, urban rail transit vehicles, construction machinery, and various electromechanical equipment, electronic equipment and their components, as well as electrical and electronic equipment and environmental protection equipment. The

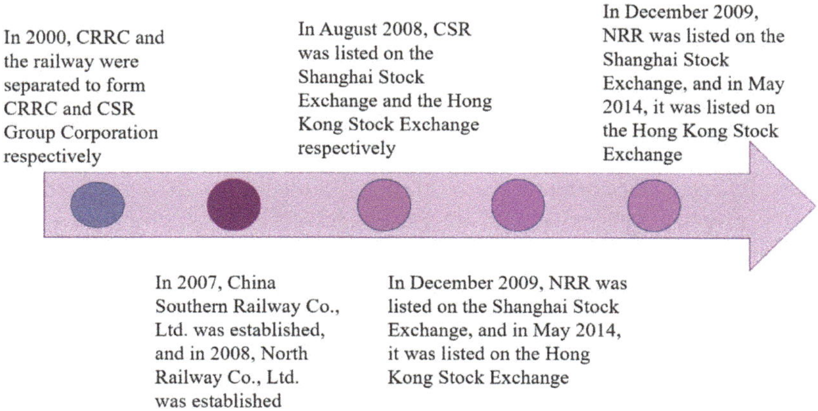

Fig. 14.2 N1 enterprise development history

company also engages in industrial investment management, asset management, and import and export operations. Since 2008, the company has led the world in new locomotive sales revenue for eight consecutive years. In recent years, the company's export turnover has shown rapid growth, with an average annual increase of over 30% from 2011 to 2015.

14.5.2 Business Layout

N1 is a global rail vehicle manufacturer and supplier with the broadest market, the most comprehensive product range, and the most advanced technology. In 2023, the company ranked 434th on the Fortune Global 500, establishing itself as a leading global rail vehicle manufacturer. Its core business focuses on the research and development, design, manufacturing, maintenance, service, and sales of locomotives, urban rail vehicles, construction machinery, environmental protection equipment, and related components. The company also provides technical services and equipment leasing, and engages in diversified businesses such as import and export, industrial investment, asset management, and information consulting. Its rail vehicle business is divided into two major segments: complete vehicles and system products. The complete vehicle business includes both rolling stock and urban rail vehicles, primarily used for mainline freight and passenger transport, intercity passenger transport, and urban and suburban commuter services. System products are used in conjunction with complete vehicle products. This demonstrates N1's near monopoly on all vehicle types in China. Unlike other manufacturers, N1 possesses a complete industrial chain, enabling it to independently manufacture rail vehicles.

14.5.3 Operating Efficiency

N1's operating revenue has maintained steady growth in recent years. According to its financial reports, N1's operating revenue showed a fluctuating trend from 2017 to 2023. From 2017 to 2019, it saw a slight increase for three consecutive years, followed by a gradual decline from 2019 to 2022. Finally, in 2023, the company's operating revenue achieved significant growth, increasing by 5.08% year-on-year, ending a three-year decline and exceeding the 2019 level. Net profit, after a slight increase from 2017 to 2018, declined continuously from 2019 to 2021, though the rate of decline gradually narrowed. In 2022, net profit achieved positive growth, increasing by 13.11% year-on-year. Net profit continued its upward momentum in 2023, albeit at a slower pace, still achieving a 0.50% increase.

The latest data shows that the company achieved rapid growth in its operating performance in the first half of 2024, with operating revenue reaching 90.039 billion yuan, a year-on-year increase of 3.13%. Net profit attributable to shareholders of the listed company reached 4.201 billion yuan, a year-on-year increase of 21.40%. Compared with the same period since 2020, the first half of 2024 saw the best performance ever, with net profit exceeding 4 billion yuan, setting the third highest record for the same period in history. Analyzing the business structure, N1's railway business revenue achieved a rapid growth of nearly 50%, becoming the main driver of the company's growth. In particular, revenue from the EMU business saw significant growth, reaching 93.16% year-on-year, accounting for 63.18% of total railway business revenue, making it a top priority for the company (Fig. 14.3).

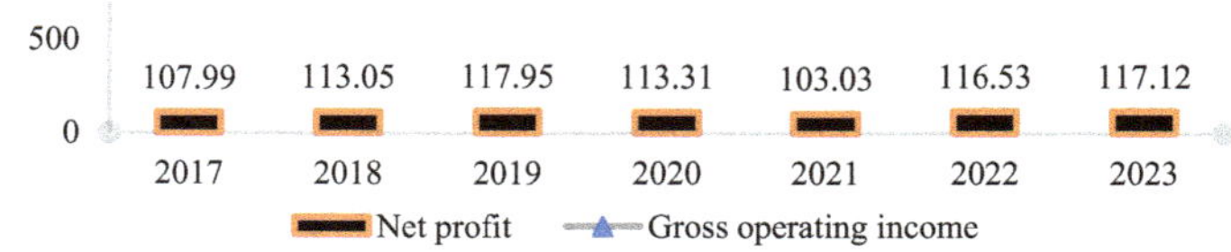

Fig. 14.3 Operating revenue and net profit from 2017 to 2023 (Unit: 100 million yuan). *Data source* N1 financial report

14.5.4 International Trade

N1 is committed to presenting China's high-quality rail transit vehicles to the world. Through diverse channels, including technology introduction, joint ventures, product exports, and after-sales service, N1 has established equal and mutually beneficial partnerships with domestic and international companies. The company's export products are distinguished by their superior quality, successfully transitioning from simple volume exports to comprehensive exports of technology, capital, and services. Its overseas target markets have expanded from developing countries to developed ones, and its export product range has evolved from a single focus to a diversified and personalized offering. Its export strategy has further evolved from "exporting" products to "deep integration" of production capacity and "high-end brand advancement." Currently, its product exports cover over 100 countries and regions worldwide, leaving a profound "Chinese imprint" on the international rail transit stage.

The company is unwaveringly committed to advancing its international operations, striving to enhance its ability to integrate global resources, navigate global competition, and meet global demands. The company is accelerating its transformation from an export-oriented enterprise to a multinational giant and even a global leader, actively expanding its international market reach and continuously raising its transnational performance index. The company is firmly focused on presenting safe, reliable, high-speed, comfortable, and environmentally friendly products and services to the world, charting the most outstanding response strategies for the sustainable development of global rail transit, and striving to become a world-class benchmark enterprise with strong international competitiveness.

14.5.5 Technological Innovation and R&D

Overall, N1's technological innovation efforts are deeply aligned with the company's strategic goals of transnational operations and global leadership, providing strong support for these goals. CRRC consistently adheres to the principle of "national needs first, industry development paramount," deeply integrating independent innovation, open innovation, and collaborative innovation models. Following a forward-thinking approach, the company is fully committed to building and continuously improving a technological innovation system aligned with its internationalization efforts. This meticulously crafts a product portfolio with exceptional international competitiveness, a cutting-edge rail transit knowledge system, and a robust international rail transit vehicle technology support framework. This comprehensive approach strengthens technological innovation and propels China's rail transit vehicle manufacturing industry up the industrial and value chains, establishing a model of "CRRC Creation" and demonstrating the power of "Created in China." This builds a solid foundation for the company's long-term, rapid development and sets a new trend in the global rail transit industry.

14.5.5.1 Technological Innovation Capabilities

N1 boasts strong technical strength and innovative capabilities, possessing numerous core technologies and patents in rail vehicle manufacturing. CRRC Corporation Limited has mastered core technologies for EMUs operating at speeds of 200–250 km/h and 300–350 km/h, and has successfully developed the Type 380 high-speed train and China Standard EMU with independent intellectual property rights. This demonstrates N1's significant achievements in heavy-haul trains, high-speed, high-power passenger locomotives, high-speed coaches, and double-decker coaches, supporting numerous major speed increases on China's railways. Furthermore, significant breakthroughs have been achieved in a number of key technologies, including suspension guidance, rolling stock, traction power supply, operation and control communications, and system integration. These breakthroughs provide a strong foundation for enhancing the performance and safety of rail vehicles.

The company's initial development is relatively complete, with a technology innovation management system covering the entire technological innovation process, ensuring sufficient funding, material investment, and a qualified pool of innovative talent. Over the past three years, N1's annual R&D funding has maintained rapid growth of approximately 8%, and the proportion of technology investment has remained above 5.6%, ensuring the continued rapid improvement of CRRC's technological innovation capabilities. The company has long been committed to building and improving an efficient and practical system of rules, regulations, and cultural concepts to ensure a well-organized technological innovation process. It has consistently established and improved management systems for scientific and technological projects, scientific and technological achievements, product technologies, knowledge management, and scientific and technological information to ensure the efficient implementation of technological innovation.

14.5.5.2 Product R&D Capabilities

The company continuously increases its R&D investment, promotes the application and dissemination of new technologies, processes, and materials, and strengthens cooperation and exchanges with universities and research institutions to jointly advance scientific and technological progress and industrial development in the rail transit vehicle industry. The company has established a product and technology R&D system centered around 11 national-level R&D institutions, including the National Engineering Laboratory for High-Speed Train System Integration, the State Key Laboratory of EMU and Locomotive Traction and Control, the National Heavy-Duty and Express Railway Freight Car Engineering Technology Research Center, and the National Railway Passenger Car System Integration Engineering Technology Research Center, as well as 19 national-level enterprise technology centers encompassing OEM manufacturers. Currently, N1 is actively producing prototypes of the next-generation Fuxing EMU train, the CR450. Its test speed can reach 450 km per hour and its commercial operating speed can reach 400 km per hour. This will further enhance the technical level of China's high-speed railway EMUs.

14.5.5.3 Talent Cultivation

According to data on N1's official website, as of the end of 2018, the group had nearly 170,000 employees, including 12,446 with a master's degree or higher, 52,420 with a bachelor's degree, and 39,992 with a college degree. Those with a college degree or higher accounted for 62.31% of the total workforce. Professional and technical personnel accounted for 61,333 employees, or 36.45% of the total workforce. Senior professional titles accounted for 18.22%, intermediate professional titles for 11,177, and junior professional titles for 20,285, respectively, accounting for 37.99% and 18.22% of the total workforce. Highly skilled personnel accounted for 56.26% of the total workforce, including 9,997 technicians and 3533 senior technicians.

14.5.6 Market Position and Brand Influence

N1 holds a significant position in the global rail transit vehicle market, forming a major supplier alongside companies such as Canada's Bombardier, France's Alstom, and Germany's Siemens. According to relevant data, N1 holds a leading position in global rail transit vehicle sales, particularly in the Chinese domestic market, where it holds a significant and near-monopoly market share. Its subsidiaries are major players in the domestic subway vehicle manufacturing market, holding over 95% of the market share. This substantial market size demonstrates N1's strength in the domestic rail transit sector, providing strong support for its international expansion and laying the foundation for its brand influence. The company actively implements its internationalization strategy, strengthens cooperation and exchanges with international markets, and promotes its products and technologies to the global stage, making significant contributions to the international development of China's rail transit industry.

In summary, N1 holds a significant position in the rail transit vehicle manufacturing industry. Its substantial capital share, domestic market dominance, technological strength and innovation capabilities, brand influence, and internationalization strategy all contribute to its strong competitiveness in this sector. With the continued development of the global rail transit market and the continuous improvement of CRRC's own capabilities, its market position is expected to be further consolidated and enhanced.

The companies represented by the letters in this article are:

N1-CRRC Corporation Limited, N2-China Railway Group Limited, N3-China Construction Bank Corporation Limited, N4-Koni Electromechanical Co., Ltd.,

N5-Jinxi Axle Co., Ltd., N6-CSR Corporation Limited, N7-CNR Corporation Limited

References

1. Yan Xi. Analysis on key points of green factory creation in rail transportation equipment manufacturing industry [J]. Machinery Industry Standardization and Quality, 2022, (05): 26–31.
2. Zhang Lili, Wang Yubing. Development opportunities and countermeasures for the rail transit equipment manufacturing industry under the new development pattern of dual circulation [J]. Market Weekly, 2021, 34(12): 43–45.
3. Long Jialing, He Zhengchu, Peng Hua, et al. Research on regional industrial agglomeration development of rail transit equipment manufacturing industry [J]. Enterprise Economy, 2021, 40(09): 146–153.
4. Fu Weichen. Building a smart enterprise for rail transportation equipment [J]. Enterprise Management, 2020, (09): 112–113.
5. He Zhengchu, Liu Yaru, Pan Hongyu, et al. Innovation network capability and innovation performance of core enterprise clusters in rail transportation equipment manufacturing industry [J]. Systems Engineering, 2018, 36(12): 22–31.
6. Wan Qifeng. Opportunities and countermeasures brought by the Belt and Road Initiative to rail transportation equipment manufacturing industry [J]. Science and Technology Innovation Herald, 2017, 4(16): 120–122+124.

Chapter 15
Innovation and Competitiveness in Civil Aviation Equipment Manufacturing Enterprises

Malin Song and Mengjie Zhu

Abstract Civil aviation equipment manufacturers form the cornerstone for ensuring flight safety and enhancing air transport efficiency and service quality, occupying an irreplaceable position within the civil aviation industry chain. This paper conducts an in-depth analysis of the current status, classification, characteristics, regional distribution, industrial chain, market competition landscape, operational management, and innovation of China's civil aviation equipment manufacturers. Using SHAYO and HHAAA as case studies, it examines the development trajectories, product lines, strategic approaches, and challenges faced by typical enterprises. The article identifies key development trends facing Chinese manufacturers, including globalisation strategies, industrial consolidation and synergy, and expansion into the general aviation market. It proposes policy recommendations to strengthen governmental support, optimise the industrial environment, and cultivate internationally renowned brands.

15.1 Current Status of Civil Aviation Equipment Manufacturing Enterprises

15.1.1 Policy Framework

The policy framework governing China's civil aviation equipment manufacturing sector has undergone significant evolution over many years. During the 1980s and 1990s, China's civil aviation industry introduced internationally leading aircraft models such as the Boeing 707 and Boeing 747SP, while simultaneously commencing independent aircraft development. Policies during this period primarily focused on introducing advanced international technologies and management expertise to enhance the overall capabilities of China's civil aviation sector. In the early 2000s,

M. Song · M. Zhu (✉)
Anhui University of Finance and Economics, Bengbu, China
e-mail: zikw2859091185@163.com

J. Wang and M. Song (eds.), *The Green Path of Chinese Transportation Enterprises*, Resources, Climate and Sustainable Development,
https://doi.org/10.1007/978-981-95-5289-4_15

following China's accession to the World Trade Organisation (WTO), the civil aviation sector underwent further liberalisation. Policies encouraged cooperation and competition between domestic and international airlines, driving market-oriented reforms within the industry. Between 2006 and 2020, the National Medium- and Long-Term Science and Technology Development Plan explicitly charted the course for aviation equipment manufacturing, emphasising the importance of technological innovation and independent R&D while providing policy support for the sector's advancement. The 14th Five-Year Plan (2021–2035) indicates that the Chinese government will continue to increase support for the aviation equipment manufacturing sector, driving the industry towards high-end, intelligent, and green development. It further underscores the importance of implementing a strategy for integrated military-civilian development. From 2023 to 2035, the Green Aviation Manufacturing Development Outline proposes establishing a comprehensive, advanced, and secure green aviation manufacturing system. It highlights that new energy aircraft will become the mainstream development direction, with domestically produced large civil aircraft achieving world-leading standards in safety, environmental performance, economy, and comfort. Concurrently, new general aviation equipment will achieve commercialisation and large-scale application (Fig. 15.1).

By 2023, China's civil aviation equipment manufacturing enterprises held the foremost position among all civil aviation enterprises, accounting for the highest proportion. Among these, MAAMA, XIANG and DADAD constitute significant components of China's aviation equipment manufacturing sector, each playing a vital role within their respective specialised fields. Subsequently, aviation maintenance enterprises, distributors and airlines also accounted for a substantial proportion (Fig. 15.2).

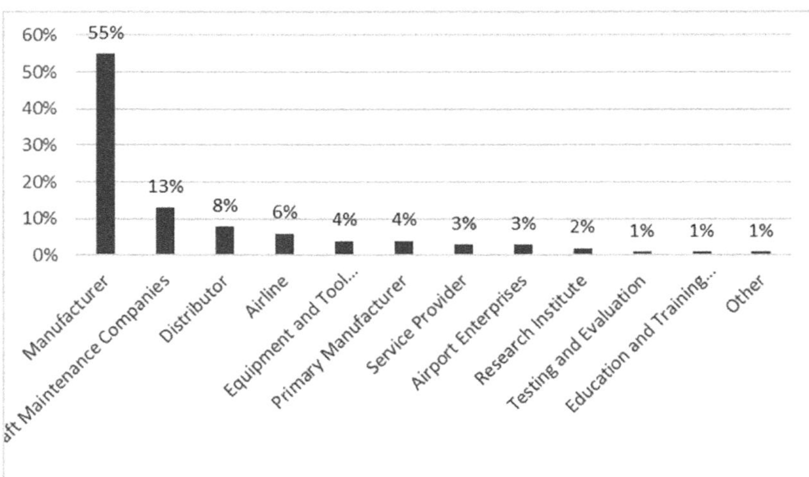

Fig. 15.1 Overview of civil aviation equipment manufacturers in China. *Source* Aviation Industry Network

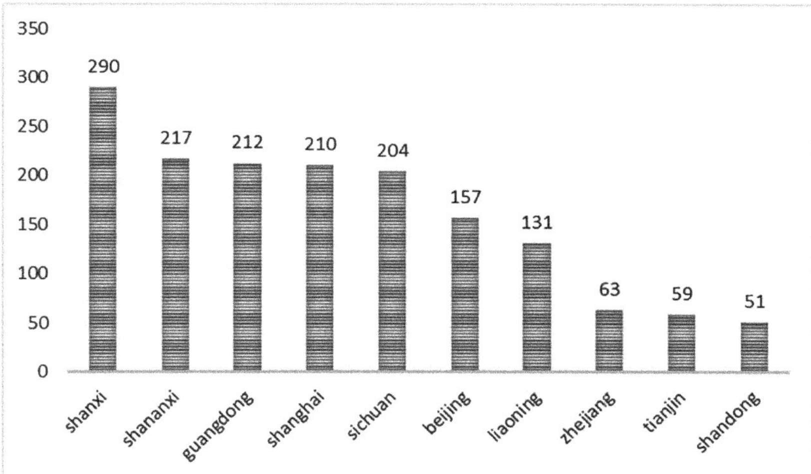

Fig. 15.2 Overview of civil aviation manufacturing enterprises in China (Geographical Distribution). *Source* Aviation Industry Network

15.1.2 The Role of Civil Aviation Equipment Manufacturers

15.1.2.1 Supporting the Development of the Civil Aviation Industry

Civil aviation manufacturing enterprises have provided a solid foundation for the development of China's civil aviation industry by supplying high-quality aerospace equipment. As a global leader in airport ground support equipment, TTAAI's product portfolio encompasses aircraft power supply vehicles and firefighting equipment, significantly enhancing operational efficiency and service standards at Chinese airports [1]. Through technological innovation and industrial upgrading, aviation manufacturing enterprises have also propelled the deep integration of civil aviation transport and aerospace manufacturing, generating synergistic effects. This has further strengthened the regulatory efficacy and nurturing role of civil aviation transport in civil aircraft manufacturing [2].

15.1.2.2 Promoting the Upgrading of the National Industrial Structure

The development of China's civil aviation manufacturing equipment enterprises has not only elevated the standards of the aviation manufacturing sector but also fostered growth across the entire upstream and downstream industrial chain. The development of the C919 large civil aircraft project has catalysed the formation of an aviation industrial cluster encompassing multiple segments, including aerospace materials suppliers, aerospace component suppliers, aircraft final assembly plants, maintenance and service providers, and aviation service providers [3]. This aviation industrial cluster not only strengthens the overall capabilities of the aviation manufacturing sector but also drives the growth of related industries, thereby contributing to the optimisation and upgrading of the national industrial structure [4].

15.1.2.3 Enhance International Competitiveness

China's civil aviation manufacturing equipment enterprises have likewise achieved remarkable success in the international marketplace. As domestic firms increase their investment in research and development, their technological innovation capabilities continue to strengthen, enabling the aviation equipment sector to secure a foothold in global markets. The establishment of SHAYO and HHHHH has propelled significant progress in China's large passenger aircraft and aero-engine sectors, enhancing the international competitiveness of the nation's aviation manufacturing industry. The government has further encouraged synergistic research mechanisms between civil aviation and aerospace industries through policy support and financial investment, thereby elevating the global competitiveness of China's aviation equipment sector [5]. Chinese civil aviation manufacturing enterprises have played a pivotal role in underpinning civil aviation development, advancing national industrial upgrading, and bolstering international competitiveness. Through technological innovation, industrial advancement, and policy backing, China's aviation manufacturing sector has not only enhanced its own capabilities but also made significant contributions to the global aviation market's evolution.

15.2 Classification and Characteristics of Civil Aviation Equipment Manufacturers

15.2.1 Classification of Civil Aviation Equipment Manufacturing Enterprises

15.2.1.1 Aircraft Manufacturing Enterprises

The primary responsibilities of aircraft manufacturers encompass the comprehensive design, assembly, and testing of aircraft. These enterprises typically possess robust research and development capabilities alongside advanced production facilities, enabling them to manufacture diverse aircraft types, including both commercial and military aircraft. Prominent aircraft manufacturers in China include SSHHH and FFFZZ. Established in 1951, SSHHH is a modern aircraft manufacturing enterprise integrating scientific research, production, testing, flight trials, and service support. Its subsidiary, SSSFF, serves as the core backbone for the serial development of China's national civil aircraft programme [6]. The latter primarily oversees integrated innovation and application in intelligent manufacturing for large commercial

Table 15.1 Aero engine corporation of China's affiliated host manufacturers and products

Automotive manufacturer	Code name	Location	Date of establishment	manufacture products
XIIHH	Factory 430	Xian	1958	WP-8、WS-9、WS-20
LIMMG	Factory 460	Guizhou	1997	WP-13、WS-13
NANFA	Factory 410	Shenyang	1954	WP-6、WP-7、WP-14、WS-10、WS-15
LIYDO	Factory 331	Zhuzhou	1951	WJ-5、WJ-6、WJ-10、WZ-8、WZ-9、WZ-10、WZ-11、WZ-16

aircraft, driving deep integration of smart manufacturing technologies within the commercial aircraft sector [7]. Additionally, XIANF—affiliated with the Aviation Industry Corporation of China—serves as a vital base for developing and producing medium-to-large military and civil aircraft in China.

15.2.1.2 Engine Manufacturing Enterprises

Engine manufacturers specialise in the research, development and production of aero-engines, providing propulsion for aircraft. These enterprises typically collaborate with globally renowned aircraft manufacturers to supply engine systems. As a leading enterprise in China's aero-engine sector, HANGD is dedicated to the manufacture, sale and maintenance of aero-engines. The company also engages in subcontracted production for export of aero-engine components, alongside the production and sale of aviation products. HANGD Power possesses four major aero-engine assembly plants: XIIHH, LIMMG, NANFA and LIYDO. These entities each specialise in distinct engine types: Xi'an Aero-Engine Corporation primarily manufactures medium-to-large military and civil aircraft engines; Liming Aero-Engine Corporation produces turbojet and turbofan engines; Nanfang Aero-Engine Corporation manufactures medium-to-small engines including turboprop and turboshaft units; while Liuyang Aero-Engine Corporation focuses on the development, production, maintenance, and servicing of small-to-medium thrust turbojet and turbofan engines [8] (Table 15.1).

15.2.1.3 Hydroelectric Equipment Manufacturing Enterprises

Avionics equipment manufacturers specialise in the research, development and production of aerospace electronic systems, encompassing navigation systems, communication systems, flight control systems and more. Shenzhen Donica Avionics Co., Ltd., established in 2003, is dedicated to the R&D, certification, production and technical support of avionics equipment. Angji Avionics, serving as the supplier and integrator for the core avionics systems of the C919 civil aircraft, has achieved remarkable progress over its decade since founding in 2012. AVIC Electronics, as the domestic leader in the avionics systems sector, develops and manufactures avionics products including lighting systems, cockpit systems, power distribution systems, and flight data recorders.

15.2.1.4 Aircraft Equipment Manufacturing Enterprises

Aviation equipment manufacturers specialise in the production of aircraft components and related equipment, including cabin interior fittings, seat cushions, and life-saving equipment. China's XINPP primarily engages in the development, production, and technical services of cabin components for civil passenger aircraft, alongside manufacturing airport ground maintenance equipment. YINYY concentrates on producing aircraft cabin interior fittings, life-saving inflatable seat cushions, and flame-retardant seat cushions. JINSH specialises in manufacturing aviation seat cushions, seat covers, and related products, serving as a Tier 1, Tier 2, and Tier 3 supplier to airlines and automotive original equipment manufacturers.

15.2.2 Characteristics of Civil Aviation Equipment Manufacturing Enterprises

15.2.2.1 High Technological Sophistication

Civil aviation manufacturing equipment enterprises belong to the high-end manufacturing sector, characterised by high technological complexity and involving cutting-edge achievements across multiple fundamental and applied sciences including physics, chemistry, metallurgy, materials science, and electronics [9]. Aviation engines operate under extreme conditions of high temperature, pressure, rotational speed, and load, representing the most technologically sophisticated and challenging power units to manufacture. The production of aerospace components also demands the machining of complex structural parts, including aircraft ribs, beams, frames, joints, and various shaped structural elements. These components feature intricate, non-standard geometries that present significant machining difficulties, requiring exceptionally high precision and product reliability.

15.2.2.2 High Capital Investment

The aerospace manufacturing sector demands substantial capital investment while simultaneously confronting high risks. Any technical or market misstep may result in significant economic losses [10] Manufacturers of aerospace components must establish proprietary technical systems encompassing materials, equipment, tooling, processes, sub-assembly, testing, and inspection. To maintain technological advancement, they require R&D personnel and skilled technicians specialising in aerospace component manufacturing techniques. The development and production of the C919 airliner involve high technological complexity and substantial capital expenditure, characterised by extended R&D cycles and prolonged return periods. To ensure flight safety, compliance with the stringent certification standards of the International Civil Aviation Organisation (ICAO) and the Federal Aviation Administration (FAA) is mandatory.

15.2.2.3 Long-Term Returns

Civil aviation manufacturing enterprises face extended R&D cycles, numerous product specifications with stringent precision requirements, and substantial fixed investments in establishing relevant production lines. The aero-engine sector is characterised by lengthy development cycles; once integrated into the dual-engine supply system, it encounters minimal competitive threats while yielding high economic returns over extended payback periods. The development of aerospace defence equipment involves high technical complexity, substantial capital requirements, and protracted R&D and production cycles, thereby establishing significant industry barriers.

15.3 Regional Distribution and Industrial Chain of Civil Aviation Equipment Manufacturers

15.3.1 Regional Distribution of Civil Aviation Equipment Manufacturers

China's civil aviation equipment manufacturing enterprises are primarily concentrated in the Yangtze River Delta region, East China, Southwest China, and other areas. Jiangsu, Shanghai, and Zhejiang in the Yangtze River Delta serve as hubs for equipment manufacturing in the region. Home to numerous universities such as Shanghai Jiao Tong University, Zhejiang University, and Nanjing University of Aeronautics and Astronautics, these areas have fostered outstanding enterprises like TUOPU and AIMEI. Certain regions in East China have also developed comprehensive aviation manufacturing industrial chains by leveraging universities and research institutions. Southwest China, including Sichuan and Shanxi, also hosts significant concentrations of aviation manufacturing enterprises. Chengdu's Xindu District collaborated with Chengdu Aircraft Industry Group (CAIG) to establish an aviation large-component industrial park. Additionally, cities like Xi'an, Nanchang, and Shenyang house substantial numbers of aviation manufacturing companies. Based

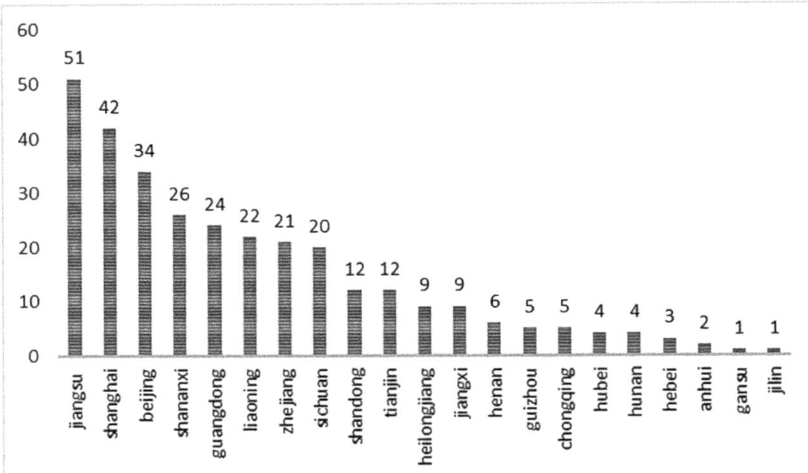

Fig. 15.3 Regional distribution map of China's aviation manufacturing equipment and tool enterprises. *Source* Aviation Industry Network

on the team's expertise and experience, and utilizing the aviation industry database, we have compiled a list of 315 aviation manufacturing equipment and tool-related enterprises across 10 major categories, all with actual industry users. The specific distribution is shown in the figure below. The six provincial-level administrative regions with the highest concentration of aviation manufacturing equipment and tool enterprises are Jiangsu, Shanghai, Beijing, Shaanxi, Guangdong, and Liaoning, collectively accounting for 64% of the total enterprises. Among these, the Yangtze River Delta region exhibits a particularly pronounced clustering effect (Fig. 15.3).

China's aviation industry clusters are primarily concentrated in regions such as the Yangtze River Delta Aviation Industry Cluster, the Chengdu Aviation Industry Cluster, and the Xi'an Aviation Industry Cluster. Among these, the Yangtze River Delta Aviation Industry Cluster, centered around Shanghai, Jiangsu, and Zhejiang, has established a relatively comprehensive aviation manufacturing industrial chain. The Chengdu Aviation Industry Cluster, anchored by the aviation large-component industrial park jointly developed by Chengdu's Xindu District and Chengdu Aircraft Industry Group (CAIG), has formed a fairly complete aviation manufacturing industrial chain. Xi'an, as a major aviation manufacturing base in China, also hosts numerous aviation manufacturing enterprises.

15.3.2 Analysis of the Industrial Chain for Civil Aviation Equipment Manufacturers

15.3.2.1 Upstream Industries (Raw Materials, Component Suppliers)

The upstream sector of the aviation industry primarily encompasses aircraft equipment design and the supply of fundamental materials. Among these foundational materials, key components include metals and composite materials such as titanium alloys, specialty steels for aircraft manufacturing, high-temperature resistant alloys, intermetallic compounds, and resin-based composite materials. Additionally, it involves the supply of critical aircraft components including electronic components, hydraulic systems, and communication facilities. China's JIALQ specializes in the development, manufacturing, and sales of aerospace composite components while providing related services. Recognized as a national "Little Giant" enterprise for its specialized, refined, distinctive, and innovative capabilities, the company has long been committed to advancing the military aviation sector.

15.3.2.2 Midstream Industry (Equipment Assembly and Integration)

The midstream segment of the aviation industry chain primarily encompasses aircraft equipment manufacturing, which includes the production of aircraft components and the assembly of complete aircraft systems. Aircraft component manufacturing can be divided into three main categories: airframes, power engines, and onboard equipment for in-flight information processing and control. The airframe constitutes the primary structural body of the aircraft, the engine provides the propulsion for flight, while the onboard equipment handles information processing and transmission during flight operations. Among these, Chinese enterprises such as Jianghang Equipment and Nanshan Aluminum are major publicly listed companies in the aviation equipment manufacturing sector.

15.4 3Downstream Industries (Airlines, Maintenance Companies)

At the terminal end of China's aviation industry, operations primarily encompass the daily management of airlines and the maintenance and repair of aircraft. With the continuous growth in the number of civil aviation carriers in China, the demand for aircraft maintenance is also rising, providing a boost to the development of the aviation maintenance sector. In China, enterprises engaged in aircraft maintenance include aircraft manufacturers, independent maintenance service providers, and third-party maintenance organizations affiliated with airlines. Among these, companies such as GEHHH, GAJIC, and WEIXI are major players in the aircraft maintenance industry. The industrial chain of China's civil aviation equipment manufacturers spans the entire process from raw material supply to equipment assembly and integration, extending to airlines and maintenance enterprises. The upstream segment handles the supply of critical raw materials and components, the midstream focuses on the manufacturing of aviation equipment, while the downstream involves the operations and services of airlines and maintenance enterprises. Through close collaboration across these segments, China's civil aviation equipment manufacturers are able to provide high-quality aviation equipment and services to meet market demands.

15.5 Market Competition Landscape of Civil Aviation Equipment Manufacturers

15.5.1 Current State of Domestic and International Market Competition

15.5.1.1 Domestic Market

The domestic aviation equipment manufacturing market is highly competitive. AVIC Group oversees over 100 member units, covering the entire industrial chain from R&D and design to component manufacturing, subsystem development, and complete aircraft assembly. Additionally, domestic aviation equipment manufacturers are continuously enhancing their capabilities in producing structural components, metal profiles, and metal parts for major international aircraft models while improving product quality.

Within the domestic market, AVIC Electromechanical Systems stands as China's primary national team in military and civil aviation electromechanical systems, occupying a pivotal position in the nation's aviation electromechanical industry. Domestic aviation manufacturers such as Aero Engine Corporation of China, Guanglian Aviation, and Lian Shi Aviation are also steadily enhancing their market competitiveness. With rapid breakthroughs in domestic technologies, the market potential for domestic substitution will be further unleashed. By 2029, the revenue of large-scale enterprises in China's aviation manufacturing sector is projected to exceed I.2 trillion yuan, achieving a compound annual growth rate (CAGR) of 13% from 2024 to 2029 [11]. This indicates that the domestic aviation equipment manufacturing market possesses immense development potential and promising prospects.

15.5.1.2 International Market

Globally, the aerospace equipment manufacturing market is virtually monopolized by Boeing of the United States and Airbus of Europe. These two giants dominate the integration of large aircraft systems, collectively capturing over 95% of the market share. Beyond this, countries such as Japan, Brazil (Embraer), Canada (Bombardier), and Russia (Sukhoi) also demonstrate leading technological advantages in specific segments of aerospace equipment. In international market competition, China's aviation manufacturers face intense competition directly from global giants like Boeing and Airbus. Boeing leads globally in aviation equipment patents with 5112 patents, while Airbus holds absolute dominance in the medium-to-large aircraft market. Against the backdrop of increasingly stringent technological restrictions imposed by Western nations on China, domestic civil aviation equipment manufacturers urgently need to achieve domestic substitution and self-reliance. In the fields of aircraft braking control systems and wheel-related technologies, products from developed Western countries boast mature technologies, widespread adoption, and formidable market competitiveness. Consequently, Chinese enterprises must continuously enhance their independent R&D capabilities to reduce reliance on foreign technologies.

China's civil aviation equipment manufacturing industry faces intense competition in both international and domestic markets. Internationally, its primary competitors are global aviation manufacturing giants like Boeing and Airbus. Domestic Chinese enterprises must persistently strengthen independent research and development alongside technological upgrades to achieve domestic substitution and self-reliance. Domestically, while competition remains fierce, companies like AVIC have secured significant market share, indicating substantial growth potential. Looking ahead, with breakthroughs in domestic technology and growing market demand, China's civil aviation equipment manufacturers are poised to assume a more prominent role in the global aviation market.

15.5.2 Market Competition Strategy

15.5.2.1 Technological Innovation

Technological innovation is central to maintaining market competitiveness for civil aviation manufacturing equipment enterprises. Recently, China has achieved significant technological breakthroughs in aviation manufacturing. China's ALLDD specializes in precision machining for high-end aviation equipment and commercial aircraft components, with production facilities capable of meeting critical requirements for aerospace part processing. GGTTT, a leading global supplier of airport equipment, holds a 100% market share for certain key products domestically and serves over 50 countries and regions worldwide. Through continuous technological innovation, these enterprises have significantly enhanced product quality and performance, thereby securing advantageous market positions.

15.5.2.2 Brand Building

Brand building plays a vital role in enhancing the market influence of civil aviation manufacturing enterprises. China has achieved significant accomplishments in brand development for its civil aviation manufacturers. AVIC has been listed on the Fortune Global 500 for 15 consecutive I years and has released its product brand naming rules for the first time, along with brand names for products across eight major aircraft categories and 18 brand series. This initiative demonstrates its commitment to integrating aviation brand development into national strategy. Additionally, enterprises such as COMAC and AECC, along with flagship products like the "Large Aircraft," "Kunlong-600," and "Taihang Engine," have evolved into globally recognized brands. These brands have established strong market reputations through their high-quality products and services.

15.5.2.3 Market Expansion

Civil aviation equipment manufacturers are expanding their business scale by increasing market share. With rapid advancements in domestic technology, the potential of the domestic substitution market has been further tapped. By 2029, the annual revenue of larger enterprises in China's aviation manufacturing industry is projected to exceed I.2 trillion yuan, with a compound annual growth rate (CAGR) of 13% expected from 2024 to 2029. Take GGTTT as an example. By offering services across six major business areas—airport maintenance, ground handling, aircraft cargo, airfield maintenance, fueling, and cabin services—and providing all necessary ground equipment, some of its key products command a 100% market share domestically while serving over 50 countries and regions globally. These enterprises have achieved rapid business growth by actively expanding both domestic and international markets.

China's civil aviation equipment manufacturers have adopted multiple strategies—including technological innovation, brand building, and market share expansion—achieving remarkable results. Through continuous technological advancement, these enterprises have enhanced product quality and performance; by establishing strong brands, they have built excellent market reputations; and through market expansion, they have realized rapid business growth. The effective implementation of these strategies has enabled China's civil aviation equipment manufacturers to secure a competitive edge in the fiercely contested market.

15.6 Operational Management and Innovation in Civil Aviation Equipment Manufacturing Enterprises

15.6.1 Operations Management

15.6.1.1 Production Management

For the aerospace manufacturing industry, production management plays a crucial role, primarily enhancing efficiency through the adoption of lean production and digital management tools. For instance, a major aerospace manufacturer drew upon advanced practices from domestic and international peers to develop a lean production management on-site control process tailored to the characteristics of China's aerospace manufacturing sector, thereby significantly improving its production management capabilities. Lean Six Sigma management methodologies have also seen widespread adoption within aviation specialty equipment manufacturers. This approach optimizes production management processes by enhancing customer satisfaction, reducing costs, and elevating product quality.

15.6.1.2 Quality Management

Quality management is a critical component in ensuring that products from aerospace manufacturing enterprises comply with international standards. SSFFF has established a standardized quality management system based on ISO 9001 and conducts research on improving the quality management system tailored to industry characteristics and the company's fundamental conditions to achieve effective implementation of quality management. AILLL has obtained certifications for international and domestic aerospace quality management systems such as AS9100D, establishing a comprehensive and controllable quality management system. Digital quality management tools like the SMART Quality inspection data management system are also widely applied in aerospace equipment maintenance. Through online data collection, automated judgment and backfilling, structured storage, and diversified analytical applications, these tools enhance quality management efficiency.

15.6.1.3 Supply Chain Management

In the operational management of civil aviation equipment manufacturers, supply chain management constitutes a complex and critical domain, encompassing the entire process from raw material procurement and production manufacturing to product delivery. Major domestic aviation manufacturers are advancing adjustments to their industrial chain structures. For instance, Aviation Industry Corporation of China (AVIC) Shenyang Aircraft Corporation has shifted certain component production to fixed-item outsourcing while implementing targeted supplier cultivation. This approach aims to achieve centralized management of outsourced parts in terms of technology, quality, and schedule. The objective of establishing a supply chain management system is to reduce management complexity and difficulty while enhancing management standards and efficiency. Its overarching philosophy emphasizes that 70% of supply chain costs and efficiency are determined by supply chain design, 20% by supply chain planning, and 10% by the execution of supply chain plans. By optimizing supply chain management, civil aviation equipment manufacturers aim to improve operational efficiency and market competitiveness.

15.6.1.4 R&D Management

Civil aviation equipment manufacturers must first conduct in-depth market demand analysis in their R&D management to determine development directions and product positioning. China has successively issued multiple plans and policies charting the course for aviation industry development. Interpreting Made in China 2025, the growth in air transport and general aviation service demand has opened up extensive market opportunities for the aviation equipment manufacturing sector. Global competition is a defining feature of the aviation industry. China's aviation sector must compete on equal footing with world aviation powers. It must analyze the global competitive landscape, formulate corresponding R&D strategies, provide a favorable R&D environment and advanced R&D facilities, and enhance the efficiency and quality of R&D.

15.6.2 Technological Innovation

15.6.2.1 Technological Innovation and R&D Investment

China's civil aviation equipment manufacturing industry places high importance on technological innovation and R&D investment, committed to driving technological advancement and product upgrades. For instance, enterprises like HAAKK have increased R&D spending, actively stimulating innovation vitality. Through continuous innovation, they propel technological breakthroughs, elevating products to new heights. Additionally, after years of sustained R&D, WAIXL has continuously enhanced its technical capabilities across multiple areas including material identification, process design, parameter selection, machining programming, fixture design, tool selection, and machining environment control.

15.6.2.2 New Material References

In the application of new materials, China's civil aviation manufacturing equipment enterprises have developed multiple advanced material applications through technological innovation and R&D. HAHAY has achieved numerous independent innovations in new material application research, focusing on the use of aerospace-grade metals such as high-temperature alloys, titanium alloys, and aluminum alloys. CHENF has also successfully developed and launched new valve well cover series products like F900 and E600, comprehensively realizing innovation in high-end airport equipment.

15.6.2.3 Smart Manufacturing

Intelligent manufacturing represents a key development direction for civil aviation manufacturing equipment enterprises. WALIH has achieved significant results in intelligent manufacturing technologies and equipment for aerospace components, digital processing technologies for special high-strength, difficult-to-machine aerospace materials, and intelligent manufacturing quality management technologies for aerospace components. WAIXL has established multiple core technology systems in areas such as intelligent CNC programming and online inspection technologies, as well as deformation control technologies for CNC machining of weakly rigid thin-walled metal structural components.

15.6.2.4 Green Environmental Technologies

Green environmental protection technology represents an important development direction for civil aviation manufacturing equipment enterprises I. HHHHH has incorporated numerous advanced energy-saving and emission-reduction green technologies in its R&D initiatives, including fuel-efficient engines, low-noise technology, and low-emission technology. The Green Aviation Manufacturing Development Plan (2023–2035) emphasizes the vigorous advancement of green design technologies, the development of green surface treatments, clean processing, and low-energy consumption processes, and the establishment of green intelligent manufacturing facilities. China's civil aviation manufacturing equipment enterprises have made significant progress in technological innovation and R&D investment, new material adoption, intelligent manufacturing, and green environmental protection technologies. Through multiple concrete measures and case studies, they have demonstrated their achievements and efforts in these fields.

Fig. 15.4 The evolution of SHAYO

15.7 Development of Typical Civil Aviation Equipment Manufacturers in China

15.7.1 A Typical Civil Aviation Equipment Manufacturing Enterprise I

15.7.1.1 Development History

See Fig. 15.4.

15.7.1.2 Product Line Analysis

Since its establishment in 2008, China's SHAYO has successfully developed multiple aircraft series including the ARJ21, C919, and C929. These series cover a broad market spectrum ranging from regional to trunk routes, extending to long-range wide-body passenger aircraft. The ARJ21, SHAYO's first civil jet aircraft I, is a new regional jet primarily targeting the domestic regional market. It has gained recognition for its outstanding economics and adaptability and has been successfully introduced into commercial service.

As SHAYO's first large passenger aircraft, the C919 aims to challenge the market dominance of Boeing and Airbus. Initiated in 2006, the project secured its airworthiness certificate in 2020 after years of development and commenced commercial operations in 2023. Its design caters to diverse domestic and international market demands, offering multiple seat configurations and range options including basic, shortened, extended, and high-altitude variants. To date, the C919 has secured 1061 orders, with two aircraft already delivered for service. Development of the C929 is progressing steadily, with expectations that it will fill China's gap in the long-range widebody market. COMAC's product portfolio now spans multiple segments—from regional jets to mainline aircraft to long-range widebodies—establishing a relatively comprehensive commercial aircraft system.

The ARJ21, as a regional jet, primarily targets the domestic regional aviation market; the C919, as a large passenger aircraft, not only meets domestic market demands but also caters to international market needs; while the C929, as a long-range wide-body aircraft, will further enhance China's competitiveness in the high-end commercial aircraft market. With the rapid expansion of China's civil aviation market, demand for domestically produced commercial aircraft continues to grow. SHAYO's product portfolio not only meets domestic requirements but also positions the company to enter international markets and challenge the market dominance of Boeing and Airbus. Looking ahead, driven by ongoing technological advancements and increasing market demand, SHAYO's product line will undergo further expansion and optimization, propelling the sustained development of China's commercial aircraft industry.

15.7.1.3 Development Strategies and Challenges

In alignment with national strategic planning and industry trends, SHAYO has established its new phase objectives—the "Three I" goals, the "Two Completions" goals, and a "Three-Step" strategic plan—prioritizing high-quality development while pursuing steady progress. Through developing regional aircraft, advancing narrow-body airliners, and expanding into wide-body aircraft, the company has formed a three-part development strategy for its full range of aircraft products, establishing a comprehensive R&D system and product portfolio. Currently, the C919 large passenger aircraft has obtained its civil airworthiness certificate and commenced deliveries, while the CR929 long-range wide-body aircraft has entered the preliminary design phase. While advancing independent development, SHAYO also prioritizes international cooperation. For instance, it engages in technical exchanges and collaborations with global aviation giants like Airbus and Boeing to enhance its technological capabilities and market competitiveness. Concurrently, it signs strategic cooperation agreements with local governments to foster the clustered development of the aviation industry. Jiangsu Province, Suzhou City, Heilongjiang Province, and other regions have signed strategic cooperation agreements with SHAYO to jointly advance joint aviation technology research and industrial chain development. SHAYO plans to list on the STAR Market, marking its official entry into the capital market. This move will not only help the company raise funds but also enhance its brand influence and market competitiveness.

Despite significant progress on the C919 project, technical gaps remain compared to international aviation giants. Talent shortages and insufficient technological reserves also constrain its development. The global aviation market is fiercely

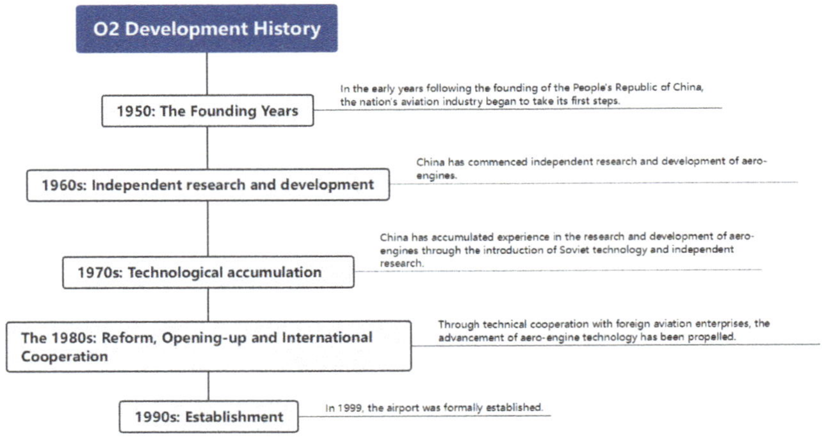

Fig. 15.5 HHAAA development timeline

competitive, with market share predominantly held by international giants like Boeing and Airbus. The C919 will face immense competitive pressure during its initial market entry. Concurrently, throughout its R&D and market expansion, COMAC requires greater legislative safeguards and policy support. The COVID-19 pandemic has dealt a severe blow to the global aviation market, presenting substantial challenges for COMAC. The company must strike a balance between pandemic prevention and model development tasks. The market introduction of domestically produced large aircraft requires time and a process of customer recognition. How to quickly gain the trust and support of domestic and international customers is a critical issue the company must address.

15.7.2 Representative Civil Aviation Equipment Manufacturers in China 2

15.7.2.1 Development History

See Fig. 15.5

15.7.2.2 Product Line Analysis

HHAAA's product portfolio is highly diversified, encompassing multiple types of aero-engines and related products. HHAAA specializes in the development, production, and delivery of commercial high-bypass-ratio turbofan engine series, establishing itself as one of the world's leading suppliers of commercial aero-engines.

Fig. 15.6 Complete industrial chain

Its product planning covers three major engine series: narrow-body aircraft engines, wide-body aircraft engines, and regional aircraft engines. LIMMH is responsible for manufacturing the Taihang and Kunlun engine series and undertaking overhaul work for a key engine model. DONGA specializes in producing helicopter gearbox transmission systems, including those for the Z-9 series and Z-10 helicopters. Additionally, HHAAA has developed and produced multiple aircraft models, including the MA60, MA600, and MA700 turboprop regional aircraft series, the Y-8 and Y-12 transport aircraft, and the Z-9 helicopter. Its product portfolio spans commercial and military aircraft engines, helicopter transmission systems, light business jets, and general aviation/special mission aircraft, demonstrating comprehensive capabilities and robust strength in aviation engine development. Through multiple restructurings, AVIC has further consolidated resources across its aviation engine manufacturing subsidiaries, establishing a complete industrial chain spanning design, manufacturing, and maintenance. AVIC Power has integrated operations from companies including Liming, Nanfang, and Liyang Power, expanding its product portfolio from large engines to encompass large, medium, and small engines (Fig. 15.6).

15.7.2.3 Development Strategy and Challenges

HHAAA has formulated an ambitious strategic plan—the "Two Integrations, Three Innovations, Five Transformations, Trillion-Yuan" initiative—aiming to establish a diversified industrial group spanning the entire industrial and value chains. The core of this strategy lies in resource integration to drive the expansion of the aviation industrial chain while focusing on the development of related financial services to achieve comprehensive value chain enhancement. HHAAA remains committed to product development as its core, emphasizing technological and management innovation. LIMMG has achieved significant breakthroughs in new engine R&D, while HHAAA has further strengthened its technical capabilities through acquisitions such as XIRUF and HAHUO. Guided by the mission of "serving the nation through aviation, strengthening the military and enriching the people," and adhering to the philosophy of "dedication, integrity, innovation, and excellence," CHUFD Co., Ltd. effectively implements the strategies of AVIC and the Engine Corporation. Its operational principles are defined by "prominent core business, military-civilian integration, people-centered approach, and coordinated development." HHAAA is also committed to building a management innovation system centered on value creation, with foundational management as the cornerstone, process management as the primary task, strategic management as the focus, and ideological management as the higher-level objective.

A primary challenge HHAAA currently faces is balancing technological accumulation with business restructuring. In industries requiring deep technical expertise, restructuring is not a quick-fix solution but demands sustained, solid efforts. During engine R&D, HHAAA has also encountered quality and performance challenges, particularly with the WS-10 engine failing to meet expected standards, indicating room for improvement in production or performance. AVIC Engine encompasses three core domains—aircraft airframes, engines, and avionics—forming an interwoven industrial chain structure. However, with the establishment of COMAC to advance large civil aircraft projects, opportunities and challenges in market competition and collaboration have emerged. In its development strategy, AVIC Engine emphasizes diversified growth across the entire industrial and value chains, prioritizes product and technological innovation, and promotes military-civilian integration.

15.8 Development Trends and Policy Recommendations for Civil Aviation Equipment Manufacturers

15.8.1 Development Trends

15.8.1.1 Globalization Strategy

China's civil aviation equipment manufacturers are progressively implementing global strategic deployments. To meet international challenges, they must formulate and execute a three-dimensional strategic model while adopting corresponding strategic control measures. Concurrently, general aviation enterprises are actively leveraging the advantages brought by the Belt and Road Initiative and free trade zone policies to promote the export of general aviation products with comparative advantages, thereby enhancing the international competitiveness of their independent brands. This globalized layout not only facilitates enterprises in further expanding international markets but also strengthens their influence within the global aviation equipment market.

15.8.1.2 Industry Integration and Synergy

For civil aviation equipment manufacturers to enhance their competitiveness, industrial integration and synergy are crucial strategies. Industrial chain integration involves adjusting and optimizing relationships across all segments of the industrial chain, aiming to achieve coordinated operations among these segments and thereby improve the overall operational efficiency of the entire industrial chain. Industrial synergy focuses on fostering cooperation and coordination among different segments within the industrial chain, aiming to maximize overall benefits. By introducing advanced production equipment, processes, and management techniques, enterprises can drive technological advancement across the industrial and supply chains, thereby strengthening their competitiveness in the global market. Furthermore, industrial chain synergy serves not only as a means of resource integration but also as a platform for innovation and value creation. It enables enterprises to achieve optimal resource allocation, collaborative innovation, and shared value creation through activities spanning the entire value chain.

15.8.1.3 General Aviation Market Expansion

According to the General Aviation Equipment Innovation and Application Implementation Plan (2024–2030), by 2030, general aviation equipment will be fully integrated into our production and daily life, with the market size expected to reach the trillion-yuan level. In recent years, its scale has shown a steady growth trend, increasing from 194.5 billion yuan in 2017 to 121.9 billion yuan in 2022. The development of the general aviation industry requires support from relevant agencies such as the Civil Aviation Administration to better serve public welfare and stimulate domestic demand. Concurrently, greater emphasis will be placed on environmental protection, promoting the adoption of clean energy solutions like electrification and hydrogen power to reduce greenhouse gas emissions.

15.8.2 Policy Recommendations

15.8.2.1 Strengthen Policy Support

Accelerate the formulation of laws, regulations, and policy measures to revitalize the aviation manufacturing industry, ensuring its sustained and healthy development. This includes promoting the formulation of regulations for the rapid development of the aviation industry, establishing legal and institutional safeguards, and improving financial support mechanisms for industrial development. Establish and refine financing systems and risk mitigation mechanisms. Strengthen airworthiness technology research, enhance capabilities in airworthiness design, verification, and certification, and comprehensively elevate airworthiness standards. Simultaneously, intensify cooperation with civil aviation authorities of developed countries and expand bilateral airworthiness agreements. Encourage civil aviation enterprises to actively pursue research, development, and innovation. While adhering to WTO rules and international management standards, promote the procurement of domestically produced aircraft and implement government procurement policies for domestic aerospace products. This includes implementing government purchases of domestically produced civil aviation products through central and local fiscal budgets, and supporting aviation manufacturing enterprises to "go global" in alignment with the national "Belt and Road" strategy. Encourage enterprises to implement smart manufacturing and green manufacturing initiatives to enhance production efficiency and product quality while minimizing environmental impact, thereby fostering a favorable development environment for China's civil aviation equipment manufacturers.

15.8.2.2 Optimize the Industrial Environment

Addressing bottlenecks, weak links, and vulnerabilities in the industrial chain, we will advance the development of key common technology platforms to enhance capabilities in market-driven product development, value-added services, and cross-industry collaboration. We will categorically promote the construction of public platforms for intelligent upgrades, green transformation, and industrial big data, reshaping new models for the transformation and development of manufacturing. National and local governments should continue to leverage policy incentives to support automation and indigenous production capabilities within the aviation equipment sector. Policy support has consistently served as a vital driver for aviation equipment industry development. Establishing market-oriented guidance systems and fostering fair competition markets—while avoiding rigid application of existing market access management models—will encourage diverse enterprises to actively participate in future industrial growth. This approach will cultivate a healthier, more efficient, and competitive industrial environment for China's civil aviation equipment manufacturers, promoting their sustained and stable development.

15.8.2.3 Cultivate Internationally Renowned Brands

Chinese civil aviation equipment manufacturers must prioritize brand storytelling and brand image building. By participating in international air shows such as the Zhuhai Airshow and Paris Air Show, they can showcase high-tech products to enhance their international brand recognition and influence. Simultaneously, leveraging global events like the Olympics and World Expos will elevate their brand's social image and reputation. Technological innovation is pivotal to enhancing brand competitiveness. Companies should increase R&D investment, particularly in emerging fields like new energy aircraft, unmanned systems, electrification, and intelligent technologies, using innovation to drive brand development. Simultaneously, they should integrate national pride and patriotic spirit into corporate culture. By fulfilling social responsibilities—such as demonstrating corporate social responsibility in major public initiatives—they can strengthen public emotional connection to the brand. This approach will gradually cultivate internationally renowned brands, elevating their competitiveness and influence within the global aviation industry.

Chapter 16
Pipeline Transportation Equipment Manufacturing Enterprises: Development, Opportunities, and Policy Implications

Malin Song and Xinmin Zhang

Abstract This paper focuses on China's pipeline transportation equipment manufacturing enterprises. Firstly, it analyzes the overall development environment of enterprises, including the impact of policy orientation, market demand, technological trends and other factors on the industry. Then, it takes JINZG, MINGS and GGKKG Co., Ltd. as the main research objects, respectively expounding the basic overview of each enterprise (such as enterprise scale and business scope), sorting out their development history and key nodes in each stage from establishment to growth, and analyzing their current development status in detail. By studying the trend chart of operating income over the years, this paper deeply explores the challenges faced in development and explores their development opportunities. Finally, based on the analysis results, relevant suggestions are put forward, aiming to provide comprehensive references for the development of pipeline transportation equipment manufacturing enterprises, and help enterprises seize opportunities, cope with challenges and achieve sustainable development.

16.1 Introduction

Pipeline transportation plays a pivotal role in the modern economic system. With the sustained growth of China's economy and the acceleration of industrialization and urbanization, the demand for transportation of energy, water resources, and other substances has become increasingly vigorous. As an efficient, safe, and environmentally friendly transportation method, pipelines have been expanding their application

M. Song
Anhui University of Finance and Economics, Bengbu, China

X. Zhang (✉)
Dongbei University of Finance and Economics, Dalian, China
e-mail: 15148635083@163.com

J. Wang and M. Song (eds.), *The Green Path of Chinese Transportation Enterprises*, Resources, Climate and Sustainable Development,
https://doi.org/10.1007/978-981-95-5289-4_16

scope. Pipeline transportation equipment is indispensable for the long-distance transportation of traditional energy sources such as oil and natural gas, the construction and improvement of urban water supply and drainage systems, and the transportation of industrial fluids like chemical raw materials. The state has vigorously promoted strategic initiatives such as energy structure adjustment and infrastructure construction, creating a favorable macro environment for the pipeline transportation equipment manufacturing industry. Meanwhile, the rapid development of science and technology has driven the continuous emergence of new materials and new processes, bringing both new opportunities and challenges to pipeline transportation equipment manufacturing enterprises. However, issues such as fierce market competition and product homogenization have also plagued the development of enterprises in the industry, which are faced with arduous tasks such as transformation and upgrading, and improving core competitiveness. This paper aims to help enrich the theoretical system in the fields of enterprise development strategy and marketing. By in-depth analysis of the problems and experiences in the development process of enterprises in this industry, it can provide practical case references for relevant academic research and further improve the theoretical framework of industry research.

16.2 Analysis of the Development Environment for China's Pipeline Transportation Equipment Manufacturing Enterprises

The Chinese government attaches great importance to the development of pipeline transportation equipment manufacturing enterprises and has introduced a series of policies to promote their sustainable progress. *Made in China 2025* emphasizes enhancing the independent development level of key core components and basic manufacturing equipment, which points out the direction of technological innovation for pipeline transportation equipment manufacturing enterprises. In the energy sector, the state has continuously increased investment in pipeline construction, directly stimulating demand in the pipeline transportation equipment market. For example, national initiatives such as the "Natural Gas Access to Every County" project and the construction of oil reserve bases all require a large amount of high-quality pipeline transportation equipment. China's current sustained and stable economic growth has created a favorable development environment for pipeline transportation equipment manufacturing enterprises. With the acceleration of industrialization and urbanization, the demand for energy continues to rise, driving the continuous expansion of the pipeline transportation industry's market scale. To meet the growing demand for energy transportation, oil, natural gas and other energy enterprises have increased their procurement of pipeline transportation equipment. At the same time, the in-depth implementation of the "Belt and Road" Initiative has brought significant opportunities for Chinese pipeline transportation equipment manufacturing enterprises to expand into the international market. From an economic development perspective, on the one hand, the stable growth of the domestic economy provides sufficient market demand for enterprises. China's energy demand will continue to grow at a relatively high rate in the next few years, which will directly drive the development of pipeline transportation equipment manufacturing enterprises. On the other hand, the exploration of the international market has brought broader development space for enterprises. With their high cost performance and good quality, Chinese enterprises are gradually highlighting their advantages in pipeline construction projects in countries along the "Belt and Road". For instance, Chinese pipeline transportation equipment has been widely used in energy cooperation projects in Central Asia, Southeast Asia and other regions. The rapid development of science and technology has brought a new technological environment for pipeline transportation equipment manufacturing enterprises. The continuous emergence of new materials and new processes has greatly improved the performance and quality of pipeline transportation equipment. The application of new materials such as high-strength steel and corrosion-resistant alloys has made pipelines more pressure-resistant and longer-lasting. At the same time, intelligent technologies are being increasingly applied in pipeline transportation equipment, such as intelligent monitoring systems and remote control technologies, which provide strong guarantees for the safe operation of pipelines. Technological innovation is a key factor driving the development of pipeline transportation equipment manufacturing enterprises. Only by continuously increasing investment in technological R&D and closely following the trend of scientific and technological development can enterprises maintain their advantageous position in the fierce market competition.

16.3 Zhejiang Jinzhou Pipeline Technology Co., Ltd.

16.3.1 Basic Overview

JINZG is an enterprise with outstanding influence in the pipeline industry. Founded in 1993 and headquartered in Huzhou, Zhejiang, it has grown into one of China's largest suppliers of galvanized steel pipes, spiral welded pipes, and steel-plastic composite pipes after years of development. JINZG's product range includes hot-dip galvanized steel pipes, high-frequency welded pipes, steel-plastic composite pipes, and double-sided submerged arc welded spiral steel pipes, with an annual production capacity of over 1 million tons of steel pipes and 20,000 tons of polyolefin pipes and fittings. Its products are widely used in numerous fields such as water supply, drainage, fire protection, gas supply, oil and natural gas transportation, construction, and communications. The sales network covers more than 20 provinces and cities across China, and products are exported to over 30 countries and regions worldwide.

In terms of technology, JINZG houses the only provincial-level technology center in Zhejiang's pipeline industry and a postdoctoral research workstation. The company continuously increases investment in technological R&D, actively introduces advanced technologies and talents from home and abroad, has successively undertaken multiple national Torch Program projects and provincial-level scientific and technological projects, and has obtained a large number of patented technologies and scientific and technological achievements. For example, in the R&D of steel-plastic composite pipes, it combines the advantages of steel pipes and plastic pipes, featuring corrosion resistance, sanitation, environmental protection, and easy installation, which meets the market demand for high-quality pipelines.

In terms of quality management, JINZG has established a strict quality inspection system to monitor quality throughout the entire process from raw material procurement to product delivery, ensuring stable and reliable product quality. The company has obtained multiple system certifications, including ISO9001 Quality Management System, ISO14001 Environmental Management System, and OHSAS18001 Occupational Health and Safety Management System, which strongly guarantee product quality.

In terms of corporate culture, the company upholds the corporate spirit of "Pragmatism, Efficiency, Perfection, and Innovation" and attaches great importance to talent cultivation and team building. Through cooperation with universities and research institutions, it has cultivated a group of professional and technical talents, providing strong talent support for the sustainable development of the enterprise. At the same time, the company actively participates in social welfare activities, fulfills corporate social responsibilities, and has shaped a good corporate image [1].

16.3.2 Development History

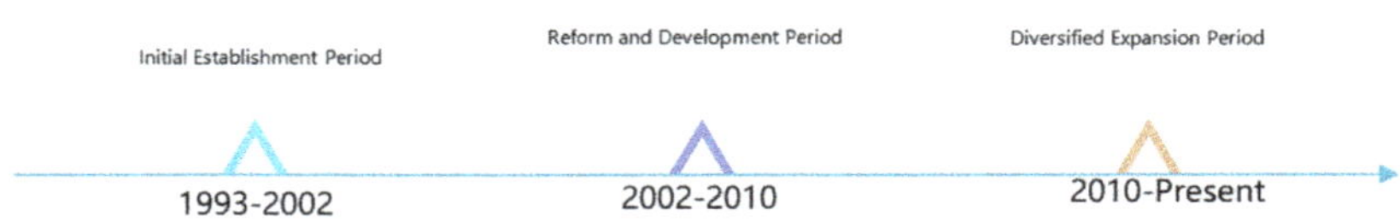

Initial Establishment and Accumulation Stage (1993–2002): In 1993, JINHU, the predecessor of the company, was established and began to engage in pipeline-related businesses. In 1994, with the advancement of national policies, JINHU integrated its enterprise resources and industries and was restructured into Zhejiang Jinzhou Group Nonferrous Metal Materials Factory, laying a certain foundation for its subsequent development. In 2000, JINHU implemented an overall restructuring to establish JINGU, further clarifying the company's development direction and organizational form.

Joint-Stock Reform and Development Stage (2002–2010): On July 24, 2002, JINGU carried out another strategic restructuring, focusing on capital operation and transforming into JINZG as a whole, marking the company's move towards a modern corporate governance structure. In 2004, the "Jinzhou" trademark was recognized as a "China Well-Known Trademark", greatly enhancing its brand influence. In 2005, "Jinzhou" brand pipes were awarded the title of "National Inspection-Exempted Product", indicating high recognition of its product quality. In 2006, JINZG established a joint venture HHSYY with ZHAIY, becoming ZHAIY's pipeline transportation base for domestic and foreign projects, thus expanding its market and business scope.

Listing and Diversified Expansion Stage (2010–Present): On July 6, 2010, the company was listed on the Shenzhen Stock Exchange, embarking on a new journey in the capital market and providing strong financial support for its scale expansion and technological innovation. After listing, the company continued to increase investment in R&D to improve the technical level and quality of its products. While consolidating its existing advantages in traditional galvanized steel pipes, spiral welded pipes, and steel-plastic composite pipes, it actively expanded the application scope of its products. For example, it supplied products for well-known large-scale domestic and foreign energy pipeline projects such as the China-Myanmar Oil and Gas Pipeline Project and the Tanzania Natural Gas Processing Plant and Transmission Pipeline Project, thereby enhancing its popularity and market share in the industry. In recent years, the company has continuously followed industry development trends and changes in market demand, actively deployed in emerging fields, strengthened cooperation with high-quality domestic and foreign customers, and strived to achieve sustainable development. It has become one of China's largest suppliers of galvanized steel pipes, spiral welded pipes, and steel-plastic composite pipes, and occupies an important position in the pipeline industry.

16.3.3 Current Development Status

According to the company's 2023 operating income report, the company achieved an operating income of 5.743 billion yuan in 2023. Although this represented a year-on-year decrease of 5.68%, the net profit attributable to shareholders of the listed company reached 283 million yuan, a year-on-year increase of 20.54%. The net cash flow from operating activities was 661 million yuan, and the net assets attributable to shareholders of the listed company as of the end of 2023 stood at 3.356 billion yuan [2]. Amid the complex economic situation and fierce competition in 2023, the company maintained a profit growth trend through rational business strategies. It was once again selected into Huzhou's "Golden Bull" Enterprise List, won the bid to be the fire pipeline supplier for Huawei's two-year new project, has been shortlisted as the steel-plastic pipe supplier for Capital Water for four consecutive years, and its subsidiary Pipeline Industry was selected as an excellent supplier for ExxonMobil's Guangdong Daya Bay Project. These achievements demonstrate its market competitiveness and recognition, and cooperation with large enterprises has further enhanced the company's brand influence and market share.

In addition, as a national high-tech enterprise, it boasts a series of scientific research platforms including a National Enterprise Technology Center, a CNAS-accredited laboratory, a provincial enterprise research institute, and a provincial engineering technology center. It currently holds more than 160 authorized patents and software copyrights, including over 50 invention patents, and has obtained multiple international and domestic certifications such as ISO. These honors and certifications are strong evidence of the company's technical strength and product quality [3].

The company's products are widely used in fields such as long-distance oil and gas pipelines, urban gas, fire protection, chemical industry, water supply and drainage, construction, communications, electricity, and structural pipes, with an annual production capacity of 1.6 million tons. It is one of the largest comprehensive pipeline production and research bases in China. With a rich variety of products that can meet the needs of different industries and customers, the company has undergone multiple technological innovations and transformations, and owns a number of advanced domestic automated production lines. It also offers a new generation of healthy and environmentally friendly "green pipeline" products, which have strong competitiveness in the market. Furthermore, the company focuses on R&D and innovation, and continuously obtains patent achievements. For example, in 2024, it obtained the patent for "A Quick Roll Change Platform and Quick Roll Change Method for Precision Welded Pipe Mills" (Authorization Announcement No. CN112058962B) and "A Welded Steel Pipe Conveying Device with Adjustable Weld Position" (Authorization Announcement No. CN221317895U). These patented technologies help improve production efficiency, product quality, and welding quality, providing support for the company to maintain a leading position in the technical field. Moreover, the "Jinzhou" trademark has been successfully registered in more than 50 countries and regions, and products are sold to over 30 countries and regions, enjoying a high reputation both domestically and internationally. The company has successively participated in national key projects such as the China-Kazakhstan Crude Oil Pipeline, Central Asia Oil and Gas Pipeline, China-Myanmar Oil and Gas Pipeline, and West–East Gas Pipeline, establishing a good brand image and gaining high customer recognition. Importantly, JINZG has become an excellent pipeline supplier for large enterprises such as SSHHH, SSYYY, and ZHAIY. The stable cooperative relationships with these industry giants not only bring stable orders and income to the company but also help it maintain a leading position in the industry, further expand the market, and enhance brand influence [4].

Overall, JINZG has achieved remarkable results in the pipeline industry. Relying on its product advantages, technical strength, brand influence, and good customer relationships, it occupies an important position in market competition. In the future, the company is expected to continue to benefit from industry development and its own competitive advantages to achieve sustained and stable development. However, at the same time, it needs to continuously respond to the challenges of market changes and technological innovation, continue to increase R&D investment, and improve product quality and service levels to maintain its leading position in the industry.

16.4 Cangzhou Mingzhu Plastic Co., Ltd.

16.4.1 Basic Overview

Founded in 1995 and headquartered in Cangzhou, Hebei, MINGS has grown into a well-known domestic manufacturer of plastic pipelines and related products after nearly 30 years of development and accumulation. The company specializes in the manufacturing of pipeline transportation equipment, with its product range covering polyethylene (PE) pipes, polypropylene (PP) pipes, gas pipes, water supply pipes and fittings, among other categories. In terms of product performance, the company's pipeline products boast multiple advantages. For instance, PE pipes feature excellent corrosion resistance and flexibility, enabling them to adapt to complex geological environments and construction conditions, thus being widely used in urban water supply and drainage systems. Gas pipes, with their strict sealing performance and good pressure resistance, provide safe and reliable guarantees for urban gas transmission. The company's products serve a wide range of fields such as urban construction, agricultural irrigation, and industrial production, making important contributions to the stable development of the social economy [5].

In terms of technology, MINGS places great emphasis on technological R&D and innovation. It has established a professional technical R&D team consisting of many experienced engineers and researchers. The company continuously explores new materials, processes and technologies, committed to improving product quality and performance. Currently, the company has mastered a number of core technologies, such as advanced pipeline forming technology and efficient connection technology. These technologies not only improve production efficiency and reduce production costs, but also help the company's products stand out in market competition. Meanwhile, the company actively collaborates with domestic and foreign research institutions and universities to jointly explore cutting-edge technologies and continuously promote the development of industry technologies.

In terms of market position, MINGS has won wide market reputation and customer recognition with its high-quality products and good services. The company has established long-term and stable cooperative relationships with large enterprises such as CNPC (China National Petroleum Corporation) and Sinopec (China Petrochemical Corporation), becoming one of their important suppliers. In addition, the company actively expands domestic and foreign markets, with its products exported to many countries and regions in Southeast Asia, the Middle East, Africa, etc., and has also gained a certain share in the international market.

16.4.2 Development History

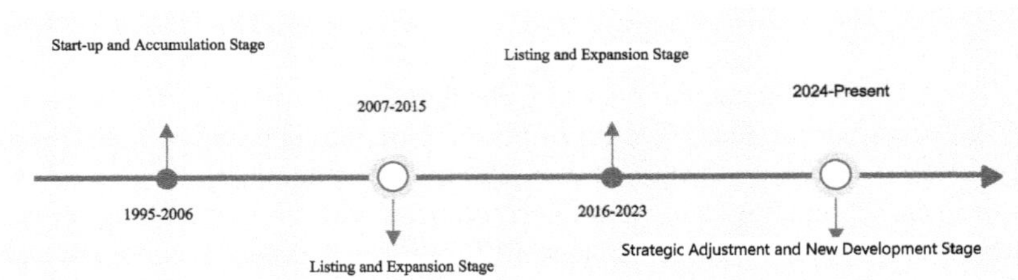

Start-up and Accumulation Stage (1995–2006): In 1995, MINGS was formally established and entered the plastic pipeline industry, mainly producing polyethylene (PE) pipes and other products, laying the foundation for the company's subsequent development.

Listing and Expansion Stage (2007–2015): In 2007, the company was successfully listed on the Shenzhen Stock Exchange, which provided capital and resource support for its further development. During this period, the company continuously expanded its production scale, established production bases in Cangzhou and other places, and improved its product supply capacity. At the same time, it kept expanding its product range—beyond PE pipes, it also developed polypropylene (PP) pipes and other products to meet diverse market demands.

Diversified Development Stage (2016–2023): Since 2016, MINGS has actively deployed in the new energy sector, focusing on the R&D and production of lithium-ion battery separator products. Its products are widely used in fields such as new energy vehicles, 3C products, electric two-wheelers, energy storage, and household appliances, opening up a new business growth path for the company. The company further strengthened technological innovation to improve product quality and performance. For example, in terms of pipeline products, it continuously improved production processes to enhance performance indicators such as pressure resistance and corrosion resistance; in terms of lithium-ion battery separators, it developed separator materials with higher safety and better electrochemical performance. It also expanded its market continuously, established cooperative relationships with many well-known domestic and foreign enterprises, built a sales network covering the whole country, and gradually extended to the international market. For instance, in the gas pipeline field, it cooperated with large gas companies to participate in urban gas transmission projects; in the lithium-ion battery separator field, it established stable cooperation with major domestic and foreign lithium battery customers.

Strategic Adjustment and New Development Stage (2024–Present): On July 19, 2024, MINGS' controlling shareholder DONSS and JJJKK reached a framework agreement on equity transfer. JJJKK acquired 184 million shares of MINGS (accounting for 11% of the total share capital) held by DONSS through agreement transfer; at the same time, DONSS entrusted the non-property rights such as voting rights and proposal rights attached to the remaining 130 million shares (accounting for

7.77% of the total share capital) to JJJKK, making the Cangzhou Municipal People's Government State-owned Assets Supervision and Administration Commission the actual controller of the company[6]. On July 4, the second production line of MINGS' "Annual Output of 200 Million Square Meters Wet Process Lithium-ion Battery Separator Project" at its Wuhu production base was officially put into operation, with a designed annual capacity of 100 million square meters. The project has realized dual-line production at full capacity, thereby improving the company's production capacity and market competitiveness in lithium-ion battery separator products. In the future, with the support of state-owned assets background, MINGS is expected to gain new development opportunities in standardized governance, resource integration, technological innovation, and other aspects, continue to play an important role in pipeline transportation equipment manufacturing and new energy materials fields, and make greater contributions to local economic development.

16.4.3 Current Development Status

In terms of financial data, its total operating income in 2023 was 1.235 billion yuan, a year-on-year decrease of 6.30%, indicating that the company has encountered certain obstacles in market expansion or product sales. The net profit was 88.2351 million yuan, a significant year-on-year decline of 46.76%, reflecting a weakening of the company's profitability. The earnings per share of 0.0528 yuan and the return on net assets of 1.72% also indicate a decline in the company's overall profitability. The operating cash flow per share was −0.0147 yuan, showing that the company's cash inflow from operating activities was relatively insufficient. The gross sales margin was 11.82%, which may also face pressure from rising costs or falling product prices compared with previous periods.

In terms of business development, different segments show different trends. The PE pipeline business has certain opportunities: the expected recovery in natural gas market demand and the quality improvement and scale expansion of rural water supply projects have brought new opportunities for it. However, the market competition situation is unclear, and there may be risks that fierce competition will affect the company's market share and profit margins. For the BOPA film business, although it has the advantage of a newly launched asynchronous stretching production line, with a good reputation for product quality and services and abundant customer resources, as production capacity continues to grow and competition intensifies, conventional specifications of products may enter a low-margin stage. The company needs to continuously innovate and optimize its product structure to enhance competitiveness. The lithium-ion battery separator business has benefited from the rapid development of new energy vehicles and energy storage, with sustained growth in demand, but it also faces challenges of fierce market competition and technological innovation. The company needs to increase R&D investment and improve product performance and quality to consolidate its market position.

16.5 Goody Technology Co., Ltd.

16.5.1 Basic Overview

Founded in 1979, GGKKG is among the earliest batch of Chinese enterprises engaged in plastic pipeline production. After its overall restructuring and establishment in 2010, it has developed into a company with eight production bases in Hubei, Chongqing and other regions over the years. The company's main business covers the R&D, production and sales of PVC, PE and PP series pipelines, whose products are widely used in municipal engineering, agricultural applications, gas supply, industrial sectors and other fields. With a marketing network spread across multiple regions nationwide, its products are sold well in various provinces (municipalities) and autonomous regions in China, and also exported to Central Asia, Southeast Asia, Africa and other places. It is one of the enterprises with relatively large scale and influence in China's plastic building materials manufacturing field [7].

In terms of technological R&D, GGKKG has a strong scientific research team composed of nearly 100 high-education and high-level professional scientific research and technical talents. In recent years, it has published nearly 400 academic papers at home and abroad, compiled three books, won two provincial and ministerial-level scientific and technological achievement awards, and holds a number of invention patents and utility model patents, boasting solid technical strength. In addition, the company has successively obtained various certifications including the ISO9001 international quality management system certification.

In terms of industry status, GGKKG is a vice-chairman unit of the China Plastics Processing Industry Association, a deputy secretary-general unit of the Pipe Professional Committee of China Plastics Processing Industry Association, and also a core member unit of the Sub-Technical Committees such as Plastic Pipes under the National Technical Committee for Standardization of Plastic Products. It enjoys high popularity and reputation in the industry. With the rapid development of China's plastic pipeline industry, GGKKG occupies an important position relying on its own technological advantages and market layout. In the future, it is expected to gain more development opportunities due to the country's urbanization construction and increased investment in water conservancy.

16.5.2 Development History

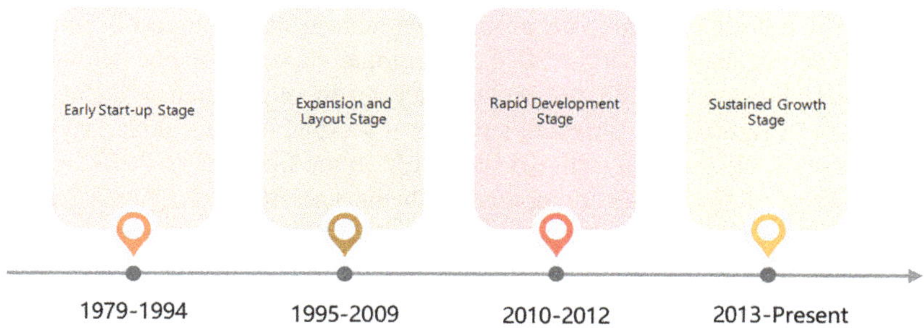

Early Start-up Stage: GGKKG was established in 1979. It is the inventor and manufacturer of China's flame-retardant PVC electrical conduits and trunking, and a pioneer in promoting the "plastic replacing steel" initiative for plastic pipelines in China. During this stage, GGKKG laid the foundation for its position in the plastic pipeline industry through technological innovation and product R&D.

Expansion and Layout Stage: In 1995, SSSJJ was established, laying the foundation for the company's business development in southwest China. In 1999, SSSGD was founded, further expanding the company's production and business scope. In 2002, BBBDD was set up, helping to explore the North China market. In 2009, MMAAA was established, enhancing the company's production capacity in east China.

Rapid Development Stage: 2010 marked a crucial turning point in the company's development. It underwent restructuring and established production bases in multiple locations including Henan, Handan, Foshan, and Gansu, completing the layout with a total of eight production bases. In the same year, based on the audited net assets of Goody Co., Ltd., the company was restructured into a joint-stock limited company by converting net assets into share capital in proportion, with a registered capital of 108 million yuan. After listing on the Shenzhen Stock Exchange in 2012, the company entered a new stage of development leveraging capital strength.

Sustained Growth Stage: Since its listing, GGKKG has continuously increased R&D investment, improved product quality and technical level, and expanded product categories. Its products are widely used in fields such as municipal water supply and drainage, agricultural water diversion and irrigation, gas transmission, and industrial fluid transportation. The marketing network covers all regions of China and extends to Central Asia, Southeast Asia, Africa, and other regions. In recent years, the company has further consolidated its industry position by participating in the drafting and revision of national and industry standards. Meanwhile, it actively responds to market changes and industry competition challenges, and makes unremitting efforts in optimizing product composition, expanding sales channels, and strengthening internal management and control to ensure its sustainable development.

16.5.3 Current Development Status

GGKKG turned losses into profits in the first half of 2023, achieving a total operating income of 454 million yuan and a net profit attributable to shareholders of 138 million yuan. However, in the first half of 2024, the company's performance declined: the total operating income was 383 million yuan, a year-on-year decrease of 15.6%, and the net profit attributable to shareholders was −40.1681 million yuan, a year-on-year decrease of 128.99%. In terms of liabilities: in the first half of 2023, accounts payable decreased by 89.73% compared with the end of the previous year, and contract liabilities increased by 12.13% compared with the end of the previous year; in the first half of 2024, short-term loans decreased by 69.87% compared with the end of the previous year, estimated liabilities decreased by 91.21% compared with the end of the previous year, accounts payable increased by 20.93% compared with the end of the previous year, other payables (including interest and dividends) decreased by 2.90% compared with the end of the previous year, and the liability structure was also adjusted. In terms of financial indicators, the gross profit margin was 14.64% and the net profit margin was 29.88% in the first half of 2023; while in the first half of 2024, the gross profit margin dropped to 8.53% and the net profit margin was −11.40%, indicating a significant decline in profitability [8].

In the first half of 2023, in response to market changes, GGKKG actively adjusted its business strategy, stripped off inferior assets and returned to its main business. For example, it transferred all equity of Mengqi Culture and Tourism to WWWYY. At the same time, the company's private placement of 468 million yuan to WWYAN was approved by the China Securities Regulatory Commission (CSRC). This measure is conducive to the company leveraging the advantages of its controlling shareholder to integrate resources, expand sales channels, exert industrial synergy effects, and further focus on its main business of plastic pipelines. However, in the first half of 2024, the company faced many challenges. The shrinkage of market demand and intensified competition led to a decline in product sales prices and gross profits. According to the data, all profit indicators have declined, and the net cash flow from operating activities is also negative, putting certain pressure on the company's operation and development.

Overall, GGKKG achieved phased profitability through strategic adjustments in 2023, but its performance declined in the first half of 2024. In the future, it needs to further optimize product structure, expand the market, strengthen cost control and risk management to improve profitability and competitiveness, cope with the pressure brought by market changes, and achieve sustainable development. At the same time, it is necessary to pay attention to the company's asset-liability structure and cash flow status to ensure its stable operation.

16.6 Challenges and Opportunities Facing the Development of China's Pipeline Transportation Equipment Manufacturing Enterprises

The analysis of the challenges faced by China's pipeline transportation equipment manufacturing enterprises starts from the following two aspects: On the one hand, from the perspective of the external environment, the instability of the macroeconomic situation has had a significant impact on the three enterprises. From 2021 to 2023, global economic growth slowed down and international trade frictions continued, making the market environment for the three enterprises complex and volatile. Under such circumstances, the demand for pipeline equipment in industries such as energy and chemical engineering decreased, directly leading to a decline in the operating income of enterprises. At the same time, the regulation of the real estate market has also impacted pipeline transportation equipment manufacturing enterprises. Pipeline products are widely used in real estate construction, and the cooling of the real estate market has reduced enterprise orders and squeezed market share. In addition, fluctuations in raw material prices are also a common problem faced by the three enterprises. The prices of raw materials such as steel and plastic required for pipeline manufacturing are volatile, increasing the uncertainty of enterprises' production costs. During periods of large price fluctuations, enterprises may face the dilemma that costs rise but product prices cannot be increased simultaneously, thereby affecting their profitability and operating income (Fig. 16.1).

On the other hand, from the perspective of enterprises' own development, the three enterprises also face common challenges in product innovation and market expansion. With the development of the industry and the continuous change of customer needs, higher requirements have been put forward for the performance, quality and environmental protection of pipeline transportation equipment. However, the investment and achievements of the three enterprises in technological innovation may not have met market demand in a timely manner. These companies are mired in product homogeneity and lack differentiated products that can break the deadlock and build

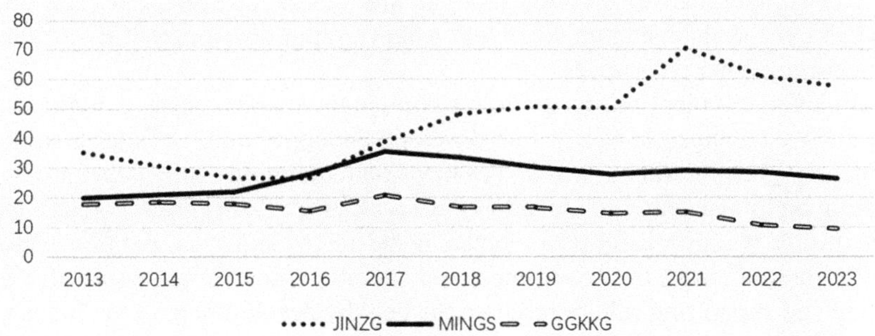

Fig. 16.1 Operating income from 2013 to 2023

core competitiveness with unique value, which affects their foothold in the market. In terms of market expansion, enterprises may face fierce competition, and it is difficult to explore new market areas. At the same time, the efficiency of enterprise operation and management needs to be improved, including production management, supply chain management, marketing and other aspects. In a fiercely competitive market, if enterprises cannot efficiently improve management efficiency and reduce operating costs, it will be difficult to maintain their competitiveness, thereby affecting the growth of operating income.

The country is vigorously adjusting its energy structure, continuously strengthening infrastructure construction, and actively implementing a new urbanization strategy, which has created opportunities and provided broad development space for pipeline transportation equipment manufacturing enterprises. For example, the layout and investment in the construction of oil and gas pipelines, urban water supply and drainage pipelines in the "14th Five-Year Plan" will directly drive the demand for related pipeline equipment. At the same time, with economic development, China's demand for energy such as oil, natural gas and coal has risen, promoting the construction and renewal of transportation equipment such as oil and gas pipelines. The development and utilization of new energy sources, such as hydrogen energy and biomass energy, have also brought new development opportunities for pipeline transportation equipment manufacturing enterprises. The application of emerging technologies such as artificial intelligence, the Internet of Things, and big data in the pipeline transportation industry provides directions and means for enterprises' technological innovation. For example, intelligent pipeline monitoring and maintenance systems and digital production management help improve enterprises' production efficiency, product quality and operation management level. The advancement of the "Belt and Road" Initiative has provided an opportunity for China's pipeline transportation equipment manufacturing enterprises to "go global" and participate in international market competition. Chinese enterprises can leverage their own technological and cost advantages to undertake pipeline construction projects in the international market, driving equipment exports and technology transfer. In addition, fierce market competition will prompt enterprises in the industry to integrate, enhance industry concentration, optimize resource allocation, and improve enterprises' scale effect and competitiveness through mergers and reorganizations. Powerful enterprises can expand their market share and achieve rapid development by acquiring and integrating other enterprises.

Overall, China's pipeline transportation equipment manufacturing enterprises are facing many challenges as well as rare development opportunities. Enterprises need to accurately grasp market trends and policy orientations, strengthen technological innovation and management innovation measures, and enhance their core competitiveness to achieve sustainable development in the fierce market competition environment. At the same time, the government and industry associations should also play a guiding and supporting role to promote the healthy development of the industry.

16.7 Countermeasures and Suggestions

In terms of technological innovation, enterprises should increase investment in research and development, and actively absorb cutting-edge technologies and high-end professional talents. They should strengthen cooperation with scientific research institutions and strive to develop high-performance and environmentally friendly pipeline transportation equipment. For example, developing pipelines made of new materials to improve their corrosion resistance, pressure resistance, temperature resistance and other properties while reducing costs. Through technological innovation, enterprises can increase the added value of their products, enhance their competitiveness in the market, and meet the ever-changing market demands.

In terms of market expansion, enterprises should actively explore markets. Domestically, they should seize the opportunity of national policies, deeply participate in infrastructure construction, energy development and other projects, and expand applications in fields such as municipal engineering, rural water conservancy, and new energy. At the same time, they should actively respond to the "Belt and Road" Initiative, expand international markets, strengthen cooperation with countries along the routes, undertake international projects, export high-quality products and technical services, and enhance their international influence.

In terms of operation and management, enterprises should optimize internal management processes to improve production efficiency and management levels. They should strengthen supply chain management to ensure the stable supply of raw materials and cost control. Meanwhile, they should attach importance to marketing and brand building to enhance corporate reputation and prestige. Through refined management, enterprises can reduce operating costs, improve profitability, and achieve sustainable development.

References

1. Zhejiang Jinzhou Pipeline Technology Co., Ltd. https://www.chinakingland.com
2. Ding Yihan. Jinzhou Pipeline's Net Profit in 2023 Reached 284 Million Yuan, with a Year-on-Year Growth of 20.54%. https://www.nbd.com.cn/rss/zaker/articles/3339440.html. April 19, 2024.
3. Wang Jiehan. "Pipelines" Create Benefits. https://cs.zjol.com.cn/kzl/202311/t20231110_26435908.shtml. November 10, 2023.
4. The Central People's Government of the People's Republic of China. Tax and Fee Policies Gather Synergy to Inject Impetus into Enterprise Development. https://www.gov.cn/yaowen/liebiao/202312/content_6918317.htm. December 4, 2023.
5. Cangzhou Mingzhu Plastic Co., Ltd. https://www.czmingzhu.com/
6. New Cangzhou (a self-media/column name, literally "New Cangzhou by the Sea"). Why Did Cangzhou Dongsu Group Led by Yu Guiting Hand Over Control of Cangzhou Mingzhu? https://baijiahao.baidu.com/s?id=1805198380746218774. July 21, 2024.
7. Goody Technology Co., Ltd. https://www.guditech.com/
8. Sina Finance. Goody Technology: Achieved a Profit of 138 Million Yuan in the First Half of 2023, Turning Losses into Profits Year-on-Year. https://m.toutiao.com/group/7359540089984369215/?upstream_biz=doubao. August 29, 2023.

Chapter 17
Digitalization and Green Transition in Modern Transport Systems

Malin Song and Xin Lu

Abstract This paper first explores the current state of service provision by transportation enterprises. By analyzing the service status of both traditional and emerging transportation companies, it identifies issues within the industry such as uneven development and insufficient capacity for sustainable development, proposing corresponding countermeasures. It then forecasts development trends in transportation enterprise services and concludes with future prospects. Transportation enterprises need to actively fulfill their social responsibilities, promote the expansion of green transportation, gradually achieve the goal of sustainable development, and build a more efficient, safe, and green comprehensive transportation system.

17.1 Introduction

17.1.1 Overview of Transportation Enterprise Services

Transportation enterprise services are not only vital to citizens' daily lives but also a crucial support for China's socio-economic development, permeating various aspects of public daily life. These services are a fundamental cornerstone of economic activity. Efficient transportation systems ensure the flow of people and goods, which is essential for maintaining trade between different regions. For instance, China's high-speed rail has significantly promoted regional economic integration and markedly improved economic efficiency. As efficient and convenient transportation services can effectively reduce commute times, their widespread availability has greatly enhanced people's time utilization efficiency and life satisfaction [1]. Simultaneously, transportation enterprise services have also greatly driven the development of China's tourism industry. Convenient transportation can attract more tourists to travel across the country, thereby boosting local economies.

M. Song · X. Lu (✉)
Anhui University of Finance and Economics, Bengbu, China
e-mail: luxinn1224@163.com

J. Wang and M. Song (eds.), *The Green Path of Chinese Transportation Enterprises*, Resources, Climate and Sustainable Development,
https://doi.org/10.1007/978-981-95-5289-4_17

In recent years, to promote the sustained development of China's transportation enterprises, relevant national departments have successively introduced a series of policy measures to support and regulate their development, creating a favorable political and developmental environment. Particularly in 2021, the Ministry of Transport officially released the *"14th Five-Year" Development Plan for Comprehensive Transport Services*. With the core goal of accelerating the construction of a leading transport nation, this *Plan* aims to build a modern comprehensive transport service system that is convenient, smooth, economically efficient, open and shared, green and intelligent, and safe and reliable [2] (Table 17.1).

Today, the transportation industry has become a crucial pillar of China's socio-economic development, with its status and role becoming increasingly prominent. As urbanization accelerates and populations continue to concentrate in cities from rural areas, economic activities have become more frequent, creating an urgent demand for convenient transportation. This requires not only strong carrying capacity from the transport system but also high efficiency, safety, and environmental friendliness of transport services. As direct providers of transport services, the quality of service from transportation enterprises directly affects the overall operational efficiency of society and the quality of life of the people. From traditional public transport to

Table 17.1 Selected recent policies related to transportation enterprise services

Release year	Issuing authority	Document name	Key interpretation
2023	Ministry of Transport	《Five-Year Action Plan for Accelerating the Building of a Leading Transport Nation (2023–2027)》	Defines the goals and action tasks for accelerating the building of a leading transport nation over the next five years and constructs a framework of ten major actions
2022	Ministry of Transport	《Comprehensive Transportation Standard System (2022)》	Aims to enhance the quality of standards in areas such as comprehensive transport facilities and transport services by 2025, promoting standardized development in the transport industry
2021	Ministry of Transport	《"14th Five-Year" Development Plan for Comprehensive Transport Services》	Optimizes the transport structure, improves transport efficiency and service quality, and builds a modern comprehensive transport service system
2019	State Council	《Outline for Building a Leading Transport Nation》	Promotes rapid and convenient travel services, builds convenient and smooth urban (agglomeration) transport networks, and accelerates the development of new business forms and models

emerging services like ride-hailing, bike-sharing, and autonomous driving, transportation enterprises are undergoing transformation and upgrading at an unprecedented pace. These changes reflect not only the profound impact of technological progress on the transportation industry but also the market's urgent need for diversified and personalized transport services.

This paper begins with an overview of the current state of transportation enterprise services, outlining the basic characteristics, operational models, and market positioning of various types of transportation enterprises, such as traditional, emerging, and smart transportation companies, and conducts further research on specific enterprises. As the market constantly changes, transportation enterprises also face numerous challenges. This paper proposes corresponding strategies and suggestions to address these challenges. It also analyzes market trends and changes in consumer demand, exploring how transportation enterprises can adjust according to market demands to enhance their market competitiveness and provide satisfactory services to consumers. Finally, the paper offers a outlook on the future development of transportation enterprises.

17.1.2 Current State of Transportation Enterprise Services

17.1.2.1 Current State of Traditional Transportation Enterprise Services

Traditional transportation enterprises, such as bus companies and railway companies, have always been important carriers of public passenger transport. However, facing rapid technological development, traditional transportation enterprises are also actively seeking transformation and upgrading. As one of China's largest bus companies, BGJIA has achieved significant success in its transformation and upgrading. Firstly, by introducing advanced intelligent dispatch systems, it conducts real-time monitoring, predicts peak passenger flow periods and popular areas, and scientifically plans departure frequencies, greatly alleviating urban travel difficulties. Furthermore, BGJIA actively promotes the use of new energy buses. This initiative not only significantly reduces exhaust emissions, effectively improving urban air quality, but also leads to a notable decrease in operational costs. As shown in Fig. 17.1, the proportion of clean energy and new energy buses has increased year by year. By the end of 2023, Beijing's public transport had 16,832 new energy vehicles, with clean energy and new energy buses accounting for as much as 94.7% [3]. In terms of customer service, BGJIA has also adopted innovative approaches. By developing online service platforms such as mobile apps and WeChat official accounts, it provides passengers with comprehensive convenient services. Passengers can easily access information on bus routes, arrival times, and fares through their mobile phones, greatly enhancing travel convenience and providing a more comfortable service experience.

In railway services, taking ZGTIE as an example, by the end of 2023, the national railway operating mileage exceeded 150,000 km, including over 45,000

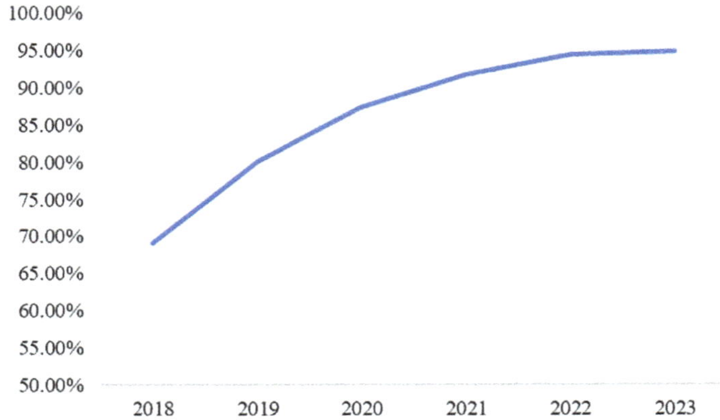

Fig. 17.1 Proportion of clean energy and new energy buses. *Data source* Corporate Social Responsibility Report

km of high-speed rail operating mileage, ranking first in the world. In 2023, ZGTIE's passenger traffic reached 3.685 billion person-trips, a year-on-year increase of 128.8%; passenger turnover reached 1471.712 billion passenger-kilometers, an increase of 123.9% [4]. As shown in Fig. 17.2, passenger traffic peaked in 2023, sufficiently demonstrating the continuously growing demand for travel. Representative enterprises include DQITE and JHGTE. DQITE primarily focuses on coal transportation, with steadily growing revenue, driving economic development along its lines, demonstrating significant advantages in freight efficiency and cost control, and achieving remarkable results in informatization construction. JHGTE is a vital artery connecting the Bohai Rim and the Yangtze River Delta. Its core business encompasses passenger transport and network services. Its revenue achieved a significant rebound after the pandemic and realized substantial profits in 2023.

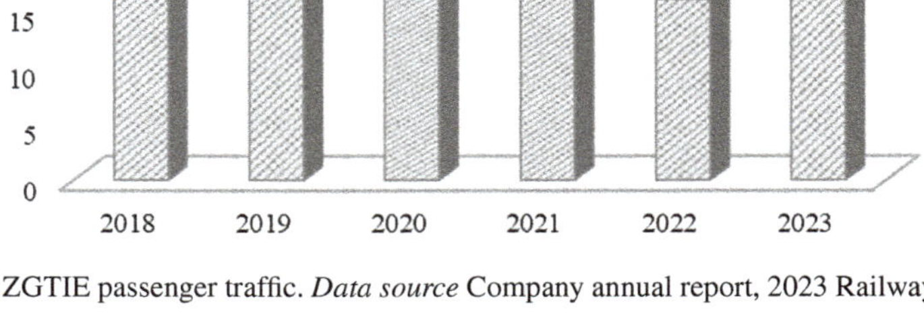

Fig. 17.2 ZGTIE passenger traffic. *Data source* Company annual report, 2023 Railway Statistical Bulletin

17.1.2.2 Current State of Emerging Transportation Enterprise Services

With the popularization of the internet, the transportation industry has witnessed the rise of a wave of emerging enterprises. These companies have innovated upon traditional service models, not only providing consumers with more convenient travel options but also meeting the demand for personalized travel experiences through efficient data processing and intelligent services. DIIII, as a leading mobile travel platform in China, achieves efficient matching of supply and demand by integrating social vehicle resources and passenger needs. Users can easily enjoy convenient travel services by downloading the Didi app and completing registration on their mobile phones. Furthermore, DIIII has established strict driver entry mechanisms and evaluation systems, ensuring all drivers pass rigorous reviews and training before registration. Simultaneously, the platform encourages passengers to evaluate and provide feedback on drivers to promptly identify and address service issues, improving overall service quality. Beyond basic travel services, DIIII continuously expands its business areas, offering diversified services such as carpooling, hitchhiking, and designated driving. These services not only enrich user choices but also reduce travel costs and improve resource utilization efficiency. By the end of 2023, the number of registered drivers on the DIIII platform exceeded several million, with average daily orders reaching tens of millions [5]. As of March 2023, DIIII offered diversified services including ride-hailing, hitchhiking, shared electric vehicles, and designated driving in 15 countries including China. It had 587 million annual active users globally and 23 million annual active drivers. The Global Transaction Value (GTV) in the first half of 2023 reached 156.7 billion yuan, a year-on-year increase of 39% [6]. This vast user base and order volume not only demonstrate DIIII's market position but also lay a solid foundation for its future development.

17.1.2.3 Current State of Smart Transportation Enterprise Construction

By applying advanced technologies such as the Internet of Things (IoT), artificial intelligence (AI), and big data, an increasing number of transportation enterprises are integrating technology into their services, achieving comprehensive perception, analysis, and control of the transportation system. JHGTE, as one of China's top high-speed rail line representatives, adopts cutting-edge intelligent dispatch and automated control system technologies, ensuring instant monitoring and precise management of train operation status. These systems can automatically adjust train speed, intervals, and arrival times, ensuring trains reach their destinations safely and on time [7]. In terms of passenger service, JHGTE also provides convenient ticket purchase, station entry, boarding, and exit services. Passengers can purchase tickets through multiple channels and enter stations with valid documents such as ID cards. Meanwhile, high-speed rail stations are equipped with self-service ticket check gates, intelligent navigation systems, and other facilities to guide passengers correctly, providing a more

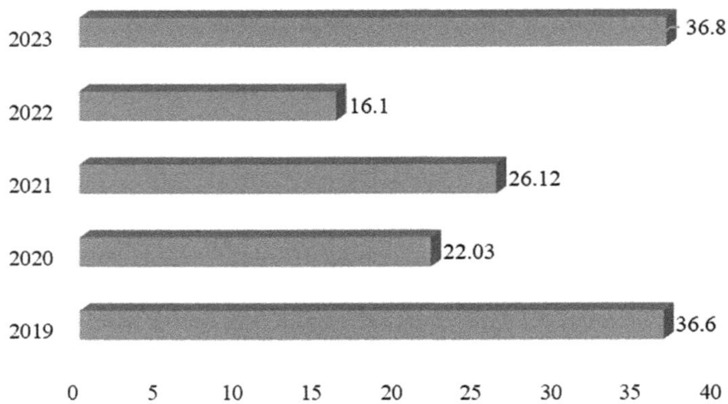

Fig. 17.3 JHGTE passenger traffic

convenient and efficient travel experience. Additionally, by constructing comprehensive transportation hubs and transfer centers, seamless connection with various transportation modes such as buses and subways is achieved. As shown in Fig. 17.3, JHGTE's passenger traffic in 2023 reached 3.68 billion person-trips [8], hitting a record high.

17.2 Problems and Countermeasures in Service Provision by Transportation Enterprises

17.2.1 Existing Problems

17.2.1.1 Uneven Development

Uneven development is one of the main problems currently faced by transportation enterprise services in China. On the one hand, major cities and developed regions, due to inherent advantages such as economy and geographical location, possess relatively complete transportation networks and advanced infrastructure, satisfying diverse freight transport demands. However, in less developed areas, problems like weak transportation infrastructure and lagging road construction frequently occur, seriously hindering the construction and optimization of transport networks. This makes it difficult for transportation services in these areas to keep pace with local economic and social development. This not only slows down logistics efficiency within the region but also restricts the potential for cross-regional cooperation and communication, hinders the effective allocation of resources, and severely limits balanced economic development. On the other hand, differences in infrastructure, policy environment, and economic development levels between regions make it difficult to achieve seamless connection of transport networks. In cross-regional freight transport processes, goods often need to undergo multiple transfers. This process not only significantly increases time costs but also raises operational costs, leading to overall low logistics efficiency. This poor connectivity of transport networks not only increases the operational costs for enterprises but also affects market response speed and supply chain stability.

17.2.1.2 Weak Capacity for Sustainable Development

With the continuous growth of transportation demand, transportation enterprises are also facing huge pressure for sustainable development, specifically manifested in the following aspects. The transportation industry, as one of the main sources of energy consumption and greenhouse gas emissions, sees extensive use of traditional fuel vehicles exacerbating air pollution and damaging environmental quality. Although new energy vehicles have received increasing attention and support in recent years, their application in the transportation field is still relatively limited, making it difficult to completely replace traditional fuel vehicles and fundamentally solve environmental pollution problems. Simultaneously, with the acceleration of urbanization, land resources are becoming scarcer. As transportation infrastructure construction requires occupying a large amount of land and cannot utilize every inch of land to the maximum extent, it leads to excessive strain on urban land use, damages the urban ecological environment, and reduces residents' quality of life. Furthermore, the transportation industry highly depends on energy, and the global energy supply situation is complex and volatile. Fluctuations in energy prices and supply shortages significantly impact the operational costs and service quality of transportation enterprises.

17.2.1.3 Technological Pressure from Accelerating Pace of Technological Updates

In the context of the informatization and intelligent era, transportation enterprise services increasingly rely on advanced information technology and intelligent transportation systems. To maintain market competitive advantage, transportation enterprises need to continuously invest sufficient R&D funds to introduce new technologies and improve service quality and efficiency. However, high technological investment, high risks, and the existence of technological barriers make it difficult for some transportation enterprises to keep up with market pace in technological innovation. The rapid development of technology also requires enterprises to possess stronger data processing and analysis capabilities. Transportation enterprises need professional technical teams to apply technologies such as big data, cloud computing, and the Internet of Things to achieve instant processing and intelligent analysis of massive data volumes. However, both sufficient R&D funds and professional technical teams remain significant challenges for many traditional transportation enterprises.

17.2.1.4 Market Pressure from Fierce Competition

On the one hand, the development of traditional transportation enterprises such as buses and taxis has been strongly impacted by emerging transportation enterprises like ride-hailing and bike-sharing. On the other hand, emerging transportation enterprises, relying on their strong technological innovation capabilities and flexible service models, occupy a large portion of the market, putting enormous pressure on traditional transportation enterprises. In the fierce market competition, enterprises need to continuously improve their service quality, optimize service processes, and reduce operational costs to win customer favor and market recognition. This requires that both traditional and emerging transportation enterprises not only have strong market competitiveness and adaptability but also possess keen market insight to respond to market changes and diverse customer needs. However, most transportation enterprises currently clearly do not yet possess these capabilities.

17.2.1.5 Diversified and Personalized User Service Demands

With the improvement of living standards and changes in consumption concepts, people's demands for transportation services are becoming increasingly diversified and personalized. Traditional transportation service models can no longer meet user needs. On the one hand, users are no longer satisfied with a single mode of travel but hope to have multiple options. For example, for short trips, they might choose shared bikes or electric vehicles; for long trips, they may prefer high-speed rail or airplanes. Secondly, users hope that transportation enterprises can provide more personalized service content. For instance, in daily tourism travel, users may want all-inclusive services including accommodation, dining, and attraction tickets; in business travel, users may desire convenient meeting venue bookings, pick-up, and parking services. On the other hand, users' requirements for the quality of transportation services are also increasing. Beyond basic transport functions, users also care about aspects such as convenience, comfort, and safety of the service. For example, when taking high-speed rail, users hope stations and trains provide convenient ticket purchase, station entry, boarding, and exit services; when flying, users wish to enjoy comfortable waiting environments and high-quality cabin services.

17.2.2 Countermeasures for Transportation Services

17.2.2.1 Narrowing the Development Gap

Increase investment in transportation infrastructure in poor areas, encourage these regions to undertake transportation infrastructure construction, improve the coverage of transportation networks such as railways, and ensure efficient and convenient transportation services are available everywhere. Promote the connection of transport networks between different regions, break down regional barriers, and achieve rapid freight transport. Simultaneously, the government should enact fair and transparent transportation policies to ensure equal opportunities and resources for all regions in transportation infrastructure construction and transport services. Encourage logistics enterprises in different regions to establish cooperative relationships, share resources such as warehousing and transportation. This can not only improve logistics efficiency and reduce operational costs but also jointly enhance service levels. Finally, increase efforts in cultivating and introducing talents in the transportation field, enhance the overall technical and management level of the industry. This is not only crucial for providing talent support for the balanced development of transportation enterprises but also key to solving the problem of uneven development in their service provision.

17.2.2.2 Promoting Green Transformation

Increase efforts to promote new energy vehicles, enhance their status in the transportation field. This can not only improve the environmental protection level of the transportation sector and reduce air pollution but also lead the automotive industry towards a more sustainable direction. Furthermore, encourage the public to use mass transit such as buses and subways more frequently to reduce the use of private cars and further alleviate environmental pressure. When constructing transportation infrastructure, fully consider the rational use of land resources, avoid excessive occupation of land resources, and ensure efficient land utilization. Optimize the layout of transportation networks, improve transportation efficiency, reduce traffic congestion, and alleviate urban land shortage. Moreover, actively explore and apply alternative energy sources such as solar and wind power to gradually reduce dependence on traditional energy sources. This helps not only to ensure energy security but also to reduce greenhouse gas emissions such as CO_2, promoting the transformation of the transportation industry towards low-carbon, green, and environmentally friendly directions. Finally, through technological innovation, improve the operational efficiency and service quality of transportation enterprises, reducing their environmental impact. For example, use technologies like big data and AI to optimize traffic management, reduce energy waste, and achieve more efficient resource allocation and service provision.

17.2.2.3 Deepening Technological Innovation and Application

Transportation enterprises should attach great importance to technological innovation. By increasing financial support for the research and development of new technologies and products, establishing specialized R&D departments or funds, they can promote the practical application of technological innovations. This not only helps enhance the enterprise's technological strength but also injects a continuous stream of technological momentum into its long-term development. Furthermore, enterprises should establish close cooperative relationships with universities and research institutions to jointly develop advanced technologies with independent intellectual property rights, injecting technological driving force for their long-term development. Since talent is the core of technological innovation, enterprises should build a scientific and reasonable system for talent introduction and cultivation to attract and nurture a team of talents with outstanding innovative abilities and profound professional skills. By providing competitive salaries, generous benefits, and broad development prospects, retain talent and stimulate their innovative potential. Finally, enterprises should closely monitor global technological changes and development trends in the transportation field to ensure timely introduction and effective absorption of new technologies. Pay particular attention to cutting-edge technologies such as autonomous driving, vehicle networking, and blockchain, exploring their application prospects in transportation services to gain a competitive edge in the market and achieve leapfrog development for the enterprise.

17.2.2.4 Enhancing Market Competitiveness

First, transportation enterprises should focus on strengthening brand building and marketing efforts. By increasing publicity, enhance their visibility and reputation to stand out in the market competition [9]. This helps not only attract more new customers but also enhance the loyalty of existing customers, laying a solid foundation for the long-term development of the enterprise. Through methods such as market research and user feedback collection, gain an in-depth understanding of users' travel needs and preferences. Strengthen cross-industry cooperation with different sectors to provide strong support for product development and service optimization [10]. Secondly, enterprises should actively explore and implement comprehensive service models like "Travel+", combining travel services with diverse scenarios such as tourism, shopping, and entertainment to provide users with a richer travel experience. Finally, by strengthening service quality and cost control, improve service quality, optimize service processes, and continuously improve through customer satisfaction surveys.

17.2.2.5 Diversified Integration and Personalized Customization

First, establish comprehensive transportation service platforms that integrate various transportation modes such as shared bikes, electric vehicles, high-speed rail, and airplanes, providing users with one-stop query and booking services to meet their travel needs in different scenarios. Utilize big data and AI technologies to deeply analyze users' travel patterns and preferences, and tailor exclusive transportation service plans. For example, provide tourism users with all-in-one services including accommodation, dining, and attraction tickets; offer business users customized services such as meeting venue bookings and professional transfers. Secondly, strengthen the construction and maintenance of transportation infrastructure to ensure the convenience, comfort, and safety of stations, airports, and vehicles. Optimize service processes, improve the quality of service personnel, and provide users with more thoughtful and professional service experiences. Also, establish user feedback channels to promptly collect and process users' opinions and suggestions on transportation services. Through continuous improvement and optimization of services, meet users' ever-changing personalized needs.

17.3 Market Trends and Changes in Consumer Demand for Transportation Services

17.3.1 Market Trend Analysis

17.3.1.1 Intelligentization and Digitalization Trend

With the continuous advancement of cutting-edge technologies such as big data, cloud computing, the Internet of Things, and artificial intelligence, intelligentization and digitalization have become important trends in the transportation service market. The deep integration and application of these advanced technologies have not only significantly enhanced the overall operational efficiency of transportation systems but also greatly optimized the travel experience for consumers. With its powerful capability for real-time data analysis, intelligent transportation systems can accurately predict traffic flow and scientifically optimize route planning. In 2022, BGJIA's average scheduled trips increased to 476, with 539 routes achieving punctuality projects. The trip fulfillment rate reached over 99%, and the departure punctuality rate reached over 98% [11]. Passenger satisfaction has also increased year by year, even reaching a high of 94.04% in 2023 [3], greatly alleviating traffic congestion and significantly reducing waiting times. Meanwhile, the widespread application of intelligent payment systems has greatly simplified the ticket purchase and payment processes, enabling consumers to enjoy smoother and more efficient travel services. These intelligent and digital innovations are gradually building a more convenient, efficient, and personalized modern transportation service system, bringing unprecedented convenience and comfort to people's daily lives.

17.3.1.2 Green Travel and Sustainable Development

With the general increase in global environmental awareness, green travel has become a growing focus for consumers. To actively respond to this trend, governments and transportation enterprises are working together to introduce a series of policies and technical means that encourage and support the development of green transportation modes. The widespread adoption of new energy vehicles not only effectively reduces carbon emissions but also provides consumers with more environmentally friendly travel options. The optimization and upgrading of public transportation systems, by improving service quality and operational efficiency, effectively guide the public towards low-carbon travel modes. The construction of cycling and walking-friendly cities starts from the level of urban planning, creating more opportunities for people to travel in ways that are close to nature, healthy, and environmentally friendly. These measures collectively push the transportation service industry steadily towards a more environmentally friendly and sustainable direction, laying a solid foundation for building a green, low-carbon future transportation service system.

17.3.1.3 The Rise of the Sharing Economy Model

The sharing economy model, with its unique advantages of efficient resource utilization and significant cost reduction, has rapidly risen in the transportation service field, gaining widespread market attention. From convenient bike-sharing and flexible car-sharing to extensive ride-hailing platforms, the sharing economy model has not only broken the time and space barriers of traditional transportation services but also provided consumers with unprecedented flexibility and diversity, greatly enriching people's travel choices. The rise of this model has not only met the public's personalized and diversified travel needs but also effectively alleviated urban traffic congestion and environmental pollution problems. Simultaneously, the sharing economy model has also stimulated cooperation and competition among transportation enterprises, promoting innovation and development across the industry. Enterprises are continuously exploring new service models and profit points under the background of the sharing economy, promoting the extension and upgrading of the transportation service industry chain, and injecting strong momentum into the sustainable development of the industry.

17.3.1.4 Cross-Border Integration and Diversified Development

As the transportation service market continues to deepen and mature, cross-border integration and diversified development have become key trends driving the transportation industry to new heights. Transportation enterprises are no longer confined to the single framework of traditional transport services but are beginning to actively expand into various fields such as tourism, logistics, finance, and economy. This strategic transformation not only broadens the business scope of enterprises but also brings unprecedented development opportunities. Through cross-border cooperation and resource integration, transportation enterprises can deeply integrate the advantageous resources of different industries, creating a series of more comprehensive and integrated travel solutions. These solutions cover basic transportation services and incorporate diverse elements such as travel planning, logistics distribution, and financial services, thereby accurately meeting consumers' increasingly diverse and personalized travel needs. Therefore, this innovative model of enterprises not only enhances the overall quality of transportation services but also injects new vitality and momentum into the sustainable development of the industry, leading the transportation service market towards a broader and more diverse future.

17.3.2 Changes in Consumer Demand

First, increased demand for personalized and customized services. With the trend towards diversification of consumer preferences, personalized and customized services have become an important demand in the transportation service market. Consumers are no longer satisfied with traditional, standardized travel services but hope to obtain more precise and personalized services based on their own needs and preferences. Therefore, transportation enterprises need to strengthen cooperation and integration with other industries, expand service areas and scope, and provide consumers with more diversified and comprehensive services. Second, safety and comfort have become primary considerations. With frequent traffic accidents and worsening environmental pollution, consumers are placing higher demands on the safety and comfort of transportation services. Transportation enterprises need not only to strengthen safety management and improve service quality to ensure travel safety for consumers but also to focus on the comfort design of vehicles and facilities to enhance the travel experience. Third, increased demand for convenience and efficiency. In the fast-paced modern life, consumers hope to complete travel tasks in the shortest possible time, reducing the time cost of waiting and transfers. This requires transportation enterprises to optimize service processes, improve operational efficiency, and provide more convenient and efficient travel services for consumers. Fourth, enhanced environmental awareness promotes green travel. As people's environmental awareness increases, more and more consumers are beginning to pay attention to green travel. They tend to choose low-carbon, environmentally friendly transportation modes to reduce environmental pollution. Therefore, transportation enterprises need to actively promote new energy vehicles, optimize transport routes, reduce empty mileage rates, and other environmental measures, provide services that align with consumer preferences, and meet their demand for green travel.

17.4 Trend Predictions and Future Outlook

17.4.1 Development Trends

After an in-depth discussion of the current state and problems faced by transportation enterprise service provision, it is clear that this field is in a period of change and innovation. With the acceleration of urbanization and the increasing diversification of people's travel needs, transportation enterprises not only undertake basic transport functions but also play an irreplaceable role in promoting socio-economic development and improving people's quality of life. Based on current technology and market trends, the development direction of the transportation service industry in the coming years can be predicted.

1. With the continuous maturation and application of digital technologies such as big data, cloud computing, the Internet of Things, and artificial intelligence, the transportation service industry will develop in a more intelligent direction. Intelligent transportation systems will achieve comprehensive perception, precise analysis, and intelligent regulation of traffic flow, improving traffic operation efficiency and safety [12]. For example, intelligent traffic signal control systems can dynamically adjust signals based on real-time traffic conditions in a specific area, reducing traffic congestion; intelligent navigation systems like GPS can provide optimal route suggestions based on real-time road conditions. This system intelligence not only enhances the travel experience but also provides powerful technical support for urban traffic management.
2. Given global climate change and increasing environmental awareness, new energy vehicles are gradually becoming a mainstream trend in the development of transportation services. Governments will increase efforts to promote new energy vehicles and build a more complete network of charging and battery swap facilities. Simultaneously, transportation enterprises will actively respond to the concept of green transportation, increase the proportion of procurement and use of new energy vehicles, reduce carbon emissions and environmental pollution, and enhance their green image and market competitiveness.
3. The sharing economy model has achieved significant results in the transportation service field, such as emerging transportation modes like ride-hailing and bike-sharing. In the future, the sharing economy model will continue to deepen and expand into more areas. For example, service models such as car-sharing and shared parking spaces will continuously emerge, providing users with more convenient and economical travel options. Meanwhile, the sharing economy model will also promote the efficient use and sharing of resources, reducing the overall operational costs of the transportation industry.
4. Facing the trend of diversified and personalized consumer demand, transportation industry enterprises should increasingly focus on providing personalized and exclusive service experiences. Leveraging the power of big data and artificial intelligence technologies, enterprises can gain deep insights into users' travel behaviors and demand characteristics, thereby providing more precise and attentive services. For instance, recommend optimal travel plans and routes based on user preferences; tailor logistics solutions for specific clients; provide detailed and comprehensive travel services for special users, etc.
5. To improve the overall efficiency and convenience of the transportation system, the future transportation service industry will pay more attention to the construction of multimodal transport and comprehensive transportation systems. By strengthening the connection and coordination between different transportation

modes, such as rail, road, waterway, and air, seamless connection and complementary advantages between various transportation modes can be achieved, enhancing the overall carrying capacity and service level of the transportation system to meet users' diverse travel needs.

17.4.2 Future Outlook

Looking ahead, the development prospects of the transportation service industry are broad, benefiting from the synergistic effects of multiple aspects including technological innovation, green development, the rise of the sharing economy, personalized services, and multimodal transport integration. Transportation enterprises need to continuously innovate service models and improve service quality to adapt to changes in the market environment and meet the upgrading of user demands. Simultaneously, enterprises must also actively fulfill their social responsibilities, promote green transportation, achieve the goal of sustainable development, always maintain a keen insight into market dynamics, and adjust strategic directions promptly. Furthermore, enterprises need to possess strong innovation capabilities and flexible market adaptability to ensure competitiveness in a rapidly changing environment. Actively respond to the global green transformation trend, dedicate efforts to developing low-carbon, environmentally friendly transportation solutions, and make important contributions to the sustained and healthy development of the socio-economy. In summary, the future transportation service industry will face unprecedented opportunities and challenges. Transportation enterprises need to comprehensively utilize the aforementioned strategies, not only making continuous efforts in technological innovation and service upgrade but also demonstrating commitment in social responsibility and environmental protection, to achieve rapid, sustainable, and high-quality development of the industry and win a brighter future.

References

1. Xie, W. High-speed rail has become the preferred mode of public travel. 2024. https://new.qq.com/rain/a/20240813A06QBM00
2. Qiao, X. F. Interpretation of the "14th Five-Year Plan for Comprehensive Transport Services" (Part II): Standing at a new starting point and implementing new requirements, China's travel services have a new direction.2021. http://finance.people.com.cn/n1/2021/1126/c1004-32293077.html
3. Jintai News. The proportion of clean energy and new energy vehicles in Beijing's public transport is 94.7%. 2024. https://www.163.com/dy/article/J6QTO0CH05346936.html
4. Lu, B. Y. China State Railway Group's 2023 revenue reached 1.25 trillion yuan with a net profit of 3.3 billion yuan, achieving the best performance in history. 2024. https://finance.sina.cn/2024-05-01/detail-inatthfu7644279.d.html
5. Yi, Y. J. Digging for gold in quality Chinese stocks series: In-depth report on DiDi Global (DIDIY). 2024. https://www.hangyan.co/reports/3292990093484098700

6. China Merchants Securities. 2023 research report on DiDi Chuxing: Returning with stable market share, the ride-sharing leader maps out a new growth blueprint. 2023. https://www.vzkoo.com/read/2023121999a3c8bc7e096892016d09b7.html
7. Jin, J., & Fei, Z. H. Optimization, upgrading, and intelligent prospects of the Beijing-Shanghai High-Speed Railway's centralized traffic control system. 2022. https://www.fx361.cc/page/2022/0518/14292182.shtml
8. The Paper. Latest | 3.68 billion passenger trips! A historical high! 2024. https://www.thepaper.cn/newsDetail_forward_25954155
9. Gao, D. R. A study on the impact of the transportation industry's development on economic growth [J]. China Market, 2023(3):161–163.
10. Zhang, P., & Wei, W., et al. New paths for enhancing urban transportation service capacity [J]. China Transportation Weekly, 2022(4):46–48.
11. Jintai News. The proportion of clean energy and new energy buses in Beijing exceeds 94%. 2023. https://new.qq.com/rain/a/20230703A02YO000
12. Wang, J., & Liu, H., et al. Development models and trends of service-oriented manufacturing in rail transit enterprises [J]. Transportation Enterprise Management, 2024, 4(3):10–12.

Chapter 18
The Development of Intelligent Transportation for Transportation Enterprises

Malin Song and Chengyue Yang

Abstract With the development of the global economy, Chinese transportation enterprises have occupied an important position internationally, and their development environment has been increasingly favorable. In the transportation field, the development of intelligent transportation has become a key trend. Intelligent transportation is of vital significance for ensuring transportation safety, improving transportation efficiency, promoting infrastructure construction and technological innovation, and helping to achieve the "dual carbon" goals. However, problems such as insufficient capital and policy support, unscientific management and operation models, unbalanced regional development, and severe information barriers have hindered the development of enterprises. Therefore, this paper conducts an in-depth discussion on the above-mentioned problems, analyzes the development status of some intelligent transportation enterprises, and draws on their valuable experience to provide relevant suggestions for the development of intelligent transportation enterprises, hoping to help Chinese transportation enterprises achieve high-quality development in the field of intelligent transportation.

18.1 Current Status of the Development of Intelligent Transp-Ortation for Transportation Enterprises

The development of intelligent transportation is of great significance. At the 20th National Congress of the Communist Party of China, General Secretary Xi Jinping emphasized the need to vigorously develop intelligent transportation. The essence of transportation enterprises realizing intelligent transportation lies in using digital and information technologies to upgrade transportation management and service

M. Song
Anhui University of Finance and Economics, Bengbu, China

C. Yang (✉)
Dongbei University of Finance and Economics, Dalian, China

J. Wang and M. Song (eds.), *The Green Path of Chinese Transportation Enterprises*, Resources, Climate and Sustainable Development,
https://doi.org/10.1007/978-981-95-5289-4_18

capabilities. Such upgrades can ensure transportation safety, improve transportation efficiency, protect the environment, and save resources. The development of intelligent transportation by transportation enterprises not only enhances their own competitiveness but also greatly improves residents' sense of well-being, making it a truly people-oriented initiative. The development of intelligent transportation by transportation enterprises has significantly improved the social operation efficiency of cities. With the application of intelligent transportation systems, the impact of transportation on a city is increasingly prominent. However, driven by factors such as technological innovation, population growth, and the explosive increase in the number of private cars purchased by residents, the traffic carrying capacity of cities is facing increasingly severe challenges. Conflicts are inevitable between the rapid development of cities and their actual carrying capacity. In addition, how to optimize their own intelligent transportation systems has become a key concern for transportation enterprises. The intelligent development of transportation enterprises is closely linked to urbanization. As urbanization progresses, the demand for transportation infrastructure continues to expand. The National Expressway Network Plan points out that the national expressway network will cover more than 1 billion people, as well as provincial capitals, large cities, medium-sized cities and other regions across the country. With the expansion of investment in expressways and urban transportation, the demand for intelligent transportation systems will also continue to grow.[1]

18.1.1 Analysis of the Development Environment for Transportation Enterprises

In the 2022 Fortune Global 500 list, 145 Chinese companies were included, ranking first in the world for four consecutive years. Among them, 12 Chinese enterprises from the engineering and construction industry made the list, and Chinese transportation construction enterprises continued to take the top six positions globally in this industry. It is evident that Chinese transportation enterprises occupy an important position worldwide [2].As the focus of global economic growth shifts from Europe and the United States to Asia, the "Belt and Road" Initiative has formed a new cooperation pattern, bringing new growth space for the development of China's transportation enterprises. With changes in people's lifestyles and major technological reforms, society is in urgent need of the development of intelligent transportation by transportation enterprises, which also brings new opportunities and challenges to these enterprises. From the supply side, the improvement of social operation efficiency and the accelerated integration of industrialization and informatization have promoted the emergence of new supply models characterized by intelligence. The development of intelligent transportation by transportation enterprises is an inevitable trend and a response to the reform and changes of the times. From the demand side, currently, not only urban transportation needs new models, but also the integrated urban-rural development has released huge market increments. This has been followed by new requirements for the intelligent development of transportation enterprises. Due to changes in social patterns, the country has focused on safeguarding and improving people's livelihoods, and the intelligent development of transportation enterprises is directly related to the convenience of residents' lives. However, demographic structure issues have gradually become prominent, such as more obvious population agglomeration in large cities and serious population aging. These multiple factors have driven transportation enterprises to improve their intelligent development level and accelerate the application of emerging technologies such as artificial intelligence. Meanwhile, market competition has put forward higher requirements for transportation enterprises.

18.1.2 The Development of Intelligent Transportation by Transportation Enterprises Is a Key Area of New Infrastructure Construction

Digitalized, networked, and intelligent infrastructure enables profound transformations in production methods, lifestyles, scientific research paradigms, and business models. The development of intelligent transportation by transportation enterprises represents an innovation in infrastructure construction, which facilitates residents' travel, improves the quality and efficiency of transportation services, and drives economic development. China will adapt to spatial change trends, optimize major infrastructure, effectively enhance guarantee capabilities, and pool urban strengths. At the local level, this can promote regional economic development, improve connectivity and interaction between regions, help drive economic growth in various areas, and boost overall economic vitality. The intelligent transportation development of enterprises cannot be separated from financial support, and the construction of new infrastructure further requires unobstructed financing channels. Sufficient capital can help enterprises advance technological innovation and lay a solid foundation for their intelligent transportation development. Intelligent transportation systems are an important component of new infrastructure construction, and the two are mutually reinforcing and interdependent. Driven by new infrastructure construction, intelligent transportation systems can achieve more efficient and intelligent management. The operation of transportation enterprises can also improve transportation efficiency and reduce energy consumption, thereby enhancing product quality and further boosting corporate competitiveness. In summary, the intelligent development of transportation enterprises and new infrastructure construction promote and influence each other: the development of intelligent transportation by transportation enterprises is a key area of new infrastructure construction, while new infrastructure construction provides technical support and infrastructure guarantees for this intelligent transportation development. In turn, the development of intelligent transportation advances the improvement and application of new infrastructure construction.

18.1.3 The Development of Intelligent Transportation by Transportation Enterprises Is Also an Important Path to Achieve the "Dual Carbon" Goals

With the increasing prominence of global environmental issues and growing attention to sustainable development, green development has become a key direction for enterprises worldwide. Improving the efficiency of infrastructure and transportation equipment, reducing resource consumption, and minimizing environmental pollution can promote the optimization of transportation production organizations, enhance passenger and cargo transportation efficiency, and drive the intensive development of the transportation industry. The development of intelligent transportation is also an important component of the national digital economy and a major scenario for digital economy construction. With the continuous advancement of urbanization, urban transportation has become an indispensable part of people's lives. However, this has been accompanied by problems such as severe urban road pollution and traffic congestion. Therefore, while developing intelligent transportation, transportation enterprises should also focus on green development and attach greater importance to the efficiency of public transportation in urban transportation planning and design. The development of intelligent transportation by transportation enterprises can reduce environmental pollution and resource consumption to a certain extent. Among other things, the construction of public transportation systems greatly improves the travel efficiency of urban residents, effectively reduces carbon dioxide emissions, and enhances transportation efficiency. The intelligent development of transportation enterprises can more effectively alleviate urban traffic pressure and improve air quality. Thus, the development of intelligent transportation by transportation enterprises is also an important path to achieve the "dual carbon" goals.

18.1.4 The Development of Intelligent Transportation by Transportation Enterprises Promotes Technological Innovation

The development of intelligent transportation by transportation enterprises cannot be separated from technological support, with cybersecurity technology, next-generation communication technology, and the Internet of Things (IoT) technology being indispensable. These technologies can be applied to data analysis and decision-making, information transmission, decision execution and operation, as well as data collection and status identification. In summary, the development of intelligent transportation by transportation enterprises is a new development model that takes user experience as the goal and improves the operational efficiency, safety, and environmental friendliness of the entire transportation system through modern information technology methods. The core of this development lies in the composition of intelligent transportation systems, whose essence relies on the application of information technology. Technological innovation is crucial to the long-term development of enterprises. With social development, cutting-edge and efficient technologies have also become an important competitive edge for transportation enterprises. These enterprises need to actively carry out technological innovation in response to changes in the market environment, improve their competitiveness, enhance product quality through intelligent means, and strengthen their own advantages. Based on technological upgrading, transportation enterprises should seek changes in marketing strategies to adapt to market environment changes. By combining their own advantages and innovating in marketing, transportation enterprises can make their products more acceptable to customers, thereby improving their economic benefits.

18.2 The Intelligent Development Course of Transportation Enterprises

18.2.1 Multi-Dimensional Strategic Layout and Practical Achievements of Intelligent Transportatio

As a leading large-scale comprehensive enterprise group in China, ZHJJJ has made remarkable achievements and formulated a clear strategic layout in the development of intelligent transportation. Market Coverage: China Merchants Xinzhi (Zhaoshang Xinzhi) covers business areas including investment and mergers & acquisitions, software development, intelligent manufacturing, IT asset performance hosting and professional operation & maintenance, scenario operation, and big data application, with its market presence spanning 26 provinces, municipalities, and autonomous regions across China. Operational Infrastructure: Through the platform of XIZHK, ZHJJJ is committed to becoming an intelligent transportation infrastructure operation service provider. Since its establishment three years ago, the company has focused on its core business of expressways and made all-round efforts with breakthroughs in multiple areas, centering on business integration, scientific research innovation, and market promotion. In 2019, XIZHK supported major livelihood projects by participating in the project of canceling provincial toll stations in more than 20 provinces and municipalities nationwide, winning multiple awards and successfully applying for national-level scientific research projects. In developing its intelligent expressway product line, XIZHK focuses on expressway safety and has developed products such as the Expressway Collaborative Command and Dispatch Cloud Platform and the Bridge Operation Safety Comprehensive Management System. Industrial Layout: With big data at its core, ZHJJJ has built an Intelligent Expressway Control and Management Platform around the product lines of intelligent safety, intelligent operation, and vehicle-road coordination. Products launched by XIZHK, such as the “Three Parties” Collaborative Command Platform and the Intelligent

Control System for High-Traffic Road Sections, have served nearly 40 expressways in 4 provinces/municipalities including Guangxi and Tianjin, covering a total mileage of over 4000 kilometers. Digital Operation Management: ZHJJJ launched the "Zhaolutong" Expressway Operation Management Platform, realizing intelligent operation management, quality travel services, and green transportation development. It also undertook the construction of the "Transportation and Logistics Cloud", accelerating the digital transformation of the transportation and logistics industry and enhancing supply chain resilience. In addition, the enterprise has the independent design and construction of high-end cruise ships to promote the development of the cruise industry. ZHJJJ's intelligent transportation development demonstrates its strategic vision and practical capabilities in technological innovation, industrial upgrading, green development, and other aspects, contributing to the development of intelligent transportation in China.

18.2.2 Development Course and Innovation Achievements of Intelligent Transportation

JJTTT's intelligent transportation development began in the 1990s, when it introduced the concept of Intelligent Transportation Systems (ITS) along with the rapid development of information technology. In the twenty-first century, JJTTT actively participated in the research, development, and demonstration application of national intelligent transportation systems, such as the Electronic Toll Collection (ETC) system for expressways and intelligent transportation management and service systems. In the innovation stage, driven by the development of next-generation information technologies such as big data and artificial intelligence, JJTTT has continuously innovated in the field of intelligent transportation, promoting the development of projects such as smart highways, smart ports, and intelligent shipping. In recent years, through technological innovation, JJTTT has promoted the in-depth integration of intelligent transportation with new technologies such as smart cities, intelligent vehicles, and the Internet of Vehicles (IoV). It has achieved a series of results in intelligent transportation, including but not limited to: Cancellation of expressway provincial toll stations and realization of "unified network operation"; Implementation of electronic tickets covering all high-speed railway and intercity railway stations nationwide; Intelligent transformation of comprehensive passenger hubs to achieve "one code (ID) for green travel"; Construction of intelligent bus dispatching systems and electronic bus stops to improve urban bus service levels; Promotion of the implementation of autonomous driving technology. JJTTT will continue to increase investment in technological innovation, promote the in-depth integration of intelligent transportation with new infrastructure construction, and contribute to the realization of transportation modernization and the construction of a transportation power.

18.2.3 Exploration in the Field of Intelligent Transportation

The development of intelligent transportation in China originated in the 1990s. ZJGGG participated in the research on the technical system framework and standard framework system of intelligent transportation, laying the foundation for its development. In the early twenty-first century, the National Science and Technology Key Project Plan set up a number of scientific and technological projects specifically to support the research on key technologies of intelligent transportation, and selected some cities to carry out the construction of intelligent transportation demonstration applications.[3]With the reform of the national science and technology system and the development of next-generation information technology, intelligent transportation has entered a stage of multi-industry collaboration and upgrading driven by new technologies. ZJGGG's key projects in the field of intelligent transportation include the National Intelligent Transportation Comprehensive Technology Integration and Application Demonstration Project, the National Road Traffic Safety Science and Technology Action Plan, and the Key Special Project on Comprehensive Transportation and Intelligent Transportation. These projects have not only promoted the development of intelligent transportation technologies but also boosted the growth of related industries. The characteristics of ZJGGG's intelligent transportation development are reflected in technology-led development, multi-industry collaborative development, high attention from both central and local governments, close integration of infrastructure construction and industry informatization, and innovative development based on China's actual conditions. Up to now, ZJGGG has formed its own technical system and standards in the field of intelligent transportation, built a series of engineering projects with international influence, and reached the world's cutting-edge level in areas such as expressway networked electronic toll collection and intelligent traffic management. At the same time, the intelligent transportation industry has begun to take shape, forming an industrial ecological system covering research and planning, design and consultation, system integration services, and professional product supply.

18.2.4 Problems Encountered in Enterprise Development

Enterprises face numerous bottlenecks and difficulties in the development of intelligent transportation. Specifically: Funding guarantee issues: The construction and maintenance of intelligent systems involve large-scale capital investment, and there is an urgent need for relevant departments to establish a sound data chain guarantee mechanism. Lack of scientific operation and market-oriented management models: This easily reduces the economic benefits of enterprises' intelligent transportation systems and hinders the commercialization process of enterprises' intelligent transportation development. Serious imbalance in regional endowments: There are significant disparities in data resources, technical endowments, and other aspects across regions in China. The intelligent transportation data, technologies, and basic conditions for developing intelligent transportation faced by enterprises have obvious regional attributes, which to a certain extent restricts the extensibility and coverage of enterprises' intelligent transportation development. Therefore, enterprises can promote the all-round development and popularization of intelligent transportation from aspects such as original capital investment, technological innovation and development, scientific management models, and data information sharing.

18.3 Outlook and Recommendations

18.3.1 Outlook on the Intelligent Development of Transportation Enterprises

With the acceleration of urbanization and the continuous growth of transportation demand, the market scale of China's intelligent transportation industry will continue to expand. It is expected that the industry will maintain a steady growth trend in the next few years. According to data from the Zhongshang Industry Research Institute, the market scale of China's intelligent transportation industry is projected to further increase to 261 billion yuan in 2024.Technological innovation is accelerating. With the continuous development of cutting-edge technologies, the intelligence level of intelligent transportation systems will be continuously improved. Autonomous driving technology, a cutting-edge field of intelligent transportation, is gradually moving from laboratories to practical applications. By continuously breaking through relevant core technologies, it aims to provide people with a comfortable, safe, and efficient autonomous driving experience. In addition, the rise of new technologies such as the Internet of Vehicles (IoV) has promoted information exchange and sharing between vehicles and between vehicles and infrastructure, building a collaborative and closely connected ecosystem. The application and promotion of these new technologies have injected new impetus into the development of the intelligent transportation industry.In terms of policy support, national-level policies such as the Outline for Building a Transportation Power and the Development Plan for the Modern Comprehensive Transportation System have laid out a grand and clear blueprint for the development of intelligent transportation. Policies encourage technological innovation and international cooperation, promote the establishment and improvement of intelligent transportation standards systems, and ensure that China maintains a leading position in global competition in the field of intelligent transportation. Currently, intelligent transportation pilot projects are mainly focused on the Beijing-Tianjin-Hebei integration region and economically developed areas along the eastern coast. Relying on their strong economic strength, advanced technological level, improved infrastructure construction, and highly concentrated market demand, these regions are expected to become pioneering demonstration areas for promoting technological innovation and application in intelligent transportation.

18.3.2 Recommendations for the Intelligent Development of Transportation Enterprises

Increase R&D Investment and Strengthen Policy Incentives. Transportation enterprises should increase investment in research and development. Meanwhile, the government should introduce a series of incentive policies to encourage enterprises to increase capital investment not only in basic technology R&D for intelligent transportation but also in cutting-edge technologies, exploring areas such as autonomous driving, IoV, big data analysis, and artificial intelligence applications to drive technological innovation and breakthroughs. At the same time, through financial and fiscal measures such as setting up special funds, providing tax reductions, and offering loan preferential policies, enterprises' R&D costs can be reduced, their innovation vitality stimulated, and the rapid development and wide application of intelligent transportation technologies promoted, laying a solid foundation for building a safer, more efficient, and greener transportation system. Establish Cross-Domain Cooperation and Build an Open Ecosystem Cross-domain cooperation has become a key driving force for the transformation and upgrading of the transportation industry. To deepen the integration of the transportation industry with information technology, communications, automobile manufacturing, and other fields, it is necessary to build a new open and cooperative ecosystem, break industry barriers, and promote resource sharing and complementary advantages. By building cross-industry exchange platforms that eliminate data silos, enterprises, universities, research institutions, government departments, and other multi-stakeholders should be encouraged to participate in in-depth exchanges and cooperation on cutting-edge topics such as intelligent transportation and smart travel. Strengthen Security Guarantees and Improve System Resilience Security guarantee work plays an important role. It is recommended to strengthen the cybersecurity protection of intelligent transportation systems and build an all-round, multi-level cybersecurity guarantee system. Efforts should be made in physical security, data security, cybersecurity monitoring and emergency response, and independent controllability capabilities to comprehensively improve the system's security protection level and ensure data security and stable system operation. Attach Importance to Standardization Construction and Solve Compatibility Problems Transportation enterprises should attach importance to standardization in the intelligent development process, as promoting standardization in the intelligent transportation field is a key measure to ensure the sustainable, healthy, and efficient development of the industry. By pooling the wisdom of experts, scholars, and enterprise representatives inside and outside the industry, unified and forward-looking technical standards and norms should be jointly formulated to solve problems such as data silos, equipment incompatibility, and difficulty in data sharing in current intelligent transportation systems. Intensify Talent Cultivation and Build a Diversified Talent Team. To address technical challenges, enterprises must attach great importance to talent cultivation for the continuous innovation and long-term development of intelligent transportation. Educational institutions should actively respond to industry needs by adjusting and optimizing curriculum systems, adding professional courses on cutting-edge technologies such as intelligent transportation, big data, artificial intelligence, and the Internet of Things, and providing comprehensive knowledge systems and practical platforms to cultivate talents' innovative thinking and ability to solve practical problems. Enterprises should establish in-depth cooperation mechanisms with universities and research institutions to jointly carry out talent cultivation projects. Through internships, joint R&D, and project cooperation, talents can have access to real working environments and technical challenges, accumulating valuable practical experience. The government should play a guiding role by formulating relevant policies to encourage and support talent cultivation in the intelligent transportation field. In the process of attracting and cultivating professionals, attention should also be paid to talent diversification and internationalization. The development of intelligent transportation requires interdisciplinary and cross-domain compound talents, as well as high-end talents with international vision and competitiveness. Focus on User Experience and Promote People-Centered Development. The fundamental of enterprise development lies in considering issues from the user's perspective. In the planning and construction of intelligent transportation systems, user experience should always be placed at the core. Through user research and data analysis, the behavioral habits and pain points of travelers should be accurately grasped. In terms of interface design and interaction logic of intelligent transportation systems, a simple, intuitive, and user-friendly operation experience should be pursued to ensure that users of different ages and technical levels can easily get started. Advanced algorithms and big data technologies should be used to provide users with accurate travel recommendations. At the same time, paying attention to users' safety experience is also indispensable. Intelligent transportation systems need to integrate multiple safety protection mechanisms, such as vehicle collision warning, pedestrian protection systems, and cybersecurity protection, to ensure users' personal safety and privacy protection during use. In addition, a sound user feedback mechanism should be established to timely collect and respond to users' opinions and suggestions, continuously optimize system performance and service quality, form a virtuous circle, and promote the continuous development of intelligent transportation systems to a higher level.

Adapt to National Conditions and Optimize System Construction.

To promote the intelligent development of enterprises, the first step is to develop intelligent transportation systems that adapt to China's national conditions. These systems should be improved with technological progress and changes in people's needs. They should not only adapt to national development at the macro level but also be tailored to local conditions—deployment and construction plans should be formulated based on actual situations and urban regional attributes to promote balanced development of intelligent transportation across regions and continuously improve the efficiency and service scope of intelligent transportation operation systems.

Improve Intelligent Dispatching and Strengthen Data Management.

To meet market demand, enterprises should improve intelligent dispatching and optimization systems, which can intelligently dispatch and manage vehicles to achieve optimal resource allocation. Such systems can automatically adjust vehicle routes based on real-time traffic conditions and passenger volume, reducing waiting time for the public and improving transportation efficiency and quality. To solve problems such as information asynchronization between enterprises, departments should break information barriers, deepen information sharing and exchange mechanisms, actively integrate data resources, and build data sharing platforms to promote efficient cooperation between enterprises. To achieve efficient transportation management, transportation enterprises need to increase investment in and construction of data centers for comprehensive data collection and storage. Enterprises should enhance their big data, cloud computing, and artificial intelligence technologies to deeply mine information such as road conditions, providing practical support for transportation management decisions. In general, enterprises need to strengthen data management and analysis of intelligent transportation systems to make accurate transportation management decisions.

Promote Multi-Party Collaboration and Achieve Win-Win Development.

Transportation enterprises and communication institutions should establish mutually trusted cooperation mechanisms, build information sharing platforms, jointly discuss development standards and measures, and actively carry out technical research cooperation and exchanges to achieve mutually beneficial and win-win development. The cooperation model between institutions and enterprises should be established to accelerate the transformation of technological innovation achievements and lay a solid foundation for the intelligent transportation development of enterprises.

References

1. 2022-2026 China Intelligent Transportation Industry Panoramic Survey and Development Strategy Consulting Report
2. Wang Jiacheng. Analysis of the environment and growth factors faced by transportation enterprises[J]. Transportation Enterprise Management, 2023, 38(03): 5-7.
3. Ministry of Transport of the People's Republic of China, Official Website of Xinhua News Agency's Economic Information Daily, Sohu.com, etc.

Chapter 19
Green and Low-Carbon Development of Transportation Enterprises

Malin Song and Chengyue Yang

Abstract This paper discusses the current situation, challenges, and future prospects of transportation enterprises in green and low-carbon development. As global warming becomes increasingly severe, the transportation industry, as one of the main sources of carbon emissions, is facing pressure to transform and upgrade. The introduction of relevant policies, diversified market demands, enterprises' practical exploration, and technological innovation have, to a certain extent, promoted the green and low-carbon development of transportation enterprises. At the same time, these enterprises also face problems in green and low-carbon development, such as excessively high costs of technological upgrading, imperfect market mechanisms, backward infrastructure, and insufficient public awareness. Therefore, corresponding measures are proposed, including the promotion of new energy vehicles, optimization of road network planning, development of intelligent transportation, strengthening of technological innovation, and improvement of public awareness. In addition, the paper demonstrates the achievements of specific transportation enterprises in green and low-carbon development through case studies, and puts forward relevant suggestions: transportation enterprises need to establish and improve an energy-saving management and environmental protection system; deepen internal environmental protection publicity and education to enhance employees' environmental awareness and sense of responsibility; and commit to the comprehensive improvement of technical levels. Thus, the dual growth of economic benefits and environmental benefits can be achieved.

M. Song
Anhui University of Finance and Economics, Bengbu, China

C. Yang (✉)
Dongbei University of Finance and Economics, Dalian, China

J. Wang and M. Song (eds.), *The Green Path of Chinese Transportation Enterprises*,
Resources, Climate and Sustainable Development,
https://doi.org/10.1007/978-981-95-5289-4_19

19.1 Current Situation of Green and Low-Carbon Development in Transportation Enterprises

As global warming becomes increasingly severe, the transportation industry stands as one of the main sources of carbon emissions. Under the low-carbon economy, its green transformation and development have become a consensus of the international community. Various ecological and environmental policies have exerted a huge impact on technological innovation and product development, prompting transportation enterprises to focus more on green innovation. According to relevant data, the proportion of petroleum consumption by global transportation vehicles has increased to over 60%, and is expected to continue growing in the next few years. In the operation process of transportation enterprises, not only a large amount of fossil fuels are consumed, but also a large quantity of waste gas, waste water and solid waste are discharged, causing serious environmental pollution. Especially in large cities and central business districts (CBDs), problems such as traffic congestion, noise pollution and declining air quality have become increasingly serious, which not only affect the quality of life of residents, but also restrict the sustainable development of cities.In recent years, governments around the world have introduced a series of policies and measures to encourage and support the green and low-carbon development of transportation enterprises. Technological progress has also provided strong support for energy conservation and emission reduction of transportation enterprises. At the same time, consumers' demand for green travel and low-carbon products continues to grow, providing new market opportunities for transportation enterprises.The goals of carbon peaking and carbon neutrality are accelerating the green transformation and upgrading of industries, reducing energy consumption, and promoting energy conservation and emission reduction. However, the green and low-carbon transformation of transportation enterprises still faces many challenges, such as excessively high costs of technological upgrading, imperfect market mechanisms, backward infrastructure, and insufficient public awareness.

19.1.1 Deepening and Refinement of Policies and Regulations

Currently, China is in the stage of high-quality development, and the principal contradiction in society has undergone changes. To support sustained economic growth, the strategy of expanding domestic demand has become a core cornerstone, while promoting coordinated development among regions heralds broad opportunities. Achieving the goals of carbon peaking and carbon neutrality will trigger comprehensive and far-reaching economic and social system reforms.[1]

On August 11, 2024, the CPC Central Committee and the State Council issued the Opinions on Accelerating the Comprehensive Green Transformation of Economic and Social Development. This document focuses on five core areas: building a green and low-carbon spatial layout for high-quality development, accelerating the green and low-carbon transformation of industrial structures, steadily advancing the low-carbon transition of energy systems, promoting green innovation in transportation, and advancing the green transformation of urban and rural development. It also covers three key links: implementing a comprehensive conservation strategy, promoting the green transformation of consumption patterns, and leveraging the supporting role of technological innovation. The document plans to accelerate the construction of a resource-saving and environment-friendly spatial layout, industrial structure, production model, and lifestyle. China's transportation enterprises should respond to national policies and maintain their green and low-carbon development. In the process of promoting the green and low-carbon development of transportation enterprises, the role of policies and regulations has become increasingly prominent. More stringent emission standards have been implemented in various countries, such as the European Union's Emissions Trading System (ETS) and China's "dual carbon" goals (carbon peaking and carbon neutrality), and legislation has been adopted to ensure the implementation of these standards. In addition, policymakers are constantly exploring innovative mechanisms, such as fiscal incentives including green funds, low-interest loans, and green tax reductions, to reduce enterprises' transformation costs and enhance the momentum for green development. At the same time, international cooperation has been strengthened, with many countries signing environmental agreements to jointly formulate cross-border emission standards and promote the green transformation of the global transportation industry. China will coordinate regional policies through fiscal and monetary economic policies, strengthen policy implementation, and promote the formation of a strong domestic market.

19.1.2 Challenges and Opportunities of Technological Innovation

Technological innovation is a key driving force for promoting the green and low-carbon development of transportation enterprises. In the field of new energy vehicles, breakthroughs in battery technology (such as the research and development of solid-state batteries), the popularization of charging infrastructure, and the integrated application of autonomous driving technology are gradually addressing issues such as range anxiety, charging convenience, and safety. In terms of intelligent transportation systems, the in-depth integration of big data, cloud computing, the Internet of Things, and other technologies has made traffic management more precise and efficient, effectively reducing congestion and emissions. Green logistics technologies reduce the carbon footprint of logistics links by optimizing distribution routes, using environmentally friendly packaging materials, and implementing intelligent warehouse management. However, the high cost of technological upgrading remains a constraining factor, requiring joint efforts from the government, enterprises, and financial institutions to lower the threshold for technology application through technological innovation subsidies, R&D cooperation, and green finance. In freight transportation, more enterprises have begun to use electric transportation tools, which are characterized by low energy consumption, low carbon emissions, and high transportation efficiency. With the rise of the sharing economy, shared logistics has also received widespread attention. For the public, the most intuitive product of transportation enterprises' green development is the popularization of new energy vehicles. In recent years, cities and enterprises have continuously improved new energy infrastructure construction, with charging stations covering public transport and social vehicles, meeting the charging needs of new energy vehicles. They have also been well-received by consumers due to their low cost and green travel benefits. To meet market demand, enterprises have continuously improved intelligent dispatching and optimization systems. Through big data and artificial intelligence algorithms, these systems intelligently dispatch and manage vehicles to achieve optimal resource allocation, thereby improving the environmental performance of public transportation. For the power support of new energy vehicles, the construction of charging facilities is also crucial. In this regard, transportation enterprises have actively invested in building charging stations to meet the energy needs of new energy vehicles. As the state has successively introduced policies to encourage new energy transportation, the new energy industry has developed steadily, and the demand for charging piles as key facilities has grown increasingly. The density of charging piles in various representative cities is shown in Fig. 19. 1.

As can be seen from Fig. 19. 1, the public charging pile density of most economically developed cities in China that rank among the top 10 is above 1.2 units per square kilometer, which indicates that economically developed cities attach greater importance to the promotion and application of new energy transportation[1]。In terms of transportation infrastructure construction, transportation enterprises widely use energy-saving and emission-reduction materials, which reduce energy consumption and environmental pollution. For example, in the Jining County-Zaozhuang-Heze Expressway Transportation-Energy Integration Demonstration Project, Jining County made full use of infrastructure such as service areas and roadside slopes to develop new energy locally, and realized the green and low-carbonization of

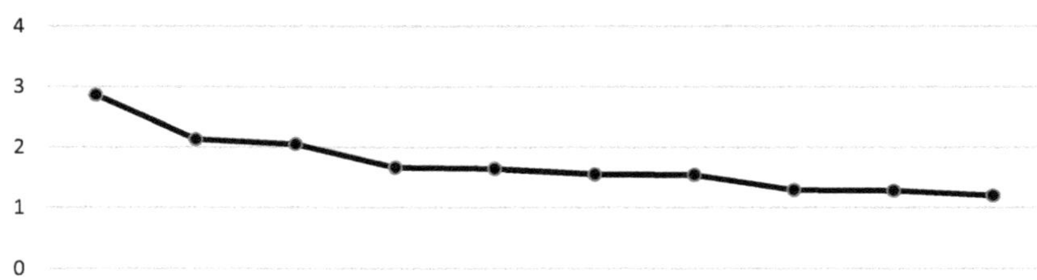

Fig. 19 1 Public Charging Pile Density in Ten Cities. *Data Source* 2019 China Urban Transportation Research Report

highway construction by adopting energy-saving and environmental protection materials and technologies. Transportation enterprises have researched and designed green tires, which have become an important means of energy conservation and emission reduction due to their low rolling resistance and high energy efficiency. The green tires launched by enterprises have significantly reduced rolling resistance and improved fuel economy by optimizing tread rubber formulas and structural design. The lightweight design of green tires uses lightweight materials and optimized design, which can reduce vehicle weight, thereby reducing fuel consumption and carbon emissions. To sum up, transportation enterprises are gradually achieving green and low-carbon development goals through various technological innovation methods. These methods can effectively reduce carbon emissions in the transportation sector, thereby promoting the sustainable development of the transportation sector.

19.1.3 Diversification and Deepening of Market Demand

With the general improvement of environmental awareness worldwide, consumers'demand for green travel and low-carbon products is growing increasingly, showing a trend of diversification and deepening. From personal travel to freight transportation, and from private vehicles to public transportation, consumers are increasingly inclined to choose products and services that can reduce environmental burdens. This shift in market demand has prompted transportation enterprises to continuously innovate and develop more products that meet green standards, such as electric buses, bike-sharing, and green freight solutions. At the same time, the popularization of green consumption concepts has also urged enterprises to pay more attention to brand image and social responsibility, and enhance consumer trust and realize brand value improvement through green marketing and sustainable development reports. To achieve the goals of energy conservation, emission reduction, and carbon reduction, deepening the adjustment of transportation structure is essential. This requires collaborative reforms on both the supply and demand sides of transportation, making full use of and optimizing the individual advantages and synergistic effects of various transportation methods, striving to promote the green transformation of port collection and distribution transportation, and accelerating the construction of a green-oriented transportation network. Enterprises and cities should vigorously promote the application of new energy vehicles in the transportation field, advance in light of local conditions and combined with local market conditions, and continue to improve the charging facilities for new energy vehicles to meet vehicle charging needs.

19.1.4 Exploration and Breakthroughs in Enterprise Practices

Facing the challenges and opportunities of green and low-carbon development, many transportation enterprises have taken active actions to explore green transformation paths suitable for their own development. On the one hand, enterprises have increased investment in the new energy vehicle field, and improved the production capacity of new energy vehicle products and expanded their share in the automobile market through self-construction or cooperation. On the other hand, enterprises should also focus on achieving energy conservation and emission reduction in operation and management, such as optimizing transportation networks, improving vehicle load rates, and adopting energy-saving equipment. In addition, enterprises have also actively participated in the carbon emission trading market, and flexibly and effectively responded to the pressure brought by emission reduction by buying or selling carbon emission rights. However, in practice, enterprises also face problems such as technical bottlenecks, capital shortages, and talent scarcity. To this end, enterprises need to strengthen internal and external cooperation, use industry-university-research cooperation platforms, introduce external intellectual resources and funds, and jointly overcome technical difficulties; at the same time, establish and improve green and low-carbon development mechanisms to ensure the sustainability and effectiveness of work.

19.2 Problems and Countermeasures in the Green and Low-Carbon Development of Transportation Enterprises

19.2.1 Problems Faced by Transportation Enterprises in Green and low-Carbon Development

Green and low-carbon development needs to be driven by scientific and technological innovation. It is an arduous task to improve energy efficiency using high-tech and develop environmental protection technologies and equipment. On the one hand, the core technology path for the green and low-carbon development of transportation enterprises is unclear, and enterprises have not yet formulated a complete and comprehensive green and low-carbon development strategy. The current relevant policies and industry standards have not fully aligned with the latest environmental protection requirements such as the "dual carbon" goals and the "ecological protection red line system". Enterprises face numerous bottlenecks in key core technologies related to pollution reduction, carbon reduction, and ecological protection, making it difficult to achieve major technological innovation breakthroughs and thus restructure the technological innovation system. On the other hand, the level of low-carbon recycling is low. Carbon emissions generated during the material production and transportation stages account for a large proportion of highway construction. However, the resource utilization of solid waste is still in the exploratory and experimental stage. In practical projects, the types of waste that can be recycled are limited, the utilization methods are relatively single, and the added value of recycling is low. For enterprises to achieve comprehensive and sustainable green and low-carbon development, they must attach importance to technological innovation. The policy problems faced by enterprises in green and low-carbon development mainly include an imperfect policy system, insufficient policy innovation, and difficulties in local implementation. On the one hand, although the state has introduced a series of policies, the implementation of policies at the local level varies, leading to unbalanced green and low-carbon development among cities. In practice, policies are not fully implemented or strictly enforced, resulting in unsatisfactory implementation effects. Insufficient policy innovation is also a problem. To address issues in the policy system, it is necessary to further deepen the construction of the policy system, strengthen policy implementation efficiency, and promote policy innovation. On the other hand, transportation enterprises lack sufficient quantitative data to support decision-making. Energy consumption and carbon emissions of highway infrastructure have not been included in the statistical accounting system, and a unified carbon emission assessment standard has not been established. These lead to a lack of basic data, making it difficult to effectively support life-cycle decision-making for engineering projects. There are deficiencies at the institutional level: energy conservation assessment for highways started late, and a complete and systematic framework of energy conservation assessment guidelines, norms, and systems has not been formed. In addition, most provinces do not set energy conservation and carbon reduction as mandatory goals when approving fixed-asset investment projects, resulting in the lack of an assessment, review, and supervision system for promoting energy conservation and carbon reduction in the planning and construction of highway projects. To promote green and low-carbon development, most transportation enterprises choose to produce new energy vehicles. Although new energy vehicles are more low-carbon and environmentally friendly for travel, there are still many problems in their production process. On the one hand, differences in subsidy policies between new energy vehicles and traditional fuel vehicles hinder the promotion of new energy vehicles. For example, the new energy transformation of rural passenger buses faces insufficient subsidies, which affects enterprise enthusiasm. In terms of purchase and operation costs, new energy vehicles have higher purchase prices, and costs such as battery life and component replacement are also relatively high, increasing the operating costs of enterprises. On the other hand, the purchase cost of new energy vehicles is higher than that of traditional fuel vehicles, which increases the investment pressure on transportation enterprises. Although the operating cost of new energy vehicles decreases year by year for consumers, enterprises still need government subsidies to produce new energy vehicles. The development of green transportation by transportation enterprises requires a complete and sound market mechanism. Currently, China's green product market is still in the primary stage of development, with a small scale. It lacks effective market supervision, and transportation enterprises have insufficient market competitiveness, making it difficult to attract more consumers. In terms of charging infrastructure construction, some remote areas and old communities lack charging infrastructure, restricting the popularization and use of new energy vehicles. The intelligent level of road infrastructure is insufficient, making it difficult to support the research, development, and application of key technologies such as autonomous driving and vehicle-road coordination. At present, highway transportation in China still mainly relies on petroleum and other petrochemical resources, and transportation by these traditional vehicles generates a large amount of carbon emissions. The promotion and application of new energy trucks face many economic difficulties. First, the purchase cost of new energy trucks is high. Second, the battery life of new energy trucks is much shorter than that of ordinary fuel trucks, and charging facilities in some areas cannot meet public demand, which restricts the promotion and application of new energy trucks.

The public's understanding and acceptance of the concept of green and low-carbon transportation need to be further improved, and publicity and education need to be strengthened. In terms of market competition and cooperation, the competition and cooperation mechanism among transportation enterprises in green and low-carbon development is still imperfect, and a more open and inclusive market environment needs to be established.

In summary, transportation enterprises face challenges in technology, policy, economy, infrastructure, and other aspects in the process of green and low-carbon development. To promote the green and low-carbon development of transportation enterprises, the government, enterprises, and all sectors of society need to work together to strengthen technology research and development, improve policies and systems, optimize economic incentives, strengthen infrastructure construction, and enhance public awareness and acceptance.

19.2.2 Countermeasures for Green and Low-Carbon Development of Transportation Enterprises

Compared with traditional fuel vehicles, electric vehicles produce less pollution and emit almost no exhaust gas in pure electric mode. Transportation enterprises should actively promote new energy vehicles for green and low-carbon development. New energy vehicles have a simple motor structure, are less prone to failure, require less frequent maintenance, and have lower maintenance costs. Additionally, they feature low travel costs, save fuel energy, and come with substantial preferential policies. As a state-supported project, new energy vehicles offer significant preferential policies for consumers, which will boost purchases.

19.2.2.1 Strengthen Policy Incentives and Infrastructure Construction

Policy Support: The government should introduce more preferential policies, such as purchase subsidies, exemption from vehicle purchase tax, parking fee reductions, and charging facility subsidies, to incentivize consumers to buy new energy vehicles through economic means. At the same time, special funds should be established to support the R&D and manufacturing of new energy vehicles, encourage technological innovation, and promote industrial upgrading.

Infrastructure Layout: Accelerate the development and deployment of supporting facilities for new energy vehicles, such as charging stations and battery swap stations. In particular, increase deployment in densely populated areas such as urban public transportation hubs, residential areas, and commercial districts to address the "range anxiety" of new energy vehicle users. Prioritize the promotion of new energy vehicles in public transportation fields such as taxis and urban buses to set industry benchmarks and drive the green transformation of the private car market.

19.2.2.2 Optimize Road Network Planning and Promote Green Travel

Enterprises should use big data, artificial intelligence, and other technologies to accurately predict and analyze traffic flow, scientifically plan road networks, improve road network efficiency, add traffic nodes, and reduce unnecessary detours and congestion. Meanwhile, optimize public transportation routes to improve bus coverage and operational efficiency. Actively apply green transportation facilities by integrating green concepts into road construction, such as building slow traffic systems including bicycle lanes and sidewalks to encourage green travel.

Promote the use of environmentally friendly materials to reduce carbon emissions during construction. Strengthen management methods such as traffic signal control and intelligent dispatching to improve road capacity and reduce vehicle waiting time and idling emissions. Road network planning optimization should also aim to narrow the urban-rural gap, improve transportation capacity in rural areas, and promote connectivity between urban and rural areas. Enterprises should integrate environmental protection concepts into road network planning, prioritize travel methods with less environmental impact, and reduce damage to the natural environment. At the same time, use modern technological means such as big data analysis and intelligent transportation systems to improve traffic management efficiency, effectively optimize urban traffic conditions, and promote the sustainable and healthy development of urban green transportation.

19.2.2.3 Intensify Exhaust Gas Treatment and Emission Supervision

The key for enterprises to strengthen exhaust gas treatment includes reducing exhaust gas generation, adopting advanced exhaust gas treatment equipment, implementing flue gas purification technologies, and optimizing combustion processes to reduce harmful gas emissions through improved processes and efficient equipment. Enterprises can also separate and recycle exhaust gas containing high-concentration useful substances. In addition, enterprises should enhance environmental protection awareness and comply with laws and regulations. Strengthen supervision over motor vehicle emission standards, implement stricter emission testing systems, and penalize or eliminate vehicles that exceed emission limits. Establish a motor vehicle emission database to realize real-time monitoring and traceability of emission information. In terms of technological upgrading, enterprises should promote the R&D and application of key technologies such as engine technology and exhaust aftertreatment technology to improve the emission purification efficiency of motor vehicles. Encourage enterprises to develop more environmentally friendly fuels and power systems. Carry out popular science education on motor vehicle exhaust emissions to raise public awareness of the importance of exhaust gas treatment. Advocate public participation in exhaust emission supervision to create a good social atmosphere of joint governance by the whole society.

19.2.2.4 Promote Technological Innovation and Intelligent Application

Enterprises should continue to invest in research, use high-tech to efficiently aggregate transportation information, provide real-time transportation data, extract key information, and support decisions on traffic management and vehicle scheduling.

Actively participate in the development of intelligent connected vehicles, accelerate industrial technology application, and save transportation time.

Accelerate the promotion of ETC technology to improve expressway traffic fluency, reduce congestion at toll stations, and cut carbon emissions from manual toll collection. Use technologies such as the Internet of Things and cloud computing to realize intelligent dispatching and optimal allocation of public transportation such as buses and taxis, improving transportation efficiency and service quality.

Promote the development and application of vehicle-road coordination technology to achieve information exchange and collaborative management between vehicles and road facilities, and enhance the overall operational efficiency and safety of the road traffic system.

19.2.2.5 Improve Innovation Ecosystem and Data Management

R&D and Talent Development: Increase R&D investment in the transportation field to promote the R&D and practical application of new technologies, new industries, new formats, and new models. Deepen collaboration among innovation entities such as enterprises, universities, and research institutions to build an industry-university-research-application collaborative innovation system. At the same time, vigorously promote talent cultivation and introduction in the transportation field to build a high-quality talent team with international vision and innovation capabilities. Strengthen vocational education and skills training to improve the technical level and professional quality of employees. Demonstration Projects and Data Systems: Establish a number of demonstration projects and pilot areas for green and low-carbon development of transportation, and form universally applicable and promotable successful examples and experiences through practical exploration and experience summary. Enterprises and cities should formulate and improve statistical standards and methodology systems for transportation energy consumption to ensure the accuracy and comparability of statistical data. Strengthen the review and supervision of statistical data to prevent data fraud and distortion, enhance the collection and collation of transportation energy consumption data, and establish a comprehensive, accurate, and timely transportation energy consumption database. Build in-depth mining models using modern information technologies such as big data and cutting-edge technical means to provide practical references for policy formulation. Regularly release reports on transportation energy consumption and carbon emissions, and disclose relevant information and data to improve policy transparency, enhance public participation, and promote the formation of a social co-governance pattern for green and low-carbon development of transportation.

19.2.2.6 Enhance Public Awareness and Encourage Participation

Enterprises can strengthen publicity and education on energy conservation, emission reduction, and green low-carbon travel through various channels and forms such as media publicity, school education, and community activities to improve public awareness of environmental protection and sustainable development. Encourage public participation in practical activities and voluntary services related to energy conservation, emission reduction, and green low-carbon travel, such as organizing green travel days and car-free days to advocate a low-carbon lifestyle. Establish a public participation feedback mechanism to timely collect public opinions and suggestions on green and low-carbon development of transportation, and continuously optimize policies, measures, and service levels based on public feedback.

19.3 Green Development Course of Transportation Enterprises

19.3.1 Actions and Explorations to Implement the "Dual Carbon" Goals

In 2022, ZSJJY released the ZSJJY Carbon Peaking and Carbon Neutrality Action Plan, which clearly sets the goal of establishing a "carbon peaking and carbon neutrality" action framework covering all businesses and processes. It aims to achieve comprehensive carbon peaking by 2030 and strive to realize the vision of carbon neutrality by 2060.The enterprise is committed to improving and perfecting the green and low-carbon governance system, enhancing the level of green and low-carbon governance and environmental governance capabilities, and promoting the development of green financial businesses to support the low-carbon transformation of enterprises and society. It actively develops the green technology industry and promotes the integrated application of green and low-carbon technologies in fields such as transportation logistics, park real estate, and green finance. Additionally, the enterprise has implemented a series of green service measures, including low-carbon shipping, low-carbon ports, low-carbon logistics, and low-carbon highways.

Through such initiatives, ZSJJY cultivates young people's awareness of green and low-carbon development and promotes sustainable development education. Responding actively to the "dual carbon" goals, the enterprise explores and layouts "transportation +" businesses such as new energy in transportation and service areas, deploys distributed photovoltaic power generation systems, explores expressway fast charging and battery swapping, and promotes the electrification of commercial vehicles. With the support of the Ministry of Transport, it has carried out projects in pilot programs such as the intelligent upgrading and transformation of container terminals, and the research and application of key technologies for smart expressways based on green energy. ZSJJY also actively explores the application of green energy in smart expressways, such as promoting the smart energy service platform in the Beijing-Tianjin-Tanggu Expressway project and building distributed photovoltaic power generation systems in expressway service areas.

19.3.2 Implementing Carbon Peaking and Carbon Neutrality to Build a New Pattern of Green Development

JJJYX has proactively aligned with the national strategic orientation of carbon peaking and carbon neutrality, integrating these goals into the company's development strategy and daily operations. The enterprise has set specific carbon peaking targets and formulated detailed action plans and implementation strategies covering industrial structure adjustment, green technological innovation, energy management, carbon asset management, and other aspects. It strives to optimize the industrial structure by reducing high-carbon emission projects, increasing the proportion of green and low-carbon projects, and developing emerging industries such as green buildings, intelligent transportation, and environmental protection engineering.

In terms of technological innovation and R&D, the enterprise has increased investment in green and low-carbon technology R&D to promote breakthroughs and innovations in key core technologies. It attaches importance to carbon asset management, participates in the carbon trading market by building mechanisms for carbon emission monitoring, disclosure, and auditing. Through participating in international cooperation projects and exchange activities, it introduces advanced international low-carbon technologies and management experience.

The enterprise focuses on enhancing employees' awareness and skills in green and low-carbon development. Through training and cultural construction, it has formed a green and low-carbon corporate culture, and established a monitoring and evaluation mechanism to regularly check the implementation effect of the carbon peaking action plan and make adjustments and optimizations based on the evaluation results. To ensure the effective green development of the enterprise, it has adopted a series of safeguard measures, including strengthening organizational leadership, ensuring capital investment, and enhancing talent team construction. It also demonstrates the company's practical achievements in green and low-carbon development through the development of typical green and low-carbon demonstration projects. In 2022, China Communications Construction Group released the JJJYX Green and Low-Carbon Action Plan, further clarifying the overall requirements, strategic goals, major actions, and safeguard measures for green and low-carbon buildings. In addition, the group has implemented a series of green and low-carbon transformation strategies, including top-level planning guidance, front-end guidance of design consulting, demonstration of typical green and low-carbon projects, exploration of green financial services, R&D and manufacturing of green and low-carbon equipment, technology innovation-driven development, carbon asset management and trading pilots, and building of green supply chains. Through these measures, China Communications Construction Group is committed to achieving comprehensive green transformation of the enterprise.

19.3.3 Building a Green Standard System to Promote Sustainable Transportation Development

Over the past decade, JJJYG has launched 143 green demonstration projects, 17 of which have been rated as typical cases. Covering fields such as expressways, urban transportation, and water transportation, these projects demonstrate the company's innovation and practice in green and low-carbon development. The company also actively participates in the R&D of green technologies and the formulation of industry standards. It has built a green and low-carbon standard system encompassing national standards, industry standards, group standards, and enterprise standards, promoting the green development of the industry. JJJYG has also achieved outstanding results in green highway construction. For example, the Yanqing-Chongli Expressway was rated as one of the "First Batch of National Typical Demonstration Projects for Green Highway Construction," and the Huashixia-Jiuzhi Expressway became the first green, circular, and low-carbon highway in northwest China. In highway construction, JJJYG has adopted recycled asphalt pavement technology, significantly improving resource recycling rates and reducing energy consumption and carbon emissions.

The company has promoted the application of new energy vehicles in urban public transport, taxis, and freight distribution, while popularizing the use of liquefied natural gas (LNG) in the water transportation industry. It has applied new materials and technologies such as ecological revetments, ecological beach protection, and artificial reefs in the regulation projects of inland waterways such as the Yangtze River, forming green waterway projects. The enterprise encourages ships calling at ports to use shore power, aiming to reduce pollution emissions within port areas. In addition, JJJYG has developed transportation modes such as rail-water intermodal transport and road-rail intermodal transport to improve transportation efficiency and reduce carbon emissions. It actively promotes the strategy of prioritizing public transport, builds a public transport network with urban rail transit as the main framework, and encourages the public to travel in environmentally friendly ways.

JJJYG also actively participates in international cooperation to promote greenhouse gas emission reduction in the global transportation field. These measures reflect the positive actions of Chinese transportation enterprises in promoting green and low-carbon development and achieving the "dual carbon" goals. Through continuous technological innovation and practice, the enterprise plays an important role in the sustainable development of the transportation industry.

19.4 Outlook and Recommendations

19.4.1 Outlook on the Green and Low-Carbon Development of Transportation Enterprises

To accelerate the promotion of new energy and the application of clean energy transportation equipment, transportation enterprises are developing products such as electric vehicles and LNG-powered ships. In response to the global initiative of green development and energy conservation and emission reduction, enterprises are actively advancing the construction of green waterways and green ports in transportation, while focusing on the ecological protection and restoration of transportation infrastructure.

To comprehensively promote the development of a green transportation system, enterprises are vigorously advancing the production of green products. In green transportation construction, enterprises promote the use of clean energy and renewable energy such as solar and wind energy in production, minimizing environmental pollution during port operations. For the protection of transportation infrastructure, enterprises adhere to the principle of "protection first" and thoroughly consider ecological protection requirements throughout the entire life cycle of the transportation system, from planning and design to construction and operation. To address the growing transportation demand and environmental pressure, enterprises are developing intelligent transportation systems to improve transportation efficiency and reduce energy consumption and emissions.

19.4.2 Recommendations for the Green and Low-Carbon Development of Transportation Enterprises

Transportation enterprises should emphasize corporate responsibility in green and low-carbon development. As an important force in social and economic development, their energy-saving and environmental protection behaviors play a pivotal role in promoting green development and achieving sustainable development goals. Firstly, enterprises must integrate energy conservation and environmental protection into their development strategies and daily operation management, and clarify their responsibilities and obligations in environmental protection and resource conservation. They should establish and improve an energy conservation and environmental protection management system, formulate and implement strict energy conservation and environmental protection standards and regulations, and ensure that production and operation activities comply with national and local environmental protection laws and regulations. At the same time, enterprises should deepen internal environmental protection publicity and education, enhance employees' environmental awareness and sense of responsibility, and create a positive atmosphere where all employees participate and work together.

Secondly, the key to promoting enterprises' green transformation lies in encouraging business owners to actively increase capital investment in green development. Enterprises should deeply recognize that green development is the future trend and the only way forward. They should actively promote the transformation and upgrading of industrial structure, optimize product portfolio structure, commit to the comprehensive improvement of technical levels, and increase investment in energy-saving and environmental protection technology R&D, equipment renewal and transformation, and clean production. Through technological innovation and industrial upgrading, enterprises can achieve the goals of reducing energy consumption, cutting pollutant emissions, and improving resource utilization efficiency, thereby realizing the dual growth and harmonious coexistence of economic benefits and environmental benefits.

Reference

1. Shu Zhong. Analysis of urban intelligent transportation and new energy industry development based on big data[J]. Times Automobile, 2020, (13): 13-15.